Impure and Worldly Geography

Tropicality is a centuries-old Western discourse that treats otherness and the exotic in binary – 'us' and 'them' – terms. It has long been implicated in empire and its anxieties over difference. However, little attention has been paid to its twentieth-century genealogy.

This book explores this neglected history through the work of Pierre Gourou, one of the century's foremost purveyors of what anti-colonial writer Aimé Césaire dubbed *tropicalité*. It explores how Gourou's interpretations of 'the nature' of the tropical world, and its innate difference from the temperate world, were built on the shifting sands of twentieth-century history – empire and freedom, modernity and disenchantment, war and revolution, culture and civilisation, and race and development. The book addresses key questions about the location and power of knowledge by focusing on Gourou's cultivation of the tropics as a romanticised, networked and affective domain. The book probes what Césaire described as Gourou's 'impure and worldly geography' as a way of opening up interdisciplinary questions of geography, ontology, epistemology, experience and materiality.

This book will be of great interest to scholars and students within historical geography, history, postcolonial studies, cultural studies and international relations.

Gavin Bowd is Reader in French, School of Modern Languages, University of St Andrews, UK.

Daniel Clayton is Senior Lecturer in Geography, School of Geography and Sustainable Development, University of St Andrews, UK.

Studies in Historical Geography
Series Editor: Robert Mayhew

Historical geography has consistently been at the cutting edge of scholarship and research in human geography for the last fifty years. The first generation of its practitioners, led by Clifford Darby, Carl Sauer and Vidal de la Blache presented diligent archival studies of patterns of agriculture, industry and the region through time and space. Drawing on this work, but transcending it in terms of theoretical scope and substantive concerns, historical geography has long since developed into a highly interdisciplinary field seeking to fuse the study of space and time. In doing so, it provides new perspectives and insights into fundamental issues across both the humanities and social sciences. Having radically altered and expanded its conception of the theoretical underpinnings, data sources and styles of writing through which it can practice its craft over the past twenty years, historical geography is now a pluralistic, vibrant and interdisciplinary field of scholarship. In particular, two important trends can be discerned. First, there has been a major 'cultural turn' in historical geography which has led to a concern with representation as driving historical-geographical consciousness, leading scholars to a concern with text, interpretation and discourse rather than the more materialist concerns of their predecessors. Secondly, there has been a development of interdisciplinary scholarship, leading to fruitful dialogues with historians of science, art historians and literary scholars in particular which has revitalised the history of geographical thought as a realm of inquiry in historical geography. Studies in Historical Geography aims to provide a forum for the publication of scholarly work which encapsulates and furthers these developments. Aiming to attract an interdisciplinary and international authorship and audience, Studies in Historical Geography will publish theoretical, historiographical and substantive contributions meshing time, space and society.

Elite Women and the Agricultural Landscape, 1700–1830
Briony McDonagh

Impure and Worldly Geography
Pierre Gourou and Tropicality
Gavin Bowd and Daniel Clayton

For more information about this series, please visit www.routledge.com/Studies-in-Historical-Geography/book-series/ASHSER-1344

Impure and Worldly Geography

Pierre Gourou and Tropicality

Gavin Bowd and Daniel Clayton

LONDON AND NEW YORK

First published 2019
by Routledge
2 Park Square, Milton Park, Abingdon, Oxon OX14 4RN

and by Routledge
52 Vanderbilt Avenue, New York, NY 10017

Routledge is an imprint of the Taylor & Francis Group, an informa business

© 2019 Gavin Bowd and Daniel Clayton

The right of Gavin Bowd and Daniel Clayton to be identified as authors of this work has been asserted by them in accordance with sections 77 and 78 of the Copyright, Designs and Patents Act 1988.

All rights reserved. No part of this book may be reprinted or reproduced or utilised in any form or by any electronic, mechanical, or other means, now known or hereafter invented, including photocopying and recording, or in any information storage or retrieval system, without permission in writing from the publishers.

Trademark notice: Product or corporate names may be trademarks or registered trademarks, and are used only for identification and explanation without intent to infringe.

British Library Cataloguing-in-Publication Data
A catalogue record for this book is available from the British Library

Library of Congress Cataloging-in-Publication Data
A catalog record for this book has been requested

ISBN: 978-1-4094-3949-3 (hbk)
ISBN: 978-1-315-58808-7 (ebk)

Typeset in Times New Roman
by Apex CoVantage, LLC

Contents

	List of figures	vi
	List of abbreviations	vii
	Interview schedule	viii
	Acknowledgements	ix
1	The tropics and the colonising gaze	1
2	Tropicalising Indochina	36
3	Romancing the tropics	87
4	Networking the tropics	137
5	*Gourou en guerre*	180
6	Affecting the tropics	206
7	Gourou's 'colonial situations'	265
8	*Fin de la tropicalité* (as we knew it)?	292
	Index	307

Figures

1.1	Cover of *Les pays tropicaux* (1947) © PUF	8
1.2	Pierre Gourou, Congo (1952) (Henri Nicolaï)	25
2.1	Indochina 1900–1945 (with the De Lattre Line, 1951) drawn by G. Sandeman	46
2.2	Orientalist imagery in Pierre-Louis Duchartre's "Imaginaire populaire et Indochine" *France Illustration* (1949)	51
2.3	Tropicalist imagery in Pierre Gourou "lÉcole des beaux-arts de Hanoi" *France Illustration* (1949)	68
2.4	Soldats de la boue © Service historique de la Défense, CHA Vincennes, 10 H 2351	71
3.1	The density of population in the Tonkin Delta. From *Les paysans du delta tonkinois. Etude de géographie humaine* © EFEO	112
3.2	Villages in the Tonkin Delta. From *Les paysans du delta tonkinois. Etude de géographie humaine* © EFEO	113
6.1	The tropical world in the era of decolonisation (drawn by G. Sandeman)	243
7.1	The density of population in Ruanda-Urundi, c. 1950. From *La densité de population au Rwanda-Burundi. Esquisse d'une étude géographique* © RAOS	272

Abbreviations

ACDF   Archives du Collège de France, Paris
ADG   Archives départementales de la Gironde
AEFEO   Archives de l'EFEO, Hanoi and Paris
AMB   Archives municipales de Bordeaux
ANF   Archives nationales de France, Paris
ANOM   Archives nationales d'Outre-mer, Aix-en-Provence
AULB   Archives de l'Université Libre de Bruxelles
BNF   Bibliotheque nationale de France
IAO   Intitut d'Asie Orientale, Lyon
PGB   Papiers Gilberte Bray (née Gourou) Private family collection
SHD   Service historique de la défense, Vincennes
UBCA   University of British Columbia Archives, Vancouver

Interview schedule

Gilberte Bray, 26 June 2008
Michel Bruneau, Paris, 13 October 2008
Paul Claval, Paris, 11 October 2008
Georges Condominas, Paris, 19 September 2008
Georges Courade, 8 September 2009
Sylvie Fanchette, Hanoi, 4 January 2009
Michel and Dominique Inguimberty, Menton, 28 June 2008
Yves Lacoste, Paris, 12 October 2008
Henri Nicolaï, Brussels, 16 June 2008
Paul Pélissier, Paris, 9 September 2009
Jean-Pierre Raison, Paris, 14 October 2008
Olivier Tessier, Hanoi, 4 January 2009
Dào Thê Tuân, Hanoi, 5 January 2009

Acknowledgements

For comments, critique, motivation, suggestions and support in connection with different aspects – and in some cases all – of the project, which is the product of a Geography-French collaboration, we would like to thank: David Arnold, David Biggs, Alison Blunt, Yann Calbérac, Julia Clancy-Smith, Paul Claval, Mark Cleary, Hugh Clout, Stuart Corbridge, Brett Christophers, Felix Driver, Jim Duncan, Hannah Fitzpatrick, Katherine Gibson, Elspeth Graham, Derek Gregory, Cole Harris, Susanna Hecht, Mike Heffernan, Rachel Hughes, Mike Kesby, John Kleinen, Audrey Koybayashi, Eric Jennings, Nina Laurie, Lisa Law, Steve Legg, the late Claude Lévi-Strauss, David Livingstone, Emma Mawdsley, Rodolphe de Koninck, James Sidaway, Pierre Singaravélou, the late Neil Smith, Matt Sothern, Paul Sutter, Gary Wilder and Charles Withers; our colleagues in the School of Modern Languages and School of Geography and Sustainable Development (and not least Graeme Sandeman for the maps); the students in our Twentieth-Century France and Colonial and Postcolonial Geographies classes, who have been brilliant enthusiasts and the most honest of critics; audiences in Belfast, Bordeaux, Cairns, Corsica, Edinburgh, Hanoi, London, Lubbock Texas, Montpellier, Paris and Rouen, for calling us out and pressing us in ever new directions; the remarkable group that gathered in St Andrews for the 'British and French Tropicality Conference' we organised in 2003 (supported by the Carnegie Trust for the Universities of Scotland and the French Embassy in London), which initially helped us on our way; and, as ever, our families for making much of what we do both possible and enjoyable. We are immensely grateful to Henri Nicolaï for his help with the Gourou material at the Université Libre de Bruxelles, his gallant responses to our numerous emails, and his permission to use the photograph he took of his teacher.

Much of the primary research for the book was undertaken between 2008 and 2011 with the help of a grant from the British Academy, which enabled us to undertake archival work in Aix-en-Provence, Bordeaux, Brussels, Hanoi, Lyon, Paris, Vancouver and Vincennes (details given in the Abbreviations), and conduct interviews with key individuals in Gourou's story (details given in the Interview Schedule). Indeed, we were in Paris interviewing during the October week in 2008 when the financial crash came to a head in Europe (and wondered whether we'd get home). Our trip to Hanoi and Gourou's Tonkin Delta was particularly poignant. We felt the need to glimpse something of what Gourou had seen in the place.

Much of our secondary research was undertaken in The British Library, London; Cambridge University Library; and the Bibliothèque Nationale de France, Paris. We are obliged to the staff of these august institutions, and as well as to those in the Interlibrary Loan Division of St Andrews University Library, for their bountiful assistance and patience in answering our sometimes arcane questions, and for tracking down often distant and obscure material.

We greatly appreciated Val Rose's and Robert Mayhew's initial excitement about the book proposal we pitched to Ashgate (its Historical Geography Series) and are thankful to the team at Routledge (especially Ruth Anderson and Kate Fornadel of Apex) for taking the project forward. Our closest and continuing debts are to Trevor Barnes, Michel Bruneau, Joe Doherty and Marie-Claire Robic, whose support over many years has on occasions gone beyond the call of duty. Bruneau, in particular, encouraged us to write this book, opened his homes in Bordeaux and Paris to us, and egged us on through some lengthy lulls in the project when we were bogged down by other things. *Michel, nous espérons que vous aimerez ce que nous avons fait et acceptez que la responsabilité de toute erreur nous incombe entièrement.* Finally, we shall miss our regular 'working lunches' at the Maisha in St Andrews (the source of the many curry stains on our Gourou material). The place withdrew its lunchtime service in the month we completed the manuscript! And we sent our story to the publisher on the day that tropical storm Isaac battered Aimé Césaire's Martinique.

G.B. and D.C.
St Andrews, 14 September 2018

Earlier versions of elements of the story and argument have appeared before in:

Gavin Bowd, "Géopoétiques de l'Indochine: Gourou, Inguimberty, Duras," in H. Velasco-Graciet ed., *Les tropiques des géographes* (Paris: Maison des Sciences de l'Homme d'Aquitaine, 2008), 15–25.

Daniel Clayton, "Militant tropicality: War, revolution and the reconfiguration of 'the tropics' c.1940 – c.1975," *Transactions of the Institute of British Geographers* 38 (2013): 180–192.

Gavin Bowd and Daniel Clayton, "Fieldwork and tropicality in French Indo-China: Reflections on Pierre Gourou's *Les Paysans du delta tonkinois* (1936)," *Singapore Journal of Tropical Geography* 24 (2003): 147–168.

Gavin Bowd and Daniel Clayton, "Tropicality, Orientalism, and French colonialism in Indochina: The work of Pierre Gourou, 1927–1982," *French Historical Studies* 28 (2005): 297–327.

Gavin Bowd and Daniel Clayton, "French tropical geographies: Editors' introduction," *Singapore Journal of Tropical Geography* 26 (2005): 271–288.

Gavin Bowd and Daniel Clayton, "Geographical warfare in the tropics: Yves Lacoste and the Vietnam War," *Annals of the Association of American Geographers* 103 (2013): 627–646.

Unless otherwise stated, all translations from French are our own.

1 The tropics and the colonising gaze

'What impure and worldly geography!'

> From Gourou, his book *Les Pays Tropicaux*, in which, amid certain correct observations, there is expressed the fundamental thesis, biased and unacceptable, that there has never been a great tropical civilisation, that great civilisations have existed only in temperate climates, that in every tropical country the germ of civilisation comes, and can only come, from some other place outside the tropics, and that if tropical countries are not under the biological curse of the racists, there at least hangs over them, with the same consequences, a no less effective geographical curse.... What impure and worldly geography![1]

So charged the Martinican writer and politician Aimé Césaire (1913–2008) in his searing 1950 *Discours sur le colonialisme*, which is one of the most incisive and influential anti-colonial texts of the post-war era.[2] He was censuring the French geographer Pierre Gourou (1900–1999), a Professor at the Collège de France in Paris and Université Libre de Bruxelles (he had a joint appointment). *Les pays tropicaux* had been published in 1947, ran to five editions to 1966, and became a touchstone of Western post-war understanding of the tropics. Césaire was concerned with how colonialism operated through language, imagery and texts like this (that is, as a 'discourse') as well as by economic, political and military means.[3] He did not dispute the notion that the world is divided into different regions, but objected to the way Gourou concluded that there was a hierarchical division between a fortunate temperate world and a perennially cursed tropical world. The way Gourou divided temperate from tropical lands had 'consequences.' Let us start this book, which is about the idea of 'impure and worldly geography' as it relates to Gourou and the tropics by delving further into this *mise-en-scène*.

Césaire contended that the "subjective good faith" of liberal Western academics like Gourou was "entirely irrelevant to the objective social implications of the evil work they perform as watchdogs of colonialism."[4] Their high-minded and well-intentioned knowledge was implicated in how the colonising West "gets into the habit of seeing the other man [sic] as an *animal*, accustoms himself to treating him like an *animal*, and tends objectively to transform *himself* into an animal." Colonisation "dehumanizes even the most civilised man,"[5] Césaire continued, and its brutal effects had rebounded on Europe in the form of Nazism

2   *The tropics and the colonising gaze*

and two devastating wars.⁶ The way colonial regions and populations had been represented, in a profoundly Eurocentric manner, as 'other' – in a twofold sense: as primitive, backward, unruly and inscrutable; and as knowable and amenable to the civilising mission of the West – had also been integral to this dehumanisation.

The expression *les pays tropicaux* (tropical lands/world) had long been used in France, and with increasing colonial purpose from the 1870s, to refer to a part of the world that was different to temperate regions (*les pays tempéré*) in environmental and racial terms.⁷ Empire created a space in which Westerners encountered and contemplated both environmental and cultural patterns, distributions and variations, chiefly on the presumption that environment and culture were somehow interconnected, and with the proviso that the nature of such links had a pivotal influence on the level of development of different regions. The 'subjective good' faith of erstwhile explorers and latter-day academic experts did not simply promote a more wise and careful (scientific and technical) understanding and management of the world. As Janet Browne reflects, it also shaped and was shaped by a "language of expansionist power" and an "expansionist geography."⁸

Gourou characterised the tropical world as "the belt of hot, wet lands" with a mean monthly temperature of over 65°F and sufficient rain for agriculture to succeed without irrigation, and as a zone which stopped at the desert (see Figure 6.1).⁹ Delimited thus, the tropics encompassed over 14 million square miles and more than one-third of "the useful portions of the earth's surface," he continued, but was sparsely populated overall because tropical nature fettered human existence and enterprise.¹⁰ "Compared to temperate countries, tropical regions are afflicted by a certain number of inferiorities," he proclaimed.¹¹ "It is generally not possible to master an elusive and difficult tropical nature, and build in hot, wet lands societies with a superior civilisation."¹² "We who live in temperate lands find it difficult to realise how baneful nature can be to humanity or to grasp that in many tropical places and regions water may swarm with dangerous germs, numerous blood-sucking insects may inject deadly microbes into the body, and the very soil may be harmful to the touch."¹³ Tropical soils were "less favourable to human existence than temperate ones," he continued, and this "basic problem of salubrity" was aggravated by "pathogenetic complexes" unique to the tropics, and "deficient" agricultural practices which "in the last analysis provide an insufficient economic basis for a bright civilisation and are not conducive to great political and intellectual achievements."¹⁴ He saw some of the rice-growing areas and cultures of the Far East as an exception to this rule, yet maintained that even there the attainment of civilisation was "without a doubt connected to cultural influences brought from outside the tropics."¹⁵ It was to the modern West that the tropical world still needed to look for help if it was to lift itself out of its "endemic bind" of poverty, disease and backwardness.¹⁶

Césaire said nothing (and perhaps knew little) about the disciplinary context out of which Gourou wrote, but he did not need to in order to castigate Gourou's outlook as a pernicious environmental Eurocentrism. The geographer's observations and generalisations were not only highly speculative, Césaire charged, but also fuelled by a double standard: for Europe would never entertain the thought of African or tropical influences on its own civilisation. Césaire pointed to the fate of Cheikh Anta Diop's *Nations nègre et culture* to hammer home his point. Diop documented the black African (Egyptian) origins of European civilisation in a book published by the radical press Présence Africaine in 1955 – the year in which it also re-issued Césaire's tract. Diop's book stemmed from a thesis he submitted for a Doctor of Letters at the Sorbonne, but he could not find a jury willing to examine his work. His treatment confirmed to Césaire that what counted as valid knowledge about Francophone Africa or Europe was the preserve of white intellectuals.[17] After World War II, and particularly in France, academic specialists operating overseas through increasingly professionalised and international networks of research and policy-making started to supplant colonial administrators and amateur collectors as the custodians of African and tropical knowledge. Their authority rested on what governments and international agencies regarded as their superior scientific know-how and objectivity, and Césaire thought it a scam.

So it was, as Césaire saw matters, that Gourou's book, which was ostensibly rational and objective (steeped in fieldwork and immense learning, and shorn of bombast), cast a "geographical curse" (*malédiction géographique*) over the tropics.[18] If, as Gourou maintained, geographers were concerned with civilisations and progress, then as Césaire continued the thought:

> it is not from the effort of these populations, from their liberating struggle, from their concrete fight for life, freedom, and culture that he expects the salvation of tropical countries to come but from the good coloniser – since the law categorically states that "it is cultural elements prepared in non-tropical regions which ensure the progress of the tropical regions towards a larger population and a higher civilisation."[19]

As Anne Gulick notes, Césaire's "true adversary" was not Gourou but "the variety of law that makes categorical statements – the law with the power to curse – and the law that is no more or less than the performance of a language of pure authority."[20] Césaire saw Gourou's primer as a malevolent quest "to reduce the most human problems to comfortable, hollow notions."[21] While Gourou eschewed the idea of 'race,' he avoided a reckoning with colonialism and Césaire saw this as a prime symptom of the duplicitous authority with which he pronounced on the problems of the tropical world.

Gourou fits the mould of what Donna Haraway, writing fifty years after Césaire, has configured as "the modest witness": the male observer who is

> the legitimate and authorized ventriloquist for the object world, adding nothing from his mere opinions, from his biasing embodiment. . . . [A witness] endowed with the remarkable power to establish the facts. He bears witness: he is objective; he guarantees the clarity and purity of objects. His subjectivity is his objectivity. His narratives have a magical power – they lose all trace of their history as stories, as products of partisan projects, as contestable representations, or as constructed documents in their potent capacity to define the facts. The narratives become clear mirrors, fully magical mirrors, without once appealing to the transcendental or the magical.[22]

Césaire's critique of Gourou anticipates elements of Haraway's treatment of "expert knowledge" and its separation "from mere opinion" as a "founding gesture of what we call modernity" – a knowledge that works "without appeal to transcendent authority" but works to give the illusion of just that.[23] Haraway also writes of the need to question "the god trick" of thinking that one can aspire to see "everything from nowhere," and to work for "politics and epistemologies of location, positioning, and situating, where partiality and not universality is the condition of being heard to make rational knowledge claims."[24] Gourou objectified the world he sought to represent, and invested both the differences between temperate and tropical worlds, and the relationship between culture and space in the tropics, with a sense of permanence that was tied to his status as an expert.

Césaire regarded *Les pays tropicaux* as just one of a much larger number of examples "purposely taken from very different disciplines" which, to his mind, "proved" that it was "not only sadistic governors and greedy bankers, not only prefects who torture and colonists who flog, not only corrupt, check-licking politicians and subservient judges" who needed to be named as "supporters of a plundering colonialism" and "enemies" of those struggling for freedom, but

> likewise and for the same reason, venomous journalists, goitrous academicians, wreathed in dollars and stupidity, ethnographers who go in for metaphysics, presumptuous Belgian theologians, chattering intellectuals born stinking out of the thigh of Nietzsche, the paternalists, the embracers, the corrupters, the back-slappers, the lovers of exoticism, the dividers, the agrarian sociologists.[25]

Gourou was one of his "lovers of exoticism" and "goitrous academicians," and Césaire drew upon his text, along with those of anthropologists, sociologists, psychologists, journalists and missionaries to declare that the denigration of non-Western peoples had been propelled by Western culture and largesse. Metropolitan knowledge about the tropical world served as a handmaiden of Western power, both in a direct (instrumentalist) sense, and more insidiously, by belittling the inhabitants of that part of the world.

This was what made Gourou's geography an "impure and worldly geography" (*géographie impure et combien séculière*). His gaze was neither inert nor detached. He had made a spectacle of the tropical world, and, for Césaire, his book was yet another hackneyed example of a "morally diseased" Western civilisation notorious for its "false objectivity," "sly racism" and "depraved passion for refusing to acknowledge any merit to the non-white races."[26] What is more, Césaire ended, this geography was imbued with "the *idea* of tropicality" (*tropicalité*) – of the tropics not simply as different, but also as inferior and subordinate.[27]

Tropicalité/tropicality

Fifty years on, historian David Arnold invoked *Les pays tropicaux* as "a high point in the discursive representation of the tropics (which will be referred to here as 'tropicality') that had formed over the preceding century of imperial expansion and control"; and he continued: "The text tells us as much about a collective (but by the 1950s already rather dated) northern world-view of the intra-tropical zone as it does about the 'tropical world' it seeks to depict."[28] Writing in a postcolonial vein – out of a body of theory that dwelt on the culture and epistemology of empire, and power of colonial discourses (systems of representation) – Arnold developed the idea of 'tropicality' as a supplement to Edward Said's hugely influential idea of Orientalism. The tropics, like the Orient, need to be regarded "as a conceptual, and not merely physical, space," Arnold ventured, and as constructed as the West's environmental foil and 'other,' in both an affirmative and demeaning fashion.[29] "The collection of ideas, representations, and experiences that can be dubbed 'tropicality'"

> was an especially potent and prevalent form of othering and one which does more to address the scientific and environmental aspects of Western imperialism than the kinds of texts Said employs. . . . [C]alling a part of the world 'the tropics' became a Western way of defining something environmentally and culturally distinct from Europe, while also perceiving a high degree of common unity between the constituent regions of the tropical world. The tropics existed in mental and spatial juxtaposition to the perceived normality of the northern temperate zone. Tropicality described the perceptions and experiences of Europeans and North Americans moving into a world alien in terms of its climate, vegetation, diseases, and human population.[30]

He distinguished between a "positive" (sensual, paradisical) and "negative" (menacing, pestilential, disparaging) tropicality – with tropical island edens encapsulating the former, and the jungle the archetype of the latter – and saw Gourou as an arch purveyor of the latter. "[T]here is scarcely a trace of the Edenic" in Gourou's writing, Arnold declared; "poverty and pathogenicity are all pervading."[31] Gourou's imagery was authoritative and his debunking of tropical fecundity continues to trail and irk environmentalists' vision of the tropical rain forest as what Philip Stott describes as "a natural moral focus" of the idealistic youth of the North.[32]

Arnold inspired a sizeable and eclectic literature on 'tropicality,' about which we shall say more shortly and throughout the book. However, he made no reference to Césaire's earlier use of the term, even though *Discours sur le colonialisme* was one of the progenitors of postcolonial theory. What is more, *Les pay tropicaux* was read in an entirely different – upbeat – way at the time it appeared.

Gourou's "little masterpiece"

In contrast to Cesaire's scolding, Gourou's primer received mostly glowing reviews in the Western academia press. The British journal *Geography* deemed it "a little masterpiece," and writing in the mainstream Paris review *Études*, François de Dainville observed that "At a time when political passions, economic interest and sentimentality regarding tropical and colonial problems may mislead, it is good to read coldly objective studies [like Gourou's] . . . providing an accurate view of reality," and to take to heart his conclusion that "nothing serious will have been achieved if the attentions of the modern world are focused solely on standards of living, which diminish appreciation of the art of living."[33] The French geographer Paul Veyret found Gourou's book "captivating yet troubling."[34] British geographer John Mogey added that it was the "most valuable account of the tropics as a whole" to date, and went on:

> The concept of tropical abundance is shown to be illusory; the soils of low fertility, the many diseases of man and beast, the geographical isolation, the practice of shifting cultivation, all result in a low density of population. South-east Asia, her civilisation adapting techniques which originated in more temperate latitudes, is the great exception.[35]

In his Preface to the book, the distinguished French ethnologist Paul Rivet noted that "The lush rainforest, exalted by writers, travelers and priests, is an optical illusion, a kind of shadow . . . hiding a fragile [tropical nature] of low vitality."[36] And the revered British scientist and director of the Imperial Agricultural Bureaux, John Russell, accepted without question Gourou's main thesis that it was from the temperate world that any hope of tropical progress towards "a larger population and higher civilisation" would need to come.[37] Even an aspiring critical theorist of colonialism, Georges Balandier, added that Gourou was one of the very few French scholars studying the peasant societies of the colonial world whose work had "real depth."[38]

One of Gourou's colleagues, Roger Lévy, a political scientist and expert on the Far East, noted that Gourou had made some "essential" findings: "The hot and rainy lands are overall sparsely populated. Their populations generally belong to backward civilisations. The insalubrity of these countries needs to be known. . . . The soil is poor. The luxuriance of the forest should not deceive. Fertile elements are mediocre and yields are weak."[39] Across the Atlantic, Jan Broek (a Dutch-born geographer based at Berkeley) noted that Gourou's "splendid exposé of the human geography of the tropics" was rich and iconoclastic in the way it counteracted a fast-growing American sociological and political science literature on "the brave new post-war world" which held that "bad government is fundamentally responsible for keeping the world's two billion people from being equally prosperous."[40]

The book was read as a fourfold advance on prevailing studies of the tropics.[41] First, it supplanted colonialist fantasies regarding the fecundity of the tropics and imagery of jungle wilderness.[42] Gourou pointed to the age-old human transformation of tropical environments, and highlighted problems of health, soil quality and disease. Second, starting the book with a chapter on population density, he supplemented geographers' zonal fixation with the physical geography of the tropics with a much-needed consideration of human geography.[43] Third, and particularly for French readers, the tropics had conventionally been represented in school textbooks and professional manuals as a realm of imperial expansion, whereas Gourou provided this part of the world with an over-riding geographical coherence.[44] And fourth, commentators applauded Gourou's fluency in multiple disciplines and literatures, and noted that this was an exception rather than the norm for a geographer.[45]

The book was translated into English as *The Tropical World* in 1953 by Edward Laborde, the Geography Master at Harrow School, and the Cambridge don B.H. Farmer (an expert on Ceylon) welcomed it with the note: "it is exactly what is required to correct the grave misapprehensions about the Tropics which … are all too common amongst University entrants," which presumably bode well for the tropics since the drivers of British overseas investment, and captains of diplomacy and the military, were drawn chiefly from Oxbridge ranks.[46] Laborde himself had in mind not only his own Harrow students and Britain's (by then beleaguered) colonial service, but also France's increasingly precarious grip over its colonial empire, when he suggested that "the book should be in the hands of every colonial administrator and planter."[47] Gourou had synthesised a vast range of material (colonial reports and statistics, and travel literature, as well as American and European academic literature), and "authoritative," "fastidious," "dispassionate," "lucid," "penetrating" and "unique" are the watchwords of reviews of the first and subsequent editions.[48]

Yet Césaire was neither the only mid-twentieth century writer to use the term 'tropicality,' and nor the only figure to censure Gourou at this time. As Susanna Hecht notes, Brazilian sociologist Gilberto Freyre could also "claim early usage [in 1944] of the terms *tropicality* (essentialisms about the tropics), *tropicalism* (ideologies that informed the practices of tropical imperialism), and *tropicology* (studies about the tropics)."[49] And in 1948 the French Marxist historian Jean-Suret Canale criticised *Les pays tropiaux* for not fully confronting the reality of colonial exploitation (the siphoning off of wealth and deliberate maintenance of the tropics in a state of backwardness through the appropriation of cheap raw materials, foodstuffs, energy and labour-power), and for focusing on pre-colonial history in much of the primer, which he found profoundly ironic given that it was published in the Presses universitaires de France's (PUF) *Colonies et empires* series.[50] Suret-Canale provided a touchstone for latter-day French tropical geographers who sought to radicalise their field and Gourou's stamp on it.

No one in these loops and cycles of adulation and critique, including Césaire, ever commented on the cover of *Les pays tropicaux*, where the idea of *tropicalité* can readily be found. The dust jacket contrasts a hilly temperate land to the north, bannered with 'colonies and empires,' and an exotic tropical land to the south, symbolised by palm trees and mangroves, and festooned with flags and a colonial outpost. The two worlds are at once set apart and connected by European ships from the Columbian age of exploration crafting a world of unequal exchange.

In spite of the sweeping and gloomy generalisations to be found within the covers of *Les pays tropicaux*, Gourou was generally lauded for his "originality, breadth of knowledge and flexibility of mind."[51] His chief American counterpart, the Yale tropical geographer Karl Pelzer, praised his "comparative studies" of the tropics, and the soundness of his overall message that "the tropics are inferior to

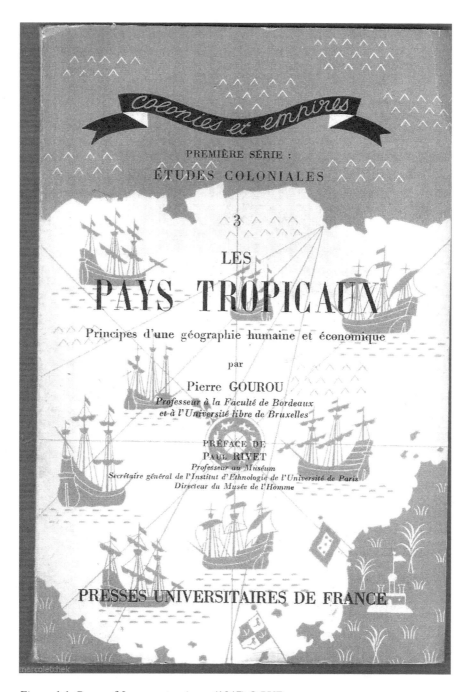

Figure 1.1 Cover of *Les pays tropicaux* (1947) © PUF

the mid-latitudes."⁵² *Les pays tropicaux* was one of the most widely used geography texts of the post-war era, and appealed to scholars and practitioners from a wide variety of disciplines and professions.⁵³ In 1983 Gourou's close friend Jean Gottmann reflected that the book had "inspired a generation of students" to study the tropical zone with better questions, more compassion, and fewer illusions than before.⁵⁴

On the other hand, no one, including Suret-Canale and latterly Arnold, ever went as far as Césaire in seeing Gourou's *tropicalité* as part and parcel of a colonising gaze, and one that was being mobilised at a time when much of the tropical world was in the throes of decolonisation (see Figure 6.1). *Les pays tropicaux* fuelled what Césaire and Frantz Fanon described as the 'inferiority complex' of the colonised, and which they saw as one of the key stumbling blocks to the attainment of true liberation. Gourou himself never acknowledged Césaire's critique in public or private. His daughter told us that her father knew nothing about it, and we are not in a position to dispute this.⁵⁵ Nor is Césaire cited in critical appraisals of tropical geography that began to appear within French geography in the 1980s, and where, one of Gourou's critics, Georges Courade, asserted, geographical research "is necessarily ideological because it is set up from the vantage point of one's own society."⁵⁶

Book reviews are an important (and much underused) resource for thinking about how post-war disciplines like geography saw themselves, and on the basis of the above comments *Les pays tropicaux* was read as a timely and unflinchingly earnest study that came at a moment when passions regarding 'tropical and colonial problems' were running high, and of which *Discours sur le colonisme* was a particularly strident anti-colonial example.

How might we account for this scene of difference between Césaire and these reviews? What connects or separates these critiques of *tropicalité*/tropicality from 50 years apart? Was Gourou's geography 'impure and worldly'? Was his tropicality entirely negative? Is the tropical world a reality or an illusion? Is there such a thing as a pure and unworldly (or objective and detached) geography? When and how does a way of looking at the tropics that is scholarly or seemingly unassuming become a colonising look (a way of appropriating that part of the world, an insidious look)? Our interest in Gourou was initially piqued by these sorts of questions, and they imbue the research that is expressed in this book. We shall say more about the above scene of difference in Chapter 6 but want to say something further here about the methodological issues raised by such questions and how we tackle them. We will do so under three headings – theory, history and biography – and use them to introduce the book, which is part intellectual biography of Gourou, part history of twentieth-century tropicality, and part theoretical rumination on the making and effects of scholarly and critical positions.

The following story is based on our work on three sets of Gourou material: first his research and writing, including an abundance of little noticed essays, reviews and unpublished reports; second, an array of archival resources pertaining to his activities and their links to wider political fields and disciplinary dynamics, including a smattering of personal correspondence and his personal dossiers at the ACDF and AULB (see Abbreviations); and third, interviews with Gourou that were conducted by others, and ones with his family and former colleagues, students, admirers and critics that were conducted by us (2008–2011; see Schedule of Interviews), which also offer insights into his career and entanglement with tropicality.[57] And we examine this primary material with a range of historical and theoretical literatures, to which we shall now turn.

Theory: orientalism, tropicality and the edges of knowledge

Both of us use Céasire in our teaching (on twentieth-century France, and colonial and postcolonial geographies, respectively), and also Edward Said's influential 1978 book *Orientalism*. Arnold's idea of tropicality is an important variant on Said's thesis, and Césaire's *Discours* is an important (if often unsung) precursor to both.[58] Let us start with Orientalism, which Said describes as a system of Western representation – style of thought, system of scholarship, and body of recurrent imagery – which has served states and empires, and buttressed Western colonial dominance. This system, or discourse on the Orient, has been centrally involved in the "mental" and "material" creation of Europe's sense of identity and superiority – with the Orient "as its contrasting image, idea, personality, experience" – and framing its colonial domination over the part of the world labelled 'the Orient' (stretching from North Africa to the Far East, although the bulk of Said's analysis is on the Arab world).[59]

Moreover, "the phenomenon of Orientalism as I study it here deals principally, not with a correspondence between Orientalism and Orient, but with the internal consistency of Orientalism and its ideas about the Orient," Said argues, "despite or beyond any correspondence, or lack thereof, with a 'real' Orientalism."[60] Western representations of the Orient shape the 'reality' they portray, he claims, and "Orientalism calls in question not only the possibility of nonpolitical scholarship but also the advisability of too close a relationship between the scholar and the state."[61] Said coined the expression "imaginative geography" to flag how images and representations "help the mind to intensify its own sense of self by dramatising the distance and difference between what is close to it and what is far away."[62]

The Orient's status as the Occident's (West's) "contrasting image" is made and sustained rather than given, prompting Said to ask: "How do ideas acquire authority, 'normality,' and even the status of 'natural' truth?"[63] While 'the Orient' is an actual material domain that has long been visited, experienced and colonised by the West, its "special place in European Western experience" also stems from imaginative geographies that construct arbitrary divisions between 'us' and 'them,' and the 'domestic' and 'foreign'; and as Said continues, that usually revolve around a "confusing [if productive] amalgam of imperial vagueness and precise detail," and spawn "representative figures, or tropes" regarding the 'other' which frame how novelty and variety are comprehended.[64]

Said had in mind "the invention and construction of a space called the Orient, for instance, with scant attention paid to the actuality of the geography and its inhabitants," and thus argued that imaginative geographies are insidious because they are fundamentally *asymmetrical*: one constituency (the imperial centre, the coloniser, Orientalist or tropicalist) controls the terms of representation, and it does not matter to the 'loftier' of the two constituencies whether the other ('lowly') one acknowledges or accepts how it is represented. Accordingly, 'native' involvement in the production of knowledge and imagination of difference is frequently trivialised and ignored, and when 'the native' resists – fights or writes back – such resistance is deemed to spring from what is innate in 'the other' rather than from the experience of colonisation.

Lastly, Said envisions the Orient as a space that "vacillates between the West's contempt for what is familiar and its shivers of delight in – or fear of – novelty."[65] The Orient is both proximate and distant, and alluring and threatening, to Europe – simultaneously "the source of its civilisations and languages, its cultural contestant, and one of its deepest and most recurring images of the Other."[66]

'The tropics' is another potent imaginative geography with a special place in Western experience, and, akin to Arnold, David Livingstone writes of the "censorial tenor" of Gourou's "zonal imaginary" in *Les pays tropicaux*.[67] "The contrast between the temperate and the tropical is one of the most enduring themes in the history of global imaginings," Felix Driver and Luciana Martins continue; and as Martin Mahoney notes, "Constructions of the local and the distant are frequently [still] painted in the primary colours of climatic difference."[68] "A thick crust of positive and negative values accreted around the tropics in Western European (or 'Northern') thought," Adam Rothman obverses; "most notably their ambivalent representation in Christian terms as either Paradise or Hell."[69] In more scientific and secular skeins of thought too, Nancy Stepan writes, the tropics "stood for many different values; for heat and warmth but also for a dangerous and diseased environment; for superabundant fertility but also for fatal excess; for species novelty but also for the bizarre and deadly; for lazy sensuality and sexuality and also for impermissible racial mixings and degeneration."[70]

Climate has been (and remains) a potent "hermeneutic device to make sense of cultural difference and to project moral categories on to global space," Livingstone argues.[71] Indeed: "In its capacity to colonise the deepest recesses of the human mind, climate must surely constitute one of the world's most successful imperial projects," and tropicality has been its principle vehicle.[72] Neil Safier and others "explore the longevity and tenacity of the torrid zone as an explanatory mechanism for describing the cultural characteristics of those populations living between the tropics."[73] This zone has been described as an uncanny domain of representation and experience where lines between desire and fear, dream and reality, and humanity and animality are blurred.[74] Nevertheless, tropicality gathers meaning as a way of thinking and mode of experience that sets up the tropical world in stark contrast to the temperate world. The palpable material differences between these two worlds should not be underplayed. They are not simply figments of the imagination. This is the point about tropicality: these environmental differences are experienced and deeply felt. But nor, Paul Sutter observes, should the potency of "the tropics as a category for defining and containing such difference" be underestimated either.[75]

Western understandings of the Orient and the tropics have been freighted through high and popular culture, and a range of artistic, imperial, literary and scientific practices and projects that whittle their 'actuality' down to a set of fixed and eternal traits and dichotomies. When Westerners looked to the East, Said argues, they did not see the cultures and inhabitants of the present, but either the great Oriental cultures of the past, upon which significant elements of Western civilisation were built, or events and processes that conformed to a set of stereotypes about the Orient and Oriental. As David Ludden remarks, in Said's thesis configurations of fact, truth and reality are at once potent and duplicitous, and it is the critic's job to reveal that "knowledge [is] constituted as truthful by the authority of a system of representation" – called Orientalism for Said, or *tropicalité*/tropicality for Césaire and Arnold – which stems from and secures Western power.[76]

Our job in this book is to delineate and evaluate Gourou's tropicality: where it came from and how it worked. A couple of passages from Said provide us with our basic (if far from straightforward) theoretical question. First, Said asks:

> Can one divide human reality, as indeed human reality seems to be genuinely divided, into clearly different cultures, histories, traditions, societies, even races, and survive the consequences humanly? By surviving the consequences humanly, I mean to ask whether there is any way of avoiding the hostility expressed by the division, say, of men [sic] into "us" (Westerners) and "they" (Orientals). For such divisions are generalities whose use historically and actually has been to press the importance of the distinction between some men and some other men, usually towards not especially admirable ends.[77]

The tropical world is a reality, but to what extent does it function as a figure of representation which creates this reality, and with what effects of power? And if imaginative geography is part and parcel of the problem Said identifies, how might we position a figure such as Gourou, whose focus, those reviewers of *Les pays tropicaux* insisted, was on 'the actuality of the geography' of the tropics and its inhabitants, and who made a concerted effort to separate fact from fiction, and reality and truth from image and myth? Or should *Les pays tropicaux* be regarding as a troubling example of geography's failure to survive the ways in which it carves up the world into different environments, spaces and landscapes, and so on, 'humanly' – a grappling with environmental difference that spawned division and hierarchy between temperate and tropical zones?

Second, Said reflects:

> I would not have undertaken a book of this sort if I did not also believe that there is a scholarship that is not as corrupt, or at least as blind to human reality, as the kind I have been mainly depicting. . . . The trouble sets in when the guild tradition of Orientalism takes over the scholar who is not vigilant, whose individual consciousness as a scholar is not on guard against *idées reçues* all too easily handed down in the profession.[78]

Can the same be said of Gourou or tropicality? We shall see that Gourou saw himself, and was regarded by others, as a single-minded scholar. But how cloistered from *idées reçues* about the tropics was he? And does tropicality revolve around a similar 'guild tradition'? Arnold suggests that different ways of studying and experiencing the tropics in different arenas – anthropology, art, architecture, botany, exploration, geography, medicine; colonial settlement, plantation production, and tourism – are linked by some common images of environmental otherness. Ideas of wilderness and jungle, and a fixation with climate and human acclimatisation, have been mainstays of the West's association of the 'torrid' zone with riot, excess and difficulty. However, and as Arnold insists, these styles of engagement and representation are not identical. Gourou was resolute in his depiction of himself as a geographer, and that, in certain ways, meant staying within specified disciplinary bounds. He did not feel duty bound to engage with literatures and debates that lay outside his disciplinary nexus. Yet he recast some of these boundaries, and grappled with ideas about the tropics that stretched beyond this nexus. It was this recasting and stretching that gave his work wider meaning and significance, and that implicates it in a broader story about tropicality. He subjected what he termed "climatic racism" (the idea that climatic differences underpinned the superiority of the West and a white race) to frequent and impassioned critique, and saw it as an ever-present danger (a point to which we shall return in the final chapter).[79] On the other hand, he was not immune from other 'easily handed down' ideas, about the 'insalubrity' of the tropics for example.

These questions about the politics and positioning of knowledge are as vital today as they have ever been, and Said, Arnold, and wider critical literatures on the histories of empire, science and geography tackle them by mobilising what feminist theorists have called a 'standpoint epistemology': a concern with the relativist embedding of knowledge in identity, difference and the body (and around positions of class, race, gender and sexuality), and the relations of power that imbue such embedding (i.e. the power to label, to curse, as Césaire opined).[80] While the tropics were – and continue to be – seen and experienced as exotic, and the zone's allure lay in the possibility of stretching the body and senses to their limits and revelling in the mystique of alterity, this part of the world has also connoted danger and fostered an impulse to contain and domesticate difference with derogatory labels and divisive categories.

As Said avers, the vexed epistemological grounds on which such vacillation between desire and fear operates is not simply a matter of theoretical and moral concern. It also has a history – and of course geography – and the historical moment that permeates the scene of difference with which we began is the moment of decolonisation, to which we shall now look and discharge as a second framing for the book.

History: geography, tropicality and the ends of empire

By 1900 there was a consistent pattern to Western dealings with the tropics. At a material level, the tropical world had been thoroughly commodified, principally through plantation complexes and processes of ecological change that were geared to the production of tropical exports – coffee, cotton, fruit, minerals, oil, rice, rubber, tea and timber – under fundamentally unequal terms of trade, and with the purported fecundity of the tropics feeding into stories of Western imperial prowess and the technological mastery of nature. The commercial exploitation and ecological degradation of the tropical world was part and parcel of what Marx characterised as the global *metabolic rift* wrought by the advent and spread of capitalism: an "irreparable rift" in the "metabolic interaction" between people and the earth, with a combination of industrial agriculture, long-distance trade, urbanisation, and monocultural production (in the above foodstuffs and raw materials) fundamentally changing the ecological relations between town and countryside, and imperial core and colonial periphery, and draining the latter (rural and colonial lands and soils) of nutrients, fertility, sustainability and wealth.[81]

Ideas about the tropics (and Orient) became dominant ideas by being propelled from the West and cast over 'the rest,' and as we shall observe in more detail in the next chapter, materialist and colonialist designs on the tropics were encapsulated in world exhibitions, which grew in ambition around the *fin de siècle*, both in terms of the scope of their displays of tropical commodities and curiosities, and their confidence in the technologies associated with the extraction and circulation of tropical things. A visual rhetoric of progress was enshrined in the pavilions (physical structures at the exhibitions) that displayed the worth of tropical colonies.

The flow of concepts, categories and understandings from the centre outwards, and the strong imperial messages emanating from these exhibitions, shaped how alien phenomena were seen both from afar and at close quarters, and infused a Eurocentric worldview in which difference was conceived in dichotomous and hierarchical terms. Texts, images and objects were prime means of constructing, capturing and reinforcing exoticism, and their credibility (truth, authenticity, or accuracy) hinged on the circulation of words and images, and objects and people, between 'there' and 'here.' Tropicality revolved around how, and from where, this movement occurred. Tropical knowledge was fashioned at particular sites of observation, examination, research, dissemination and display – ships, laboratories, research field sites, universities, conferences, museums and exhibitions. In turn, these procedures of "translocation and transference" in the production and circulation of knowledge, as Livingstone puts it, became associated with particular textual, visual and rhetorical formats and genres.[82] Said writes of "the journey, the fable, the stereotype, the polemical confrontation" as "typical encapsulations" of the Orient, and Stepan similarly writes of the scientific (geographical, agronomic, medical) treatise, the picture (painting, engraving, sketch and photograph of tropical flora, fauna, people and terrain), and botanical gardens and zoos (which inspired the French artist Henri Rousseau, who never left Paris, to produce vivid paintings of jungle scenes) as intoxicating 'encapsulations' of the tropics.[83]

At a visceral level, Westerners responded to the tropics in a variety of ways, and some of which one might not seem blatantly colonialist. Some wished to immerse themselves in this otherness and 'go native,' so to speak; and from afar, as in Rousseau's art, the tropics were represented as exotic and to be preserved as such. However, many recoiled from the exoticism they encountered upon arrival in the tropics, often through physical separation, for fear of engulfment, contamination and degeneration, and reinstated barriers between 'us' and 'them,' and the trappings of a 'temperate' way of life. Elements of what was new and alien were also made less threatening by being studied and being brought within the folds of reason, science, spectacle and colonial governance. Colonial interactions with the peoples of the tropics sometimes fostered toleration and generated cross-cultural liaisons, but in many tropical places they did not. What Gourou typically dubbed 'the colour bar' (the spatial separation of races, and reservation of the best land and resources for colonists) was common to myriad imperial formations over the centuries, and, as Fanon argued, the colonialist urge to separate and distance a Western self from the colonial other became a perennial source of resentment and envy among the colonised.[84] These cognitive and corporeal processes of 'othering' both spurred imperial expansion, by rendering distant lands and people as 'backward' and beseeching 'civilisation,' and sustained an illusion of imperial righteousness by regarding the peoples of the tropics, and the colonised world in general, as unable to represent and civilise themselves.

Arnold's empirical work on tropicality focuses on science (especially botany and medicine) and landscape in colonial India. Tropicality there was expedited through what he calls a "travelling gaze": the deployment and reworking of Western ways of visualising, observing and sojourning in exotic landscapes. On the one hand, tropicality was a discursive and classificatory project bound up with the ambitions of Western science, technology and medicine, and one, Driver and Martins note, that was "more often characterised by typification than generalisation . . . [with] very particular views and visions, represented, for example, by iconic images of tropical forest or desert island scenes, stand[ing] in for tropical landscapes as a whole."[85] As Arnold surmises,

> External to the cultural and physical landscape through which the European travelled, this scientific and scenic gaze was itself an ordering, even disciplining, mechanism that edited as well as elicited information and actively meddled in the construction of the knowledge it sought to shepherd and cajole into meaningful shapes and approved scientific forms.[86]

Tropical reality became a function of this ordering process.

On the other hand, tropicality was (and remains) a form of colonial exoticism and primitivism that enacts a Romantic quest for, and return to, primeval conditions and experiences. In the late eighteenth century, and again the late nineteenth and early twentieth centuries, primitivist discourse was bound up with Western aesthetic reaction to the anomie of modern life and what Max Weber termed "the disenchantment of the world" (the loss of the "great enchanted garden" of traditional societies, and of the spiritual and emotional qualities of life, to increasing rationalisation, secularisation and bureaucratisation).[87] The tropics became a privileged site at which to experience and contemplate the sublime majesty of nature, and fulfil a modern Western yearning for more 'natural' – authentic, primitive, organic, 'back to the land' – ways of being in the world (i.e., a place where one could kick back against the artifices of modernity).

Such tropical visions and travels made the Caribbean, Indian Ocean and East Indies, then the South Pacific, and by the mid-nineteenth century equatorial Africa, crucibles of Western imperial fascination and image-making, and concomitantly tropicality propelled modernity's veneration of sight over the other senses. However, this system of representation and body of experience was multisensory and never simply occularcentric. The tropics became associated with both personal reverie and an assault on the body and all the senses, and, to return to Césaire, tropicality thus amounted to a colonising gaze that unleashed a 'boomerang effect' (*choc en retour*) on the temperate north and the coloniser rather than just a 'travelling gaze.' Visual practices were embodied and enacted, and the startlingly alien environments of the tropics scrambled the presumed hierarchy of the senses. Journeying into and sojourning in the tropics inspired awe, brought enlightenment, induced trances, and effected personal transformation. As John Lindsay-Poland notes, for the American tropical adventurer (and later President) Theodore Roosevelt "through contact with the wild natural world, men were tested and became stronger."[88]

From the earliest days of Spanish exploration in the New World, tropical environments also induced dread and fears of degeneration and engulfment – a fear, as Eric Jennings puts it, "that Europeans denaturalized in the tropics," that they would become someone else in mind and body.[89] Pathogens, flora and fauna that thrived in the heat of the tropics had their own colonising – disturbing, invasive, slithering, frightful and debilitating – properties. This flip side of colonisation weighed heavily on the Western body and mind, and, as James Duncan observes, was thought to bring out a wild and brutal streak in human nature, and placed large question marks over the prospects of fruitful Western investment and settlement in the tropics.[90] In his 1944 *The Amazon: A New Frontier?*, for instance, the otherwise optimistic American geographer Earl Parker Hanson wrote of "vast jungle wildernesses, filled with poisonous insects and unpleasantly savage Indians"; and the year before the well know American tropical naturalist William Beebe, who otherwise sought to delight his readers with jungle yarns, noted how for American soldiers serving in Southeast Asia and the Pacific the jungle presented "unknown horrors."[91]

Contemplative detachment from the tropics might be seen as a fraught reaction to such fear, and as Arnold suggests, it gave the travelling gaze a contradictory hue:

> Aided by increased European mobility and observational opportunity, the seemingly contradictory role of the traveling gaze was to render the novel and exotic more familiar by attaching it to the cultural norms and epistemological systems of Western Europe while simultaneously emphasizing what was alien. . . . And while part of the function of the gaze was to disaggregate – to break down larger entities (patterns of vegetation, disease environments, racial categories) and so enable them to become objects of more precise

scrutiny – it was also deemed necessary to view and interpret such entities as a whole, as "tableaux," as "spectacles," or "scenes."[92]

The same contradiction can be found in Orientalism. In a similar vein, Driver and Martins warn against seeing tropicality as an "already fully formed, ready-to-be-projected" discourse positing essential and eternal differences between natures and cultures. Such a view overstates the unanimity and consistency (or 'guilded' quality) of Western representations of the tropics, and can obscure how tropicality works as a process of syncretisation (the stupefaction, proliferation and transformation of matter, meaning and identity). As Monique Allewart relates, tropicality involves the movement of animal and plant life, and microorganisms, and the transformation of ecologies, and human bodies and psyches, through their disaggregation. The tropical invasion and derangement of the Western self should not necessarily be seen as catastrophic, as it generally was in colonial times, she suggests, but as generative of "an alternate materialism of the body." Her focus is on slave and maroon groups in the American plantation tropics that evaded and resisted Anglo-European power and knowledge, and materialised "at the interstices between human and nonhuman life."[93]

However, tropicality also exemplified the power of visualisation that lay at the heart of modernity – a power, Haraway observes, wherein "The eyes have been used to signify a perverse capacity – honed to perfection in the history of science tied to militarism, capitalism, colonialism, and male supremacy – to distance the knowing subject from everybody and everything in the interests of unfettered power."[94] The academic discipline of Geography has been an instrument of visualisation *par excellence*, and Césaire saw this 'perverse capacity' in the way Gourou's imagery bestowed a "geographical curse." We shall see that Gourou was an accomplished visualiser, and part of our reason for seeking to re-situate his work in these critical terrains of tropicality and visuality is to suggest that we still know remarkably little about geography's early-to-mid twentieth-century performance of what Haraway characterises as "the god trick of seeing everything from nowhere"[95]

Said makes an intriguing point in this regard. His survey of Orientalism includes two twentieth-century figures very much like Gourou – the ethnologists Louis Massignon and Hamilton Gibb – and he suggests that their work points to a significant shift in Western knowledge about the Orient during the course of the twentieth century from an "academic to an *instrumental* attitude."[96] By the middle decades of the twentieth century, Orientalists such as Massignon and Gibb were no longer individuals belonging to the "guild community" of Orientalists we have already encountered; rather, they became "representative" figures of Western culture, expediting through their work "a major duality of which that work (regardless of its specific form) is the symbolic expression: Occidental consciousness, knowledge, science taking hold of the furthest Oriental reaches as well as the most minute Oriental particulars."[97] Herein lies another meaning of 'worldliness' for Said: Western consciousness which encompasses the world, albeit in highly selective terms. It is Said's sensitivity to the precise connection between structure and form in the history of Orientalism that intrigues us, and for the following reason.

Arnold developed the idea of tropicality at a moment when geographers were probing the complicity of their discipline in empire and posing broader questions about the inherent spatiality of colonialism. Imperial expansion and colonisation had been propelled by geographic perceptions of distant and exotic land, and geographic practices of exploration, mapping, surveying, and territorial administration and control. This 'postcolonial' turn in geographical inquiry – as it became dubbed – was integral to the prosecution of the wider claim that geography is a partial, shifting and power-laden realm of knowledge and understanding rather than an autonomous, immutable or detached discipline.

Arnold's survey of tropicality appeared in a special issue of the *Singapore Journal of Tropical Geography* entitled 'constructing the tropics,' and one of the editors of that publication, Driver, who was a leading figure in the 'geography and empire' literature, had earlier called for more comparative work on how different national (American, British, French and so forth) traditions of geographical inquiry were implicated in empire. In their introduction to this special issue, Driver and Brenda Yeoh gloss Arnold's notable contribution by referencing French geographers' earlier criticism of Gourou, and thus intimated (at least to us) that there was more to the Gourou story than we initially imagined.[98]

We also recognised that Gourou was an outlier in this critical liaison between geography and empire, in that geographers' energies were largely trained on the nineteenth century and prior imperial times and places, seldom made it past 1945, and almost wholly avoided the post-war era of decolonisation. This bias did not simply pertain to geography. Much of the postcolonial theory upon which geographers drew had its eyes firmly trained on the colonial past too – and often the quite distant colonial past. In 1992 Driver coined the term 'geography's empire' to capture both the discipline's nineteenth-century service to empire and how such assistance helped to shape how geography became a professional discipline, and empire became a pivotal site of theoretical rumination within geography more generally – a testing ground for theories of power and subjectivity, and a potent arena in which allegories of disciplinary objectivity might be exploded. Or as Frederick Cooper noted in a wider register, the ideas and methods that Western observers and scholars, including geographers, brought to foreign and colonial regions were "less a natural means of analysis of bounded societies located elsewhere than part of a process of intellectual pacification and ordering of the world."[99]

However, there was a basic missing link in this literature: it paid scant attention to the era of post-war decolonisation. By decolonisation we mean the processes that led to independence (both anti-colonial struggle, and attempts to rejig and reform empire), and the first steps taken by new post-colonial states.[100] That the only essay in the geography and empire literature that latched on to Césaire's 'impure and worldly geography,' by Clive Barnett deals with the nineteenth-century Africanist (racial) discourse of the Royal Geographical Society, and that Arnold's survey of tropicality stops at 1950, was telling in this regard.[101] That *Les pays tropicaux* was written and reviewed on the cusp of independence in many parts of the world was of great interest to us. While Césaire's critique was driven by questions of decolonisation, and especially the need to decolonise knowledge, reviews of the first edition of the book were written at a juncture when the complete end of empire did not seem a foregone conclusion. By 1960, however, the days of empire were by and large over, and we thus wondered: How were geographers involved in this dramatic shift in power and thought? Did they defend empire or support independence? Did they look on from afar or look to help or intervene? Were they aware of the imperial harm that their discipline had done in the past? Did they look to make amends? Was recognition of geography's imbroglio with empire subsequently buried, ignored or forgotten? Was the discipline's complicity subsequently exaggerated by late twentieth-century 'critical' human geographers who were seeking to distance themselves from this imperial heritage?

Gourou seemed important to us because his life and work spanned the twentieth-century story of imperial apogee, and then decolonisation, which was concentrated in tropical Africa and Asia (see Figure 6.1). The literature on tropicality has grown considerably over the last 20 years. Some of it reaches into the twentieth century, and we shall interact with relevant portions of it in ensuing chapters. It is also important to observe that much other secondary literature we use deals with 'the tropics' without engaging explicitly with the idea of tropicality. This serves as a reminder that this discourse does not tell us everything about northern views and experiences of the tropics, just as Orientalism does not tell us everything about the Orient and the West's handling of it. Indeed, in the last third of the twentieth century critical appraisal of Gourou's work was couched in a language of development and under-development rather than post-colonialism or tropicality.

Lastly in this section, the issue of decolonisation is important because it re-invests our scene of difference with further theoretical significance. *Discours sur le colonialisme* is a profound indictment of how Eurocentrism works as a fount of modernity (a declaration of what is right, normal and true) and why, without concerted critical effort, it was likely to outlive decolonisation.[102] There has been no let-up in the critique of Eurocentrism since 1950, and yet it persists, is good at changing its spots, and continues to marginalise alternative ways of knowing and explaining. Gennaro Ascione makes a vital observation: if "any theoretical standpoint within the theoretical framework of modernity [now] has little chance of avoiding the question of what stances it takes on Eurocentrism," how and why, during the course of the twentieth century, have so many standpoints successfully sidestepped or ignored this very issue (as the reviews of Gourou's book do), and made themselves seem obvious or right?[103]

In this sense, Cesaire's *tropicalité* flags both a *blasé* (or unthinking) Eurocentrism that is complacent about its own standpoint, and a knowledge that always has to work to sustain its hierarchies of meaning (superiority, normality), and keep the alternatives out. 'The real' – in our case the 'reality' of the tropics – has and continues to be a harbinger of truth through which this carefree attitude is simultaneously articulated and disguised. Haraway puts the problem thus: "my problem, and 'our' problem, is how to have *simultaneously* an account of radical historical contingency for all knowledge claims and knowing subjects . . . and a no-nonsense commitment to a faithful account of a 'real' world, one that can be partially shared and that is friendly to earthwide projects of finite freedom, adequate material abundance, modest meaning in suffering, and limited happiness."[104] Edmund Siderius adds:

> What this position entails is the recognition of an ever-present 'epistemologically impure' agent observing the world without the rejection of that agent's observations as purely 'subjective' or constructed. It thus accepts a certain degree of indeterminacy in regards to an absolute frame of reference while accepting an internal or contextual 'rightness' and 'wrongness' from within its context, given various degrees of epistemic quality. In such a scheme there are a plurality of justified truths in various area of human life which have no need for an ultimate hierarchy of facts for their collective validity.[105]

We see this position as germane to tropicality, which, we have noted, is a profoundly ambivalent discourse with positive and negative traits, and works via epistemologically impure agents. We seek to understand Gourou's "plurality of justified truths" yet without arriving at an "ultimate hierarchy" – at least at the outset.

In recent decades matters have also moved on with respect to the postcolonial tongue in which much of the literature on tropicality has endeavoured to speak. The disciplinary project of postcolonialism is concerned with the ongoing sway of the imperial/colonial past in the present, and the need to recall this past and hold it to account in the present remains strong. As Jane Jacobs argued 15 years ago, geography cannot "effectively decolonize its practices" without historical scholarship. To foster alternative, subversive or more vigilant or ecumenical geographies – be they postcolonial, anti-capitalist, ant-racist or indeed post-tropicalist – we need to look back, for such geographies cannot emerge or be discerned "outside of the histories . . . of the geographies that preceded them."[106] Yet what these colonial histories and geographies 'stand for' – planetary exploitation, social exclusion, ethnic division, authoritarianism, jingoism, racism, misogyny, corruption and cronyism, and spiralling inequality – are still rife today: so much so that talk of what Derek Gregory calls the "colonial present" seems more apt, and geography now envisions itself as needing to talk truth to power.[107] Accordingly, while this book is focused chiefly on the middle decades of the twentieth century, we end with some reflections on tropicality today.

Biography: 'lives lived' and 'lives told'

We now seek to outline and justify our focus on Gourou – the main character in this story. In recent years there has been a resurgence of interest among geographers (and historians of science and postcolonial scholars too) in biography as a means of thinking about intellectual change and the complex intersections between the 'lives' of projects, their authors, and the times and places in which they are made.[108] There are some noteworthy studies by geographers in this vein, and as Jake Hodder usefully explains, the concern is less "with knowing a life per se than how those experiences can cast light on the wider social and cultural worlds that a life inhabits." Furthermore, biography can be used "as a 'sampling' device . . . to navigate abundant archive collections without compromising on conceptual ambition. It allows consideration of the widest range of material by restricting the relevant volume of material to tell stories that are wider, deeper and more revealingly complex within the existing time and financial constraints of humanities research."[109] The twentieth-century history of tropicality can be – we think in good measure – be told through Gourou and how he positioned himself in relation to wider pressure points, tendencies and shifts in this discourse.

Conceived thus, a biographical orientation towards how Gourou both experienced and wrote about the tropics provides a window on to the twentieth century. It also allows us to explore how geographical ideas and practices are transacted between what Trevor Barnes very usefully describes as "lives lived" and "lives told": between what we know of geographers from their work, and how the discipline (its practitioners and historians) and commentators from farther afield reflect on what they did (or did not do).[110] Barnes develops a "lives told" approach to geography and the so-called 'quantitative revolution' that transformed Geography during the 1950s and 1960s. A significant element of this approach is drawn from pioneering work in the history of science and connects individual endeavour to wider realms of scientific output and development. Barnes sees this approach as a more revealing means of interpreting how science or geography is done (made and performed) than a more narrowly conceived "lives lived" approach, which tends to remove the inquirer from wider historical and institutional dynamics.[111] The "lives told" approach helps us to see the complex relations between working and recounting, intention and effect, self-understanding and reputation, and stresses that geographical thought has developed in neither a historical vacuum nor a tightly defined disciplinary milieu, but through in a range of sometimes concentrated, sometimes expansive, and sometimes contradictory encounters, institutions and networks.

Extant treatments of Gourou's life and work are framed by two interpretative motifs which veer in a 'lives lived' direction – even if, for us, they provide a starting point for a 'lives told' approach. The first is to whittle his eventful career and eclectic interests down to a 'system of thought,' identify different periods in its development, and separate his work out from his life. For instance, Gourou's "thought" and "life" are discussed separately in Paul Claval's essay on him in the *Geographers Biobibliographical Studies* series, and the former discussion of Gourou's thought traces four "steps" in his intellectual development, culminating in his attempt at "systematisation" in his 1973 *Pour une géographie humaine*.[112] Similarly, a number of Gourou's students and colleagues sought to distil his 'system of thought'; one of them, Gilles Sautter, judged it to be a "highly personal and highly coherent system" and described his teacher as "the anti-Pickwickian of French geography" – a single-minded scholar "set against all forms of contrived intellectualisation."[113] One of Gourou's critics, Michel Bruneau, reflects that Gourou's "production followed a consistent line, with the recurrent problematic of the relations between people and their natural milieu, refracted through the prism of civilisation . . . [with] societies considered in the light of the landscapes they create and their ability to occupy space in varying densities of population," and with Gourou "pushing social and political history to the background."[114] We shall more about Gourou's 'system' later in the book, but suffice it here to note that such packaging was also encouraged by Gourou himself, who, especially after his retirement, reflected more and more on the 'unities' of his work.

This first motif, which tends to suspend Gourou's work above the everyday and shifting circumstances in which he worked, is connected to a second: one of personality. Commentators (and by no means just Sautter) have portrayed Gourou as the humble and disinterested scholar who was driven by curiosity about the human elements of landscapes and their physical frameworks.[115] Again, this view of the detached and inquisitive scholar was encouraged by Gourou himself, who was also fond of deploying epigraphs from the scientific and humanist influences upon him, and including Montaigne's maxim "*Le monde, école d'inquisition*" (The world, the school of inquiry).[116] In a 1984 interview with the radical geography journal *Hérodote*, for instance, he proclaimed (partly, we shall see later, in frustration): "The term judge of the world expresses my attitude to what I see . . . I try to understand without approving or condemning."[117]

Those who query whether Gourou's work should be arraigned for what we are calling its tropicality, and seek to defend 'the tropics' as a coherent, credible or apolitical object of geographical inquiry, invoke these two motifs, and the adjectives 'kind,' 'self-effacing,' 'dedicated' and 'detached' have frequently been used to describe him.[118] Our point, however, is that while the world 'is there' in many commentaries on Gourou's life and work, and his work has not received doggedly in-house reviews that invest it with a singularly disciplinary meaning, the world is not made to do much – or as much as it might. We are not reaching for a contextual approach in which the capaciousness and integrity of Gourou's arguments and texts melt into context and are diminished accordingly. Rather, we seek what might be termed a more 'circumstantial' way into and through his intellectual journey. Geographers' works and lives cannot easily be distilled into 'systems of thought' or 'programmes of research.' Yet nor are they without rhyme or reason. Positions are cultivated and defended, and habits of mind and research inclinations are formed, defended and critiqued. Césaire caught Gourou in an anti-colonial moment. But we shall see that other – romanticist, internationalist, conflictual and developmentalist – moments count too.

In Gourou's case there are some significant bumps on the road to this 'circumstantial' understanding. *Impure and Worldly Geography* is not a biography of Gourou that seeks to tell his life story from memoirs and by sticking close to personal testimony. It is not such a study because Gourou left very little in the way of personal papers (fieldwork notes, journals, diaries or correspondence), shied away from autobiographical writing, and gave only a handful of interviews (and three of them when in his nineties).[119] Little is known about his early life in Tunis, Lyon and French Indochina because whatever private papers he possessed were seized by the Gestapo during a May 1940 raid on his Brussels apartment block (as he and his family were fleeing the city), and were presumably destroyed.[120]

Much of what we might know about Gourou comes from what he left on his academic sleeve – in books, essays and reviews – and on this 'lives lived' showing he had a much more oblique relationship with institutions of power, the field over which he presided, and the *tropicalité* that his work emitted, than Césaire knew or acknowledged. Yet as we seek to show, we discover much else about Gourou via a 'lives told' approach: through the interviews he gave; his semi-autobiographical *Terres de bonne espérance*, and from the interviews we conducted (which variously corroborated, extended and refuted what we thought we knew about him from his writing).

We find much of Gourou's professional persona in the way he portrayed his friend and contemporary, Roger Dion, in a 1982 obituary: they both had a "culture, spirit of inquiry, passion for landscapes, [and a] historian's skill and elegance of expression."[121] Nevertheless, we have also sought to build a personal and professional impression of Gourou through our second cache of material (interviews and correspondence) and obituaries and reflections on his life and work. This research showed us – perhaps unsurprisingly – that Gourou did not live in a solitary ivory tower and was not cocooned from the wider and turbulent historical forces and political debates of his day, and that lives are complex and sometimes contradictory, as well as unique. But this work revealed that he had cultivated the aura of a 'life lived' – a solicitation to judge him by what he wrote. In many respects, it is difficult to know Gourou except by his writing, and we sensed that he worked to remove himself from what he put down on paper. He did not talk about himself, and as Césaire surmised his "subjective good faith" was "highly problematical" because of the aura of scientific objectivity that was so central to what he wrote, and to how he assumed the mantles of scholar and expert.

As this suggests, and Barnes acknowledges, lives lived are entangled with lives told; and Gourou's aura was never just 'scientific.' 'Culture' is a word that looms large in his 'life told.' His daughter described her father as private, single-minded, cultured, self-effacing, idiosyncratic and paradoxical: in a nutshell, she said, "an unbeliever who liked order."[122] Gourou's wartime student Paul Pélissier viewed his teacher as "ferociously attached to the intellectual independence of himself and others. He was solitary, but his culture and courtesy also kept him from indulging in loud polemic."[123] For much of his life, Gourou "observed a disciplined colonial lifestyle," his daughter continued, starting work at 5 A.M. each day and locking himself in his study until 9.30, and retiring religiously at 8 P.M. He was born a Catholic and became a moderate socialist; and we will see that like other French intellectuals in this political mould (from Jules Michelet to Lucien Febvre and Claude Lévi-Strauss), he found and venerated symbolism in the things he studied, and was suspicious of linear and teleological models of history and progress. One of Gourou's admirers, Jean-Pierre Raison (one of Pélissier's students), was convinced that he was a freemason, although had no proof of this (and we have none either).[124]

Gourou played a prominent political role in the French Resistance but was happy to see the fall of communism in 1989. The radical geographer Georges Courade pegged him as a centre-left, secular, republican and anti-Communist scholar who argued for the reform rather than abandonment of the French empire after World War II. And Courade and others observe that Gourou's influence – as large as it undoubtedly was – emanated more from his writing than his teaching, chiefly on account of the fact that he was geographically and intellectually marginal to the political bastions of geography in post-war Paris – the Sorbonne and Institut de géographie.[125]

Gourou seldom wrote letters, was reluctant to give interviews, and never owned a television. He was a prolific academic researcher, but was more drawn to baroque poetry, philosophy and theatre than dry scientific prose, and many (including his sternest critics) have lauded the literary quality of his writing.[126] Descartes and Molière were among his favourite writers. He also embraced the rational and meritocratic "culture of politeness" and republic of letters of the French Enlightenment.[127] Like Denis Diderot, he believed that literary style and wit were integral to the development of civilisation.

He liked epistolary literature, yet spoke and wrote in a lapidary style. His top-floor apartment at 14 Place Meunier, off Avenue Winston Churchill in suburban Brussels (a short walk from the Université Libre de Bruxelles), was adorned with his friend Joseph Inguimberty's paintings of traditional peasant life in French Indochina. While accepting of empire, he was critical of French colonial policy, and his daughter added that while he reminisced fondly about his years in Indochina he never became nostalgic about the loss of empire. Roger Lévy described him as "a man of vast culture and learning" who rose above intellectual dogma and fads; and while fêted as an expert by a number of governments and international organisations, he shied away from the intellectual and political limelight.[128] At the same time, Lévy continued, Gourou was prone to making sweeping generalisations.

Lévy remarked at length on his friend's deep affinity with sceptical and satirical traditions of French thought. Gourou was particularly inspired by Michel de Montaigne's famous account of the three Tupi Indians (captured by conquistadors and shipped back to Spain) he encountered in Rouen in 1563, which inspired his well-known essay and auto-critique of empire, *Des cannibales*; and Molière's 1672 satire, *Les femmes savantes*, which deals with academic pretension and cultural posturing. But he was also drawn to French rationalists such as Descartes and Bernard le Bovier de Fontenelle, who taught him to doubt childhood impressions, and contemporary authors such Victor Segalen, who placed exoticism under the critical spotlight.[129]

In these ways and more, Gourou admired writers who grappled rigorously and sensitively with the human condition, and he was as approving of this Western literary and rationalist tradition as he was dismissive of Montesquieu's "climatic racism."[130] In 1947, the American geographer John K. Wright lamented that his discipline's "deep-seated distrust of our artistic and poetic impulses too often causes us to repress them and cover them over with incrustations of prosaic matter, and thus to become crusty in our attitude toward anything in the realm of geography that savors the aesthetic."[131] Much French geographical writing betrays a tension between literary and scientific values, and Wright's remarks left a deep impression on us and how we read Gourou.

The tropics and the colonising gaze 25

In short, we think that biographical ambiguity and paradox are vibrant ways of situating Gourou, and we get a glimpse of what his daughter fondly recalled as the mischievous glint in her father's eye in Figure 1.2. This photo of Gourou, which was taken near the city of Luluabourg in southern Congo by his Belgian student Henri Nicolaï in 1952.[132] Gourou was visiting Nicolaï and fellow student Jules Jacques, who were in the midst of their fieldwork, investigating the impact of the railway on the Congolese landscape. With a sense of spectacle and occasion, Gourou posed for the picture with a pith helmet and in khaki drill – the classic garb of Europeans in Africa and the tropics (it was thought to diminish the effects of heat and humidity on the body) – along with the latest travel accoutrements, a Leica camera and Polaroid sunglasses.[133] And Nicolaï told us that it was his supervisor's suggestion that they ventured from the roadside into the bush to have the picture taken. The scene conjures up

Figure 1.2 Pierre Gourou, Congo (1952)
Source: taken by Henri Nicolaï

26 *The tropics and the colonising gaze*

wider and longer histories of imperial exploration and conquest – etchings, paintings and photographs of explorers, soldiers, missionaries, scholars and colonial administrators of yesteryear – from Henry Morton Stanley to Marcel Griaulle – in similar poses. Such garb was part of what we shall call the exhibitionism of empire (see Chapter 2). But Nicolaï regarded the whole set-up as curious, even preposterous – a get-up. For by the 1950s such colonial attire was clichéd, and Nicolaï had not seen his mentor wearing it before and did not again thereafter. It was as if the iconography of empire called for it. Or perhaps Gourou was poking fun at the amateurism of a bygone era of exploration and contrasting it with a new era in which a better world would spring from a more rigorous social science understanding and new breed of experts. Etienne Gilbert noted that Gourou had "a great sense of humour and nonconformity," and this photograph might be viewed in this light.[134]

Times, tropicalities, themes

These three elements of theory, history and biography are woven into the chapters that follow, and a short overview of them might usefully be offered before we jump back into the story. The chapters are organised in a threefold way, and each of them provides specific ways of thinking about when and how 'trouble' sets in, to use Said's words, with respect to tropicality.

First, we take a chronological path through Gourou's career, and starting, in Chapters 2 and 3, in Indochina, where he began his research career. We proceed from there, in Chapters 4 and 5, back to Belgium and France between 1936 and 1945. Chapters 6 and 7 consider his post-war career, and we end, in Chapter 8, with an overall assessment of Gourou's tropicality and some reflections on this discourse today. In this register, the 'trouble' with Gourou's tropicality is gauged in relation to the particular metropolitan, colonial and international settings in which he worked, and the interplay between individual consciousness, public events and political processes.

Second, our research taught us not to endorse a single script about tropicality. This discourse is made – projected and negotiated, imagined and experienced, visualised and embodied, asserted and resisted, varied and fractured, and comprised of fragments and remains – rather than fixed or given, and it behoves us to account for this making and these tensions. It is important to think about the different expressions of tropicality in different times, places, projects. Our focus is on geography, and more particularly French geography. The story of twentieth-century tropicality would not come out exactly the same if it was traced through, say, the disciplinary fields of ecology or medicine, or the public realms of film, literature and tourism. However, Gourou's geography of the tropics was positioned in relation to, and often interacted with, other fields and wider realms of tropical experience and representation, and this positioning is pivotal to how the nature and distinctiveness of his work and influence might be read. Our French interviewees also made us mindful of how we were seeking to link Gourou to the idea of tropicality. This idea (as articulated by Arnold and with the links we have discussed back to Said and Césaire) had not crossed the radars of the French geographers – a number of them leading tropical geographers – with whom we spoke. They said we brought 'news' and an Anglophone perspective. Said and postcolonial theory had seemingly made limited headway into French geography.

In short, tropicality might be deemed a fragmented discourse, and one in which not everything that is thought, experienced or written about the tropics might have an insidious element to it. However, following Said and postcolonial theory, and as a notable essay by Stuart Hall impressed upon us, with tropicality, as with colonialism, "we cannot afford to forget the over-determining effects of the colonial moment, the 'work' its binaries were [are] constantly required to do re-present the proliferation of cultural and forms of life, which were always there, within the sutured and over-determined 'unity' of the simplifying, overarching binary, 'the West and the rest.'"[135] While there are dangers in thinking that all knowledge pertaining to the tropics is ultimately caught within the clutches of – or 'over-determined' by – a colonising gaze, there are dangers too in underestimating the persistence and persuasiveness of tropicality as a mode of othering that works in a binary fashion.

Third, each of the chapters also has a thematic focus, and, for this reason, a number of them have overlapping time frames. Gourou's influence on twentieth-century tropicality is appraised in terms of a string of motifs, each of which casts tropicality, and our worries about it, in a specific light. We start the next chapter by exploring a French publication on Indochina that appeared the year before Césaire's critique. We use it as a way into thinking about a fifty-year period (roughly 1900–1950) over which a particular region, Indochina, became 'tropicalised' – identified, experienced and constructed as 'tropical' – and into exploring how, in the context of the Indochina War, tropicality was seized and inverted by the 'tropicalised' (in this instance Vietnamese communists resisting the French, and later the Americans). While Said and Arnold explore different genres and 'encapsulations' of Orientalism and tropicality, and volumes like Driver and Martins' *Tropical Visions in An Age of Empire* consider different 'sites' and 'scenes' in the making of this discourse, we pursue Susanna Hecht's argument with respect to the Amazon and work of Euclides da Cunha that it is important to think about how tropicality was assembled in particular regions.[136] At the regional scale, she shows, the complacencies and complexities surrounding the 'tropicalisation' of particular parts of the world, and resistance to that process, come into sharp relief. As Frances Aparicio and Susanna Chávez-Silverman also observe, "to tropicalize" means not only "to trope" but also "to imbue a particular space, geography, group, or nation with a set of traits, images and values."[137] In Chapter 2, French Indochina, and the beginnings of modern Vietnam, is our space.

Chapter 3 tracks back to the 1920s and 1930s and explores Gourou's contribution to the tropical make-over of Indochina in more depth. We explore his fieldwork in the Tonkin Delta and how it was indicative of a broader Western penchant during this period (and stretching farther back into the annals of tropicality) to romanticise exotic cultures and landscapes. Our exploration of this 'affirmative' tropicality also serves as a foil to the more commonplace portrayal of Gourou as the arch purveyor of a negative tropicality. In other words, there are important historical shifts and geographical variations in Gourou's entanglement with this centuries-old discourse. Chapters 4 and 6 consider how, upon his return to France in 1936 and through the rest of his career, he worked in an increasingly 'networked' world. While he regarded himself as a lone scholar, and never as serving the state directly, his work and influence was bound up with networks of academic expertise and their political utility to national governments and international organisations. During the middle decades of the twentieth century Gourou became part of what Timothy Mitchell has dubbed a "rule of experts," and the tropical world became a realm of Western expertise *par excellence*.[138]

Chapters 5–7 explore a final theme that is writ large, we think, in twentieth-century tropicality, and to which Gourou made a decisive contribution: engagement with the tropics as an 'affective' domain – a domain *moved* by reason and emotion; one *affected* by economic, social, political, ideological agendas, and geographical and environmental conditions and transformations; and one that was impinged upon by war and the destruction it wrought. These chapters return to the scene of difference with which we started and alight on (the barely noticed) significance of wartime events and experiences on how Gourou and other Westerners viewed the tropics and the processes affecting it. To date, nothing has been made of the war element of Gourou's story, and that story is interesting because it provides a counterpoint to the prevailing narrative about geography and war, which is about how geographers served their nations and how the discipline was put on a war footing.[139] Gourou, on the other hand, eschewed military aggression and used his geographical skills in a clandestine way, serving in the French Resistance.

Notes

1 Aimé Césaire, *Discourse on Colonialism* trans. Joan Pinkham (New York: Monthly Review Press, 1972), 34, 37.
2 Robin D. G. Kelley, "A poetics of anticolonialism," *Monthly Review* 51, no. 6 (1999), http://monthlyreview.org/1999/11/01/a-poetics-of-anticolonialism/. Initially published by Réclame, a small Parisian press linked to the French Communist Party, Césaire's tract was re-issued in 1955 by Présence africaine (based in Paris's Latin Quarter, and which supported francophone black African writers) and became its best ever selling book. See Imorou Abdoulaye ed., *La littérature africaine francophone: Mesures d'une présence au monde* (Dijon: Editions Universitaires de Dijon, 2014).
3 Robert J.C. Young, *Postcolonialism: An Historical Introduction* (Oxford: Blackwell Publishers, 2001), 274 and *passim*; Gary Wilder, *Freedom Time: Negritude, Decolonization, and the Future of the World* (Durham, NC and London: Duke University Press, 2015).
4 Césaire, *Discourse on Colonialism*, 34.
5 Césaire, *Discourse on Colonialism*, 41, 20.
6 Césaire, *Discourse on Colonialism*, 34, 41, 20. On Césaire and the *choc en retour* (boomerang effect) of colonisation, see Michael Rothberg, *Multidirectional Memory: Remembering the Holocaust in the Age of Decolonization* (Stanford, CA: Stanford University Press, 2009).
7 See, for example, Octave Saint-Vel, *Hygiène des Européens dans les climats tropicaux, des créoles et des races colorées, dans les pays tempérés* (Paris: A. Delahaye, 1872),

1–25 and *passim*; Julien Brault, *Traité pratique des maladies des pays chauds et tropicaux* (Paris: J-B. Baillière et fils, 1900); Paul Carton, *La météorologie et ses applications dans les pays tropicaux* (Paris: Exposition Coloniale, 1931).
8 Janet Browne, "A science of empire: British biogeography before Darwin," *Revue d'histoire des sciences* 45, no. 4 (1992): 453–475, at 457, 467.
9 Pierre Gourou, *Les pays tropicaux: Principes d'une géographie humaine et economique* (Paris: Presses Universitaires de France, 1947), 1. We quote here from the original French edition as this was the edition Césaire read.
10 Gourou, *Les pays tropicaux*, 2–4.
11 Gourou, *Les pays tropicaux*, 173.
12 Gourou, *Les pays tropicaux*, 174.
13 Gourou, *Les pays tropicaux*, 7.
14 Gourou, *Les pays tropicaux*, 28, 49–50, 157. Gourou drew the term 'pathogenic complexes' from Maximilian Sorre's *Les fondements de la géographie humaine, Vol. I: Les fondements biologiques: Essai d'une ecocologie de l'homme* (Paris: Armand Colin, 1943). He was particularly drawn to Sorre's argument that disease was not determined solely or ultimately by environmental conditions, but also by human factors of population density, culture, subsistence and lifestyle, and the diffusion of animals, plants, insects and microbes. Gourou, *Les pays tropicaux*, 6.
15 Gourou, *Les pays tropicaux*, 175.
16 Gourou, *Les pays tropicaux*, 178–179.
17 Césaire declared the work "the most daring book yet written by a Negro": *Discourse on Colonialism*, 35; Sheikh Anta Diop, *Nations nègres et culture* (Paris: Présence africaine, 1954). Diop's thesis work was supervised by the distinguished French ethnographer Marcel Griaulle. See François-Xavier Fauvelle-Aymar, *L'Afrique de Cheikh Anta Diop: histoire et idéologie* (Paris: Karthala, 1996), 15–22. During the 1940s and 1950s the empowerment of Africans through education was weaker in the francophone world than the Anglophone world. Fewer francophone African scholars were trained to PhD level or attained university positions. See Catherine Coquery-Vidrovitch, "Colonial history and decolonization: The French imperial case," *European Journal of Development Research* 3, no. 1 (1991): 28–43. However, Britain's post-war aim of providing higher education to Africans, in preparation for self-government albeit as a means of keeping the elites of newly independent countries within Britain's orbit, handed the University of London, where many Africans studied, a new imperial remit and sway. See John Flint, "Planned decolonization and its failure in British Africa," *African Affairs* 82, no. 328 (1983): 403.
18 Césaire, *Discourse on Colonialism*, 35.
19 Césaire, *Discourse on Colonialism*, 36–37. One of Césaire's students, Frantz Fanon, developed a similar 'discursive' outlook, and with inflections of *tropicalité*, in his 1952 *Peau noire, masques blancs*: "To speak means to be in a position to use a certain syntax, to grasp the morphology of this or that language, but it means above all to assume a culture, to support the weight of a civilization. . . . Every colonized people – in other words, every people in whose soul an inferiority complex has been created by the death and burial of its local cultural originality – finds itself face to face with the language of the civilizing nation. . . . The colonized is elevated above his jungle status in proportion to his adoption of the mother country's cultural standards. He becomes whiter as he renounces his blackness, his jungle." Frantz Fanon, *Black Skins, White Masks* trans. Charles Lam Markmann (New York: Grove Press, 1967), 17–18.
20 Anne Gulick, "A universal rich in all its particulars: Aimé Césaire's negritude and mid-twentieth century discourses of justice," in Isabel Constant and Kahiudi C. Mabama eds., *Negritude: Legacy and Present Relevance* (Newcastle Upon Tyne: Cambridge Scholars Publishing, 2009), 116.
21 Césaire, *Discourse on Colonialism*, 35.

22 Donna J. Haraway, *Modest_Witness@Second_Millennium: FemaleMan©_Meets_Onco-Mouse: Feminism and Technoscience* (New York and London: Routledge, 1997), 24.
23 Haraway, *Modest_Witness*, 24.
24 Donna Haraway, "Situated knowledges: The science question in feminism and the privilege of partial perspective," *Feminist Studies* 14, no. 3 (1988): 575–599, at 581, 589.
25 Césaire, *Discourse on Colonialism*, 33–34, 43 (emphasis in original).
26 Césaire, *Discourse on Colonialism*, 36, 10, 35.
27 Césaire, *Discourse on Colonialism*, 43 (emphasis in original).
28 David Arnold, "'Illusory' riches': Representations of the tropical world, 1840–1950," *Singapore Journal of Tropical Geography* 21, no. 1 (2000): 6–18, at 6–7.
29 David Arnold, *The Problem of Nature: Environment, Culture and European Expansion* (Oxford: Basil Blackwell, 1995), 142.
30 David Arnold, "India's place in the tropical world, 1770–1930," *Journal of Imperial and Commonwealth History* 26, no. 1 (1998): 1–21, at 4. Around this time similar arguments were developed in: David Arnold and Ramachandra Guha, eds., *Nature, Culture, and Imperialism: Essays on the Environmental History of South Asia* (New Delhi: Oxford University Press, 1996); Richard Drayton, *Nature's Government: Science, Imperial Britain, and the "Improvement" of the World* (New Haven, CT: Yale University Press, 2000); Roy MacLeod ed., "Nature and empire: Science and the colonial enterprise," special issue, *Osiris* 15, no. 1 (2001); and Michael A. Osborne, *Nature, the Exotic, and the Science of French Colonialism* (Indianapolis: Indiana University Press, 1994).
31 Arnold, "'Illusory riches'," 10.
32 Philip Stott, *Tropical Rain Forest: A Political Ecology of Hegemonic Myth Making* (London: IEA, 1999), 35.
33 François de Dainville, "Revue des livres," *Études* 255, October–December (1947): 126–127.
34 Paul Veyret, "Review of *Les pays tropicaux*," *Revue de géographie alpine* 36, no. 1 (1948): 178.
35 J.M. Mogey, "Review of *Les pays tropicaux*," *Man* 49, no. 2 (1949): 22.
36 Paul Rivet, "Préface," *Les pays tropicaux* 3.
37 E. John Russell, "Review of *The tropical world*," *International Affairs* 29, no. 4 (1953): 489. Russell was the founder of the Imperial Agricultural Bureaux.
38 Georges Balandier, "The colonial situation: A theoretical approach (1951)," in Immanuel Wallerstein ed., *Social Change: The Colonial Situation* (New York: John Wiley & Sons, 1966): 34–61, at 41.
39 Roger Lévy, "Review, *Les pays tropicaux*," *Politique étrangère* 12, no. 2 (1947): 228–229, at 229.
40 Jan O.M. Broek, "Review of *Les pays tropicaux*," *Pacific Affairs* 20, no. 3 (1947): 337–338. This thesis about bad government was pronounced in post-war American thinking about agricultural and rural development in tropical Africa. It was given a new lease of the life in the 1980s by the economist Robert H. Bates in his *Markets and States in Tropical Africa: The Political Basis of Agricultural Policies* (Berkeley: University of California Press, 1981), and is reassessed in British colonial contexts in Matthew Lange's, *Lineages of Despotism and Development: British Colonialism and State Power* (Chicago: University of Chicago Press, 2009). We shall see in the last chapter that in recent years new connections have been drawn between 'bad attitude' (government) and 'bad latitude' (environmental conditions).
41 A scientific assault led by botanists, agronomists, anthropologists and medical scientists as well as geographers. On American efforts, see Kelly Enright, *The Maximum of Wilderness: The Jungle in the American Imagination* (Charlottesville and London: University of Virginia Press, 2012).
42 See Elsa Viellard-Baron, "La jungle entre nature et culture: un imaginaire socio-spatial de l'antimonde," Doctorat de géographie, Université Paris 7-Denis Diderot, 2011;

Mark Harrison, *Climates and Constitutions: Health, Race, Environment and British Imperialism in India* (Oxford: Oxford University Press, 1999); David N. Livingstone, "Tropical climate and moral hygiene: The anatomy of a Victorian debate," *The British Journal for the History of Science* 32, no. 1 (1999): 93–110.

43 See Emmanuel de Martonne, "Géographie zonale: la zone tropicale," *Annales de Géographie* 55, no. 1 (1946): 1–18; Karl J. Pelzer, "Geography and the tropics," in Griffith Taylor ed., *Geography in the Twentieth Century: A Study of Growth, Fields, Techniques, Aims, and Trends* (New York: Philosophical Library, 1957), 311–344. On zonality, climate and regional classification, see, for example, "Classification of regions of the world: Report of a committee of the Geographical Association," *Geography* 22, no. 4 (1937): 253–282; Alfred Hettner, *Die Klimate der Erde* (Leipzig: Tuebner, 1930); Wladimir Köppen, *Handbuch der Klimatologie* (Berlin: Borntraeger, 1936); and Karl Sapper, *Die Tropen: Natur und Mensch Zwischen den Wendekreisen* (Stuttgart: Strecker und Schröder, 1923).

44 Hervé Théry, "Les pays tropicaux dans les livres de géographie: manuels de l'enseignement secondaire entre 1925 et 1960," *L'Espace Géographique* 17, no. 4 (1988): 299–306.

45 See, for example, Rodolphe Rougerie, "Trois ouvrages sur l'agriculture tropicale," *Annales de Géographie* 73, no. 398 (1964): 469–472, bemoaning the postwar chasm between Anglophone and francophone literatures on similar tropical problems.

46 Benjamin H. Farmer, "Review of *The tropical world*," *Geography* 38, no. 4 (1953): 337.

47 Pierre Gourou, *The Tropical World: Its Social and Economic Conditions and Its Future Status* trans. E.D. Laborde (London: Longmans, Green and Co., 1953), v.

48 In addition to the above, see J.H. Boeke, "Review of *Les pays tropicaux*," *Geographical Review* 37, no. 3 (1947): 513–514; L.P. Mair, "Review of *Les pays tropicaux*," *International Affairs* 23 (1947): 391; Charles Robequain, "Review of *Les pays tropicaux*," *Annales de Géographie* 57 (1948): 70–73; Robert Woolbert, "Review of *Les pays tropicaux*," *Foreign Affairs* 26 (1948): 113; G.R.C., "Review of *Les pays tropicaux*," *The Geographical Journal* 110, no. 1 (1947): 126; R.W. Steel, "Review of the Tropical World," *The Geographical Journal* 120, no. 2 (1954): 242–243; O.H.K. Spate, "Review of *The Tropical World*," *Economic Geography* 31, no. 2 (1955): 185–186; B.H. Farmer, "Review of *The Tropical World* 2nd Edition," *Geography* 44, no. 3 (1959): 213–214; P.R.C., "Review of *The Tropical World* 4th Edition," *Geography* 52, no. 2 (1967): 224–225.

49 Susanna B. Hecht, *The Scramble of the Amazon and the 'Lost Paradise' of Euclides da Cunha* (Chicago and London: University of Chicago Press, 2013), 421 (emphases in original).

50 Jean Suret-Canale, "Les géographes françaises face à la colonisation," in Michel Bruneau and Daniel Dory, eds., *Géographies des colonisations XV–XX siècles* (Paris: Harmattan, 1994), 155–169, reflecting on his earlier review, and how geographers largely ignored it. The relationship between the four cheap assets – or 'cheap natures' – of the colonial periphery flagged by Suret-Canale, and the logic of capitalist accumulation, have been theorised in greater depth within Marxist theory since he wrote. See, recently, Jason W. Moore, *Capitalism Within the Web of Life: Ecology and the Accumulation of Capital* (London: Verso, 2015), 17 and *passim*.

51 B.W. Hodder, "Review of *Leçons de géographie tropicale*," *Bulletin of the School of Oriental and African Studies* 37 (1974): 285.

52 Karl Peltzer, "Review of *The Tropical World*," *Pacific Affairs* 28, no. 4 (1955): 383–384.

53 For an overview of Gourou's influence on economics, see Joseph A. Tosi Jr. and Robert F. Voertman, "Some environmental factors in the economic development of the tropics," *Economic Geography* 40, no. 2 (1964): 189–205.

54 Jean Gottmann, "The bounty of the tropics," *Times Literary Supplement* no. 4163 (1983): 41–43, at 41.

55 Gilberte Bray. Interview with the authors.
56 Christian Taillard, Georges Courade, Gilles Sautter, François Durand-Dastès, Alain Durand-Lasserve and Michel Bruneau, "La géographie tropicale de Pierre Gourou et le développement," *Espace géographique* 13, no. 4 (1984): 329–337, at 332.
57 ACDF: Pierre Gourou, personal dossier – CV.
58 Although Said wrote a good deal about Césaire in his *Culture and Imperialism* (New York: Alfred A. Knopf, 1993) and subsequent work.
59 Edward W. Said, *Orientalism* (New York: Random House, 1978), 1–2.
60 Said, *Orientalism*, 5.
61 Said, *Orientalism*, 326.
62 Said, *Orientalism*, 55. For overviews of the impressions left by Said's Orientalism and Arnold's tropicality, see, for example, Daniel Clayton, "Critical imperial and colonial geographies," in Kay Anderson, Mona Domosh, Steve Pile and Nigel Thrift eds., *Sage Handbook of Cultural Geography* (London: Sage, 2003), 354–368; Derek Gregory, "Imaginative geographies," *Progress in Human Geography* 19, no. 4 (2005): 447–485.
63 Said, *Orientalism*, 325–326.
64 Said, *Orientalism*, 1–2, 71, 50.
65 Edward W. Said, "Invention, memory and place," *Critical Inquiry* 26, no. 2 (2000): 181; Said, *Orientalism*, 59.
66 Said, *Orientalism*, 1.
67 David N. Livingstone, "Race, space and moral climatology: Notes towards a genealogy," *Journal of Historical Geography* 28, no. 2 (2002): 159–180, at 160.
68 Felix Driver and Luciana Martins, "Views and visions of the tropical world," in Felix Driver and Luciana Martins eds., *Imperial Visions in an Age of Empire* (Chicago and London: University of Chicago Press, 2005), 3. And see Martin Mahoney, "Picturing the future-conditional: Montage and the global geographies of climate change," *Geo: Geography and Environment* 3, no. 2 (2016): 1–18, at 13; Felix Driver, "Imagining the tropics: Views and visions of the tropical world," *Singapore Journal of Tropical Geography* 25, no. 1 (2004): 1–17.
69 Adam Rothman, "Lafcadio Hearn in New Orleans and the Caribbean," in William Boelhower ed., *New Orleans in the Atlantic World: Between Land and Sea* (Oxford and New York: Routledge, 2013), 114.
70 Nancy Stepan, *Picturing Tropical Nature* (Chicago: Reaktion Books, 2001), 21.
71 David N. Livingstone, *Science, Space and Hermeneutics*, Hettner Lecture Series (Heidelberg: University of Heidelberg, 2002), 43 and *passim*.
72 David N. Livingstone, "Review of Jessica Howell, *Exploring Victorian Travel Literature*," *Social History of Medicine* 28, no. 4 (2015): 937–938, at 937.
73 Neil Safier, "The tenacious travels of the torrid zone and the global dimensions of geographical knowledge in the eighteenth century," *Journal of Early Modern History* 18, no. 1 (2014): 141–172.
74 On the tropical uncanny see Barbara Creed, "Tropical malady: Film and the question of the uncanny human-animal," *etropic* 10, no. 1 (2011): 131–140.
75 Paul S. Sutter, "The tropics: A brief history of an environmental imaginary," in Andrew C. Isenberg ed., *The Oxford Handbook of Environment History* (Oxford: Oxford University Press, 2014), 178–198, at 179.
76 David Ludden, "Orientalist empiricism: Transformations of colonial knowledge," in Carol A. Breckenbridge and Peter van der Veer eds., *Orientalism and the Postcolonial Predicament: Perspectives on Southeast Asia* (Philadelphia: University of Pennsylvania Press, 1993): 250–278, at 250.
77 Said, *Orientalism*, 45.
78 Said, *Orientalism*, 326.
79 Most trenchantly in Pierre Gourou, "Le déterminisme physique dans 'l'Esprit des lois'," *L'Homme* 3, no. 3 (1963): 5–11. See Roger Lévy, "Review of Pierre Gourou *Pour une géographie humaine*," *Politique étrangère* 38, no. 5 (1973): 647–648 for a broader summary of Gourou's "principled position."

80 See Sandra Harding ed., *The Feminist Standpoint Theory Reader* (New York and London: Routledge, 2004).
81 Marx cited in John Bellamy Foster and Brett Clark, "Ecological imperialism: The curse of capitalism," *The Socialist Register* (2004): 186–201, at 188–189; they deploy the expression "metabolic rift" as a shorthand for what Marx was saying.
82 David N. Livingstone, *Putting Science in Its Place: Geographies of Scientific Knowledge* (Chicago: University of Chicago Press, 2003), 140.
83 See Christopher Green and Frances Morris eds., *Henri Rousseau: Jungles in Paris* (London: Tate – Harry N. Abrams, 2006).
84 Frantz Fanon, *The Wretched of the Earth* trans. Richard Philcox; orig. pub. Eng. 1963 (New York: Grove Press, 2004); and see, e.g., Pierre Gourou, "Les Kikuyu et la crise Mau-Mau: une paysannerie au milieu du XXe siècle," *Les Cahiers d'Outre-Mer* 7, no. 3 (1954): 317–341, at 328.
85 Driver and Martins, "Views and visions of the tropical world," 16.
86 David Arnold, *The Tropics and the Traveling Gaze: India, Landscape and Science, 1800–1856* (New Delhi: Permanent Black, 2005), 31.
87 Max Weber, cited in Richard Jenkins, "Disenchantment, enchantment and re-enchantment: Max Weber at the millennium," *Weber Studies* 1, no. 1 (2000): 11–32. The theme of disenchantment imbues critical theories of modernity.
88 John Lyndsay-Poland, *Emperors in the Jungle: The Hidden History of the U.S. in Panama* (Durham, NC: Duke University Press, 2003), 191.
89 Eric Jennings, *Curing the Colonizers: Hydrotherapy, Climatology, and French Colonial Spas* (Durham, NC: Duke University Press, 2006), 23. And see Daniel Clayton, "Militant tropicality: War, revolution and the reconfiguration of 'the tropics' c.1940–c.1975," *Transactions of the Institute of British Geographers* 38, no. 1 (2013): 180–192; and Nicolás Wey Gómez, *The Tropics of Empire: Why Columbus Sailed South to the Indies* (Cambridge, MA: MIT Press, 2008).
90 James S. Duncan, *In the Shadows of the Tropics: Climate, Race and Biopower in Nineteenth Century* (Aldershot: Ashgate, 2007).
91 Earl Parker Hanson, *The Amazon: A New Frontier?* (New York: Foreign Policy Association, 1944), 5; William Beebe, "How lost jungle fighters can survive," *Popular Science Monthly* 142, no. 3 (1943): 49.
92 Arnold, *The Tropics and the Traveling Gaze*, 31–32. Also see Paul Carter, "Tropical knowledge: Archipelago consciousness and the governance of excess," *etropic* 12, no. 2 (2013): 92.
93 Monique Allewaert, *Ariel's Ecology: Plantations, Personhood, and Colonialism in the American Tropics* (Minneapolis: University of Minnesota Press, 2013), 4–7.
94 Haraway, "Situated knowledges," 575–599, at 579.
95 Haraway, "Situated knowledges," 581, 589.
96 Said, *Orientalism*, 246 (emphasis in original).
97 Said, *Orientalism*, 246.
98 Taillard, Courade, Sautter, Durand-Dastès, Durand-Lasserve and Bruneau, "La géographie tropicale de Pierre Gourou et le développement," 329–337.
99 Frederick Cooper, *Colonialism in Question: Theory, Knowledge, History* (Berkeley: University of California Press, 2005), 15.
100 Jean-François Klein and Marie-Albane de Suremain, "Clio et les colonies retour sur des historiographies en situation," *Romantisme* 139, no. 1 (2008): 59–80; Ruth Craggs, "Postcolonial geographies, decolonization, and the performance of geopolitics at Commonwealth conferences," *Singapore Journal of Tropical Geography* 35, no. 1 (2014): 39–55.
101 Clive Barnett, "Impure and worldly geography: The Africanist discourse of the Royal Geographical Society," *Transactions of the Institute of British Geographers* 23, no. 2 (1998): 239–252. Barnett does not mention Gourou in this article, although does mention that Césaire was critiquing Gourou at the start of his Oxford PhD dissertation, which has this title too.

34 *The tropics and the colonising gaze*

102 Michael Burrawoy, "Provincializing the social sciences," in George Steinmetz ed., *The Politics of Method in the Human Sciences: Positivism and Its Epistemological Others* (Durham, NC: Duke University Press, 2005), 508–525; Carola Dietze, "Toward a history on equal terms: A discussion of provincializing Europe," *History and Theory* 47, no. 1 (2008): 69–84.
103 Gennaro Ascione, *Science and the Decolonization of Social Theory: Unthinking Modernity* (London: Palgrave Macmillan, 2016), 61–62.
104 Haraway, "Situated knowledges," 579.
105 Edmund Siderius, "Donna Haraway and Hilary Putnam on god's spectacles," *Starry Messenger*, January 3, 2011, https://en.gravatar.com/edmundsiderius.
106 Jane Jacobs, "Introduction: After empire?" Section 6 in Kay Anderson, Mona Domosh, Steve Pile and Nigel Thrift eds. *Sage Handbook of Cultural Geography* (London: Sage, 2003), 345–353, at 350–351.
107 Derek Gregory, *The Colonial Present: Afghanistan, Palestine, Iraq* (Oxford: Basil Blackwell, 2004).
108 See, for example, Hecht, *The Scramble of the Amazon*; Gerry Kearns, *Geopolitics and Empire: The Legacy of Halford Mackinder* (Oxford: Oxford University Press, 2009); Innes M. Keighren, *Bringing Geography to Book: Ellen Semple and the Reception of Geographical Knowledge* (London: I. B. Tauris, 2010); Robert J. Mayhew, *Malthus: The Life and Legacies of an Untimely Project* (Cambridge, MA: Harvard University Press, 2014); Neil Smith, *American Empire: Roosevelt's Geographer and the Prelude to Globalization* (Berkeley: University of California Press, 2004).
109 Jake Hodder, "On absence and abundance: Biography as method in archival research," *Area* 49, no. 4 (2017): 452–459, at 453.
110 See Trevor Barnes, "Lives lived and lives told: Biographies of geography's quantitative revolution," *Environment and Planning D: Society and Space* 19, no. 4 (2001): 409–429. Also see Trevor Barnes, "Obituaries, wars, corporal remains, and life," *Progress in Human Geography* 31 (2009): 693–701.
111 Barnes, "Lives lived and lives told."
112 Paul Claval, "Pierre Gourou," *Geographers: Biobibliographical Studies* 25 (2006): 60–80, although it should be noted that this, at the time, was generally the approach taken to these biobibliographical studies.
113 See Jean Gallais, "L'evolution de la pensée géographique de Pierre Gourou sur es pays tropicaux (1935–1970)," *Annales de Géographie* 90 (1981): 129–149; Olando Ribeiro, "La pensée géographique de Pierre Gourou," *Annales de Géographie* 82, no. 449 (1973): 1–7; Jean-Pierre Raison, "L'œuvre de Pierre Gourou," (avec Sylvain Allemand et Paul Pélissier) Géoblog-géographie, http://geoblog.canalblog.com/archives/2011/11/index.html. The quotation is from Gilles Sautter, "On a recent book: The geographical system of Pierre Gourou," *Social Science Information* 15, no. 2/3 (1976): 227–228, at 227.
114 Michel Bruneau, "Pierre Gourou (1900–1999): Géographie et civilisations," *L'Homme* 153 (2000): 7–26, at 7, 20.
115 See, especially, Ribeiro, "La pensée géographique de Pierre Gourou," 3.
116 Pierre Gourou, "Deux geographes: Paul Pelissier, Gilles Sautter," *Tropiques: Lieux et liens: Florilège offert à Paul Pelissier et Gilles Sautter* (Paris: Editions de 'ORSTOM, 1989), 23.
117 Pierre Gourou, "La géographie comme 'divertissement'? Entretiens de Pierre Gourou avec Jean Malaurie, Paul Pélissier, Gilles Sautter, Yves Lacoste," *Hérodote* 33, no. 1 (1984): 50–72, at 68.
118 See, for example, Jean-Pierre Raison, "Nommer, c'est créer un peu. De 'tiers-monde', à 'tropicalisme': les avatars d'un vocabulaire," *Autrepart* 41, no. 1 (2007): 57–68.
119 Gourou's *Terres de bonne espérance, le monde tropical* (Paris: Pion, 1982) is a semi-autobiographical work, and one, Yves Lacoste told us, which was produced at the bidding of the French explorer Jean Malaurie. Yves Lacoste, interview with the authors.

120 John Kleinen, Correspondence with the authors, 10 September 2005.
121 Pierre Gourou, "Obituary: Roger Dion, 1896–1981," *Journal of Historical Geography* 8, no. 2 (1982): 181–182, at 182.
122 Gilberte Bray, Interview with the authors.
123 Paul Pélissier, "Pierre Gourou, 1900–1999," *Annales de Géographie* 109, no. 612 (2000): 212–217, at 216.
124 The Grand Orient de France was France's largest masonic organisation, and was, for many, a progressive secular force that clung to order, stability and ritual. Jean-Pierre Raison, Interview with the authors.
125 Georges Courade, Interview with the authors.
126 Olivier Orain, "Le plain-pied du monde: Postures épistémologiques et pratiques d'écriture dans la géographie française au XXe siècle," PhD Geography, Universite Pantheon-Sorbonne – Paris I, 2003.
127 See Gourou's reflections, pp. 111–114, in "La bibliothèque imaginaire du collège de France," *Bulletin des bibliothèques de France* [en ligne], n° 6, 1990 http://bbf.enssib.fr/consulter/bbf-1990-06-0408-001. The quoted phrase is from Martin Jay, *The Virtues of Mendacity: On Lying in Politics* (Charlottesville and London: University of Virginia Press, 2010), 32, who glosses this posture in politics and academia.
128 Lévy, "Review, *Pour une géographie humaine*," 647–648.
129 Gilbert Etienne, "Pierre Gourou (1900–1999): Le terrain, le vent du large, l'homme," in Christophe Gironde and Jean-Luc Maurer eds., *Le Vietnam à l'aube du XXIe siècle* (Paris: Karthala, 2004), 18. Gourou's starts each of the chapters of his last major work, *Riz et Civilization*, with epigraphs from these and other (mainly French) authors.
130 Lévy, "Review, *Pour une géographie humaine*," 648.
131 J.K. Wright, "Terrae incognitae: The place of imagination in geography," *Annals of the Association of American Geographers* 37, no. 1 (1947): 1–15, at 7.
132 Henri Nicolaï, Interview with the authors.
133 Gourou noted the important use he made of a "small Leica" during his fieldwork in the Tonkin Delta – probably a Leica I (released in 1930). He bought a new and "simpler" model in 1946 – probably the widely used Leica Standard (from 1933). In the 1950s French colonists and troops used a mixture of American, British and French khaki and 'green jungle' drill. Nicolaï, Interview with the authors; and see Marianne Hulsbosch, *Pointy Shoes and Pith Helmets: Dress and Identity Construction in Ambon from 1850 to 1942* (Honolulu: University of Hawai'i Press, 2014).
134 Étienne Gilbert, "Review of *Pour une géographie humaine*," *Tiers-Monde* 14, no. 56 (1973): 890.
135 Stuart Hall, "When was 'the post-colonial'? Thinking at the limit," in Ian Chambers and Linda Curti eds., *The Postcolonial Question: Common Skies, Divided Horizons* (New York and London: Routledge, 1996), 242–260, at 249.
136 Hecht, *The Scramble of the Amazon*. Cf. Candice Slater, *Entangled Edens: Visions of the Amazon* (Berkeley: University of California Press, 2001); Arnold takes a similar approach, with respect to India, in *The Tropics and the Traveling Gaze*.
137 Frances R. Aparicio and Susana Chávez-Silverman, *Tropicalizations: Transcultural Representations of Latinidad* (Dartmouth: University Press of New England, 1997), 8.
138 Timothy Mitchell, *Rule of Experts: Egypt, Techno-Politics, Modernity* (Berkeley: University of California Press, 2002).
139 See, recently, Daniel Clayton and Trevor J. Barnes, "Continental European geographers and World War II," *Journal of Historical Geography* 47, no. 1 (2015): 11–15.

2 Tropicalising Indochina

The tropics in an age of war and empire, 1900–1950

In June 1949 the Parisian weekly magazine *France Illustration* published a special issue on Indochina featuring 20 lavishly illustrated articles by a distinguished cast of politicians, administrators, military figures and scholars.[1] The publication sought to raise public awareness about this distant and beleaguered colonial region, where French troops were locked into a bloody and perilous conflict with the Viet Minh, the political and military front forged by Ho Chi Minh in 1941 which had precipitated North Vietnam's declaration of Independence as the Democratic Republic of Vietnam (DRV) in September 1945 and prompted France to mobilise 50,000 troops (the bulk of them from its empire rather than France itself) to defend its possession and national honour. The special issue appeared at a critical juncture in the Indochina War and points to a longer colonial history and turbulent process of decolonisation.[2] Between 1900 and 1950 war touched the masses on an unprecedented scale, chiefly in the form of two world wars, and the tropics were brought home to the general public of Western countries like never before in an array of commodity forms, and at a string of colonial and world exhibitions (35 in total) that extolled national and imperial prowess, and attracted many millions of visitors to "fleeting cities" of their own within Western and colonial capitals.[3]

In this chapter we use this *France Illustration* special issue to discuss how, during the first half of the twentieth century, Indochina became 'tropicalised' – identified and experienced as distinctly (and in many ways troublingly) tropical. One of our touchstones in what follows is the French geographer Pierre Gourou, who was born in 1900, lived in Indochina between 1926 and 1935, became an authority on the peasant culture and landscape of the Tonkin Delta – the 15,000 square mile area in northern Vietnam stretching from northwest of Hanoi to the coast – and contributed to this publication. In the next chapter we shall probe how Gourou aestheticised the Tonkin Delta and constructed an 'affirmative tropicality' and romantic geography, but we will start to divulge elements of this geography later in this chapter. Our broader aim here, however, is to reveal how Gourou's imagery serves as a foil to the largely negative (downbeat) tropicality of French settlers, travellers and soldiers, and thus to treat tropicality as a fluid and fractured site of difference with a range of meanings and intensities, and to think about how, during the first half of the twentieth century, this discourse operated within a public age of war and empire.

This leads us to a final, perhaps coincidental but we think telling, introductory point. 1949 was also the publication date of Fernand Braudel's *chef d'ouevre*, *La Méditerranée et le Monde Méditerranéen à l'Epoque de Philippe II*, which some see as the most important historical work of the twentieth century, and where Braudel conjures with three levels of history: *événement* (pertaining to the short term – days and months – the history of occurrences, battles and political events); *conjoncture* (the medium term – from decades to centuries – which he associated with the economic and social history of capitalism); and *structure* (the *longue durée* – from hundreds to thousands of years – the large scale and long-term history of environmental and social change). Braudel famously used an ocean metaphor to capture the operation of his three levels, with *structure* the almost imperceptible movement of water in the ocean's depths; *conjoncture* the movement of tides and currents; and *événement* the froth formed by waves.[4] He was particularly drawn to the second and third of these levels, and his sense of them was marked deeply by his experience of war and captivity. It was from World War II German prisoner of war camps, and writing largely from memory, that he sent "school copy book after school copy book" to his mentor Lucien Febvre, which formed the basis of his study. "All those occurrences [*événements*] which poured in upon us from the radio and the newspapers of our enemies, even the news from London," he recalled, "I had to outdistance, reject, deny. . . . Down with the occurrences, especially the vexing ones! I had to believe that history, destiny, was written at a much more profound level. . . . So it was that I consciously set forth in search of a historical language . . . in order to present unchanging (or at least very slowly changing) conditions."[5]

It is possible to detect the operations of each of Braudel's levels in the diverse contributions to the *France Illustration* special issue. Many of the contributors were concerned with the short and medium terms: what was happening in the summer of 1949; and how the military conflict had been precipitated by decades of colonial turmoil. Gourou's contribution – which we shall get to later in the chapter – is interesting because it operates at the level of *structure* and in so doing (and in disguising the other two levels) raises important questions about how or whether tropicality works as a colonial discourse of "anti-conquest," to borrow Mary Louise Pratt's term: as a passive but nonetheless culpable intellectual project of othering involving "strategies of representation whereby European bourgeois subjects seek to secure their innocence in the same moment as they assert European hegemony."[6] As Nancy Stepan has shown in a similar vein, tropicality was – is – envisioned and built with texts, images and spectacles as well as with political economy and colonial violence (slavery, plantations, plunder and profit), and has been shaped by artists, travellers and scientists as well by entrepreneurs, settlers and militia. Artistic and scientific connoisseurs of the exotic produced and exploited distinctions between the temperate and the topical, and Stepan insists that their textual and visual outputs were "not just *illustrations* of tropical nature but its *argument*."[7] Travel narratives, scientific treatises, paintings, sketches and maps were part and parcel of the West's imperial quest to unlock the mysteries and riches of foreign realms, and to arrogate to the Western observer or pundit the power to decide on what counted as right, normal and true (and what counted as aberrant and outlandish).[8]

Gourou and Braudel were colleagues at the Collège de France and good friends.[9] During the three-month stint each year that Gourou commuted from Brussels to Paris (usually staying between a Tuesday and Friday) to give his annual lecture course at the Collège, he would have lunch with Braudel most Wednesdays at a Vietnamese restaurant just off the Rue St Jacques.[10] Braudel referenced Gourou's work in many of his books, and Gourou developed an antipathy towards 'occurrences' too, and for some of the same reasons as Braudel.[11] But *structure* had different connotations in colonial Indochina than it had in the Mediterranean and history of European civilisation. In this colonial context, we shall see, structure was linked to exoticism, and Gourou's use of the idea was more akin to the structuralism of Claude Lévi-Strauss (another of his Collège de France colleagues) and his interest in 'societies before power.'[12]

'Jungle capitalism'

With regard to a medium-term understanding of tropicality, Indochina became tropicalised in the process of being colonised, commodified and beset by war. These drivers stretched back to what Ricardo Padrón describes as the sixteenth-century "invention of tropics" as a "rich, populous, fertile zone ripe for the taking by Europeans," and a zone that was taken violently.[13] From then on, this invention – what we are calling tropicality – became yoked to capitalist and imperialist expansion, and machinations of conflict and disease, and was trumpeted as the 'conquest of nature.' Western consumers' growing taste for, and ultimately dependence upon, a suite of tropical commodities – bananas, coffee, cocoa, copper, cotton, oil, rice, rubber, sugar, tea, timber, tin and tobacco– implicated them (however unwittingly) in imperial projects. Tropical colonies and dependencies, and the extensive and exploitative trading and labour systems tying them to the temperate West, lay at the heart of what Corey Ross describes as "a world of goods" based on an "ecology of colonial extraction."[14]

By 1900, Ross continues, "The European claim to mastery over nature was a central legitimising prop of modern imperialism, one that not only resonated with contemporary notions of racial hierarchy and societal evolution, but that also nourished a belief in the right, even duty, of Europeans to govern those who were less capable [in Europe's estimation] of controlling the world around them."[15] The tropics became turned into a scientific object and integrated into the West's "calculative hold" over the world, Christophe Bonneuil continues, and with the *Jardin d'Agronomie Tropicale*, established in 1899, and other botanical gardens serving as "laboratories of a domesticated tropicality."[16] Over the next 50 years there emerged what Ross construes as "a distinctly interventionist form of colonialism that spanned the realms of agriculture, forestry, resource extraction, conservation, and public hygiene."[17]

This promethean ability to exploit the tropical world, along with the capacity of Europeans to ameliorate the hitherto deleterious impact of tropical diseases, was facilitated by late nineteenth-century advances in medicine, science and technology. In 1898, for instance, the American writer Trumbull White introduced his account of *Our New Possessions* (Cuba, Puerto Rico, Hawaii and the Philippines) by representing them as tropical and noting: "It has been my special pleasure, as well as a care, in all my tropical journeys in the West Indies and in the islands of the south and west Pacific alike, to study the methods of tropical agriculture and the care and profit of crops strange to farmers of the United States. In all of these islands one finds sugar and coffee and tobacco, with a host of products of lesser importance but equal interest and sometimes great profit."[18] Those 'lesser' products included oil and rubber, and White saw a direct connection between their exploitation and the extension of political and economic "liberty" (democracy and trade) to the tropics. With respect to the Philippines, the largest of these 'possessions,' White downplayed the problems that climate and disease posed to white settlers and steered the conversation back to the archipelago's "magnificent virgin forests" and "astonishingly fertile soil."[19] One of the keys to reaping such rewards was the advent of cheap land (rail and road) and sea (river and oceanic) transportation, which allowed 'trade' in tropical commodities to be rendered on an industrial scale.

Another American, Frederick Adams, wrote in 1914 of the "conquest of tropics" by the United Fruit Company (formed 1899), which built a banana empire in Central and South America, making the fruit an affordable mass market staple.[20] Peter Chapman dubs this and other companies that targeted the tropics and made its produce part of daily life in the West – companies such as the Firestone Rubber Company – "jungle capitalists" that recognised yet largely ignored the deleterious environmental consequences of their practices – soil erosion, deforestation and habitat destruction.[21] For these and other tropical enterprises, and including those attempting coffee, rice and rubber export production in Indochina, scholars have shown that 'the tropics' were produced and placed in a tug of war between capitalist and conservationist, and colonial and humanitarian, interests and ideals.[22] "It is clear today," Richard Grove argued in 1995, "that modern environmentalism, no longer exclusively a product of European and North American philosophy, has emerged as a direct response to the socially and ecologically destructive conditions of colonial domination."[23] Colonial experience and local knowledge played a much more significant (if often haphazard and incomplete) role in the constitution of global science than metropolitan scholarship, until quite recently, acknowledged. And, as Richard Tucker sees matters, histories and critiques of Western consumers' "insatiable appetite" for tropical staples led not only to the domestication of tropical ecosystems (the capitalist and colonialist production of tropical nature) but also precipitated their degeneration and an unprecedented (and some areas irreversible) loss of biodiversity.[24]

Latter-day debates about the benefits, gains and sins of this 'jungle capitalism' are still imbued with agendas and perspectives surrounding development, decolonisation and environmentalism that came to fruition in the late 1960s and 1970s.

For instance, in his 1971 study of French colonialism in tropical Africa during the first half of the twentieth century, the historian Jean-Suret-Canale questioned how the "obligation to produce more without being able to use any supplementary acreage, and often over an area reduced by concessions to European settlers and classified forests, led to a speeding-up in [crop] rotation, a decline in yields, and often the permanent ruin of the soil along with a reduction in pastoral land."[25] He had also witnessed the Dakar-Niger railway strike of 1947, which involved Senegalese railway workers demanding the same rights as their French counterparts and prompted the French to accept that the colonial labour system needed to "involve African agency as much as imperial design," as Frederick Cooper sees matters.[26] All of this was important to Suret-Canale too, but the strike also alerted him to the significance of transportation and port development in tropical development under imperialism and following independence.

On the other hand, the Saint Lucian Nobel Prize–winning economist Arthur Lewis, writing around the same time as Suret-Canale, bemoaned the "sombre" tone cast in the literature on tropical development. "Terms which occur most frequently," he noted, "are 'colonialism,' 'monoculture,' 'exploitation,' 'low wages,' 'sources of raw materials,' 'drain,' 'peripheral,' 'unstable.' These terms are appropriate and true, but they represent only one aspect of what took place," he continued, referring to the period 1880–1913. Then, and in a subsequent era of independence, tropical countries that were in a position to respond to the demand for exports "did well" – "their modern sectors grew as rapidly as the modern sectors of Western Europe" – and the availability of fertile agricultural land as well as mineral resources was key. But the subsistence sector was so large in many tropical countries that these nations necessarily became "late starters" in the field of economic development, Lewis went on, and continued to be beset by a colonial "enclave" mentality by which "developed" spaces and sectors became isolated from, and did not benefit, the vast majority of the population.[27]

However, Suret-Canale and Lewis, and other advocates and critics of tropical development alike, recognised that the exploitation of the tropics had natural and human limits, and that in places capitalist and colonial schemes of development became counter-productive. During the 1920s and 1930s British and French researchers and administrators in tropical Africa and the Indian subcontinent became concerned with environmental problems arising from short-sighted agricultural and engineering projects (and as we show in Chapter 6, Gourou raised similar concerns in the post-war era).[28] In the *France Illustration* special issue, one of Indochina's leading economic advisers, Paul Bernard, warned that the communist "revolutionaries'" promise to deliver rapid development and improve peasant standards of living rested on a weak agricultural, industrial and financial platform. Indochina was rich in rice and timber, ores and minerals (especially coal), and Bernard noted that if the conflict abated France's large rubber plantation on the red soils of Cochinchina and Cambodia might be restored to profitability. But the region remained an overwhelmingly peasant, rice-growing domain, and "a weak people" thus defined, he snipped, "is incapable of controlling its own destiny."[29]

Empire and exhibitionism

Nor, for either older or newer imperial powers such as the United States, did this 'conquest' of nature dissipate concerns about the tropics as 'the white man's grave.'[30] While British visitors to the African and Asian pavilions at the 1924 British Empire Exhibition enjoyed "spectacles of exotic people and landscapes while seeing their imagined self-image reflected in representations of progress . . . [and new] travel possibilities," Daniel Stephen notes, the spectre of disease (especially cholera, dysentery, malaria and yellow fever) spoiled the narrative and prompted organisers to recoup the idea of progress with elaborate spectacles of scientific strength and achievement regarding the medical conquest of the tropics.[31]

This world, ecology and self-image did not operate in a bubble of trade, profit and progress. Frantz Fanon argued that it was also integral to the core "relationship . . . of physical mass" between colonisers and the colonised the world over, and to the deep-seated anxieties and violence of empire.[32] In many parts of the colonial world, and particularly in tropical Africa and Asia, colonists were vastly outnumbered by the indigenous population, and colonial conquest inevitably fomented native anger and resentment. Colonisation was marked by the constant threat and myriad instances of violent rebellion and imperial reprisal. The conquest of the tropics involved not only the plunder of natural resources, but also the exploitation of labour, the coloniser's disdain for the colonised, and elaborate (and for Fanon always bogus) defences of colonialism as a 'civilising mission' that would at one and the same time unleash the economic potential of 'backward' regions and justify the use of violence against 'savage' and 'superstitious' peoples. The political economy of tropical development, and the brutal trading and labour regimes by which it operated, was situated in a wider culture of empire based on division, hierarchy and derision.

Fanon urged that apprehension and vulnerability turned the colonist into "an exhibitionist" who needed constantly to remind the colonised "Here I am master," and, consequently, kept the colonised in "a state of rage."[33] This dynamic – or double inferno of mastery and anxiety, resistance and reprisal – was aided and abetted by disciplinary and governmental (ordering and securitising) practices of inquiry, survey, display, and divide and rule, and with many of them buttressed by military power and scientific organisation. "By deploying technosciences such as military organization and modern agriculture in its colonies," Milton Osborne shows with reference to Algeria and Indochina, "France gained the skills and knowledge which allowed it to accomplish the task of military recruitment. Intent on extracting the wealth of the colonies, France experimented with various methods of farming, and organizing and disciplining colonial labor."[34] In Indochina, as in Algeria, the French grafted these colonial technosciences on to indigenous forms of hierarchy and rule.

It was partly on account of the nature and difficulty of this grafting – i.e., the question of how or whether colonisers were able to wield power on their own terms – that Fanon's 'exhibitionism' was also refracted through an array of media – including, by 1900, grandiose world exhibitions – that solicited the public to witness, participate in, consume and ultimately condone empire as a mass spectacle. The world exhibitions became powerful metropolitan sites at which superiority and anxiety were conveyed, colonial guilt was deflected, and the vulnerable expanses of empire were recouped into a more orderly and manageable scene.[35] They embodied "the exhibitionary spirit" of the *fin de siècle*, Emily Rosenbery observes: "They presumed to present tours of the world, but each was a tour confined to a constricted space." Western cities vied to stage them, and in spite of their fleetingness – they opened for just a few months – they were "major cultural enterprises of global significance because their representations projected powerful imaginaries about the world, its diverse cultures, and its interconnectivity and divisions."[36] One of the most potent ways in which these binds and tensions of empire were negotiated was through food. Culinary figureheads, such as Escoffier in France, did their imperial bit by flagging the exotic taste and nutritional value of colonial ingredients (although far less often dishes). Rice was one such foodstuff that the great Escoffier praised as a nutritious household and restaurant supplement to bread and potatoes, and the colonial lobbyist Pierre Cordemoy made the chef's endorsement central to his 1928 pamphlet *L'alimentation nationale et les produits coloniaux*, which proposed that a more varied – by which he meant colonial and tropical diet – was a remedy to many domestic social ills.[37]

However, by 1949 the empire was eating France and a new era of decolonisation and development was dawning. Over the preceding 50 years, colonial exhibitionism had involved a great range of investigations, schemes, fantasies, projections, objects, texts and images, and the world exhibitions were not the only spaces in which the profits and ruses, bonds and fissures, and splendours and sins of empire were displayed, debated and deflected. Texts such as the 1949 *France Illustration* special issue, to which we now turn, were also important in this regard.

Illustrating colonial crisis

This publication appeared at a pivotal moment in the Indochina War and was a spectacle – performance – in its own right. Diplomatic negotiations over the future of the region had broken down, and French government ministries were at loggerheads about the way forward. The DRV had denounced the rule of Bao Dai, the last emperor of the Nguyen Dynasty, who had been installed as a colonial monarch (albeit a titular figurehead) by the French, had abdicated in 1945, and in the summer of 1949 was in the process of being restored to power by his old colonial masters as the head of a new Associated State of Vietnam. The Viet Minh was consolidating its grip on the north with the help of land reforms and the promotion of the Romanised Vietnamese script *quoc ngu* as a national language, and extending its liberation efforts in the south with a campaign of guerrilla sabotage, particularly around the capital, Saigon.[38] Its pursuit of the French and their southern Vietnamese collaborators was made possible by hundreds of thousands of civilian porters.[39] However, this push was compromised by the lingering effects of the devastating famine of 1944–1945, which sent rice production plummeting

Tropicalising Indochina 43

to pre-war levels and left between 400,000 and 2 million people dead from starvation and disease (mainly in the north).⁴⁰

In a wider international sphere, Chinese communists were supplying the Viet Minh with arms and military advice, and in May 1949 the French Army's chief of staff, General Georges Revers, was dispatched to the region, both to evaluate the military situation and morale of French troops, and to gauge the strategic threat posed by Mao Zedong's impending defeat of Chiang Kai-shek's nationalist army. Revers' mission was earmarked 'secret' but was soon leaked to the press, and the Viet Minh quickly turned the general's downbeat assessment of France's military prospects into useful propaganda.⁴¹ There was growing concern in Paris that France's hefty military force in Indochina was being "whittled away ingloriously and futilely in the kind of war they are being forced to fight" against the DRV and the military mastermind of its People's Army of Vietnam (PAVN), Võ Nguyên Giáp. The problem facing the French, Giáp reflected, was that they never fully ascertained "whether it would be easy for them to return to North Vietnam by force."⁴²

What had started, in 1946, as a low-intensity and scattered guerrilla insurgency was morphing into a fully fledged conventional war between well-equipped regular armies, and with the United States funding France's war effort.⁴³ It has been estimated that by the end of the war, in 1954, around 100,000 lives had been lost on the French side and 300,000 on the Vietnamese side.⁴⁴

In the summer of 1949 Ho Chi Minh eschewed too close an association with the Chinese communists, lest it alienate moderate DRV nationalists who feared that their hard-won independence might herald by a new form of foreign domination. By October 1949, however, the DRV Council of Ministers was making a formal approach to China for economic and military assistance, and Giáp was furnishing the PAVN with conventional divisions (although it never abandoned guerrilla tactics) and planning a General Counter Offensive (in accordance with the third phase of Mao Zedong's strategy of People's War) against the French.⁴⁵ In the estimation of the American Central Intelligence Agency, which kept close tabs on these developments, "the major underlying motives and justifications of the French in refusing the demands (from both the Resistance and Bao Dai) to deliver full sovereignty are, in order of importance: (1) a desire to perpetuate French prestige; (2) a wish to protect local French interests; (3) a desire to assist in containing Communism in Asia; and (4) a belief that the Vietnamese are not yet ready to govern themselves."⁴⁶ In March 1949 the U.S. State Department had advised President Truman that "the choice confronting the United States is either support the French in Indochina or face the extension of Communism over the remainder of the continental area of Southeast Asia and possibly farther westward."⁴⁷ It was prophetic advice.

In short, the situation was fluid, fast-changing and by no means just regional in meaning or scope. "So few problems [in the Far East] are isolated," U.S. Secretary of State Dean Acheson reflected in a widely reported speech of 21 October 1949. "Most are part of a very complicated mosaic."⁴⁸ Mao had proclaimed victory in China just a few weeks before, and the Soviet Union completed its first successful nuclear weapons test in late August. In October 1949 it was still unclear whether China and the Soviet Union would adopt the same view of Hanoi's struggle to oust the French.[49] However, Acheson saw an urgent need to make Indochina both a bulwark against the further spread of communism, and an economically secure region (one rich in coal and rice) that might aid the post-war recovery of Japan.[50] Meanwhile, Revers would propose that the increasingly worn down and demoralised French Expeditionary Corps in Indochina retreat from its dispersed mountain outposts along the Chinese border, where it had been particularly exposed to Viet Minh guerrilla attacks, and regroup in the densely populated and watery Tonkin Delta. A defensive line of fortified posts, artillery and checkpoints, supported by air cover, might be established there while the French sought a diplomatic solution and American aid, Revers advised.[51]

Revers hinted that American support might be decisive, that there was too little concern about the conflict in France, and that it was likely to be protracted and inconclusive. Revers pointed up a military dilemma for the French that would last until the war's end: either try to defend remote mountain regions in the north of Vietnam, where the Viet Minh were at their strongest, and thus weaken the ability of the French force in the delta to interdict DRV exports of coal and rice, and its movement of supplies between the north and south; or cede the mountains to Giáp in order to secure the delta but then suffer the consequences of the Viet Minh's unhampered opportunity to recruit, and to develop what, in time, became called the 'Ho Chi Minh Trail' – the elongated strategic supply route in the mountainous borderlands of Vietnam, Cambodia and Laos (mostly in Laos) connecting north and south Vietnam (which became the most bombed military passageway in history).[52] Giáp noted that from 1949 to France's cataclysmic defeat at Điện Biên Phủ in 1954, he preyed on his enemy's predilection to "concentrate their forces," hitting them at "points they had left relatively unprotected," and "oblig[ing] them to scatter their troops all over the place in order to ward off our blows."[53] Moreover, and as we shall soon see, the longer the conflict dragged on the more the alien tropical environment of Indochina weighed on the minds and bodies of French officers and troops.

It was against this complex regional and international backdrop that parts of the French press, including the founder of *France Illustration*, Georges Oudard, who was close to Charles de Gaulle, sought to rally public support for the war. The special issue was bookended by adverts for Indochinese products and companies, pronouncements from the former Governor-General of Indochina (1912–1914, 1917–1919) and incumbent President of the Assembly of the Union Française, Albert Sarraut, and the High Commissioner of la Fédération indochinoise, Léon Pignon, who had been pivotal to the formation of the Fédération.[54] Between 1945 and 1949 both Sarraut and Pignon worked hard to keep Indochina in the French imperial fold. Sarraut began his article by underlining the history and significance of "Franco-Vietnamese symbiosis," which he traced back to creation of the Union Indochinoise (the political name for French Indochina) in 1887, which integrated the northern, central and southern regions of what was then known as Dai Nam – Tonkin, Annam and Cochinchina, respectively – and combined them with Cambodia following a series of military conquests between 1858 and 1883. Laos was added in 1893.

The lie of imperial association

French power in the region had its roots in imperial prospecting. Nguyen Thi Dieu writes of the dream of nineteenth-century Western colonisers "to locate and penetrate the mythical China market with its millions of potential consumers" and how it was the French, upon searching for a way into China, who "stumbled inadvertently on the Mekong River, and then on the Red River, leading them to conquer Vietnam."[55] Even so, Sarraut began his piece in the special issue with the idea of symbiosis and by linking it to the colonial doctrine of *mise en valeur*, which he forged during his time as Governor-General and then as France's Minister of Colonies (1920–1924).[56] He coined this term – literally meaning a duty to and spirit of improvement – to flag France's commitment to develop rather than simply exploit its colonies and "civilise" their colonial populations.[57] The doctrine was both a practical and ideological response to the realisation that expansive colonial empires like France's struggled to gain the acquiescence of their colonial subjects to their imperial plans without the participation of native leaders and go-betweens.[58] As Martin Thomas notes, *mise en valeur* sprang from "a republican vision of imperialism predicated on collaboration with indigenous elites in the day-to-day management of colonial affairs."[59]

In Indochina and elsewhere in the French empire, however, the hope was that programmes of economic, educational and governmental reform and modernisation would dissipate anti-colonial liberation movements and create a more productive (albeit largely *corvée* – unfree) labour force and efficient colonial administration. Sarraut genuinely believed in an enlightened colonial humanism that was based on collaboration with Indochinese elites (both a traditional mandarin class, and Indochinese civil servants recruited and trained by the French), and expedited it through what became called the "Bao Dai solution": he resurrected a titular colonial monarchy under French suzerainty as a means of winning the hearts and minds of the Indochinese population.[60] Yet as the French Orientalist Paul Mus noted in 1949, the village-oriented political arrangements of rural Vietnam, where over 80 per cent of the population lived, had seemed particularly inscrutable to the French and intractable to manage by 'association.'[61]

In practice, 'association' and its governmental corollary, indirect rule, did not reach nearly as far as Sarraut envisioned, never amounted to equal partnership or greater political participation for the vast majority of the population, and historians have queried its difference from earlier colonial policies of 'assimilation' (direct rule and often limited and brutal conversion to French ways).[62] Ultimately, distinctions between colonisers and the colonised were defended rather than disavowed, and the French frequently failed to recognise differences between different colonial populations and ethnic groups. Sarraut's vision of Franco-Indochinese symbiosis was "nothing more than a bluff," Christopher Goscha concludes: "As in Algeria, the French in Indochina were largely opposed to according metropolitan

46 *Tropicalising Indochina*

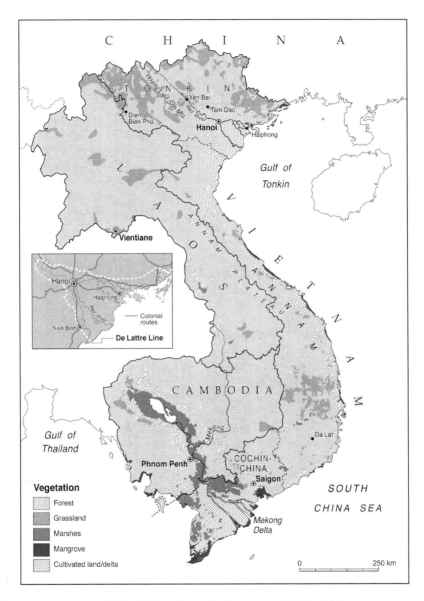

Figure 2.1 Indochina 1900–1945 (with the De Lattre Line, 1951) (G. Sandeman)

citizenship to members of the native elite or to broadening the native electorate."[63] *Mise en valeur* was a potent and flawed example of what postcolonial theory invokes as the contradictory double-time of post-enlightenment modernity – extolling equality as an "elementary virtue" of progress on the one hand, but suspending and constricting its attainment by the colonised on the other, as Leela Gandhi puts it.[64] Educational, health care, and road and rail projects that were much touted by the French as evidence of 'improvement' had limited reach and served the French first; and Indochinese understanding of the difference between what was promised and what was delivered stoked resentment. Hugues Tertrais sees the hesitancy with which the French unrolled its plan to electrify Indochina beyond French urban enclaves as emblematic in this regard.[65]

In the eyes of the small but vocal French settler community in Indochina, Goscha continues, "the extension of even limited voting rights to the most 'civilized' Vietnamese raised the specter of the 'native masses' running the Europeans out of the colony – legally."[66] By 1940, around 80 per cent of the Indochinese population remained illiterate, and there were only 2 doctors for every 1,000 people. This lack of 'improvement' was particularly stark in the countryside, where French rule had helped to consolidate the power of Vietnamese landlords and elites, and create a new class of tenant and landless peasants. Talk of reform rang hollow, and rebellion against colonial land, tax and *corvée* regimes, particularly in the peasant rice, plantation and mining regions of Vietnam, was met with violent repression by the French. In any event, the French were vastly outnumbered in Indochina. At the time of the first systematic census, in 1937, the French community consisted of just 36,000 "nationals by right of birth" out of a total population of 23 million, and with around 19,000 of these nationals born in France and 15,000 in Indochina. As the French geographer Charles Robequain reported, the French were concentrated in Cochinchina and Tonkin, but even there they comprised just 0.35 per cent and 0.2 per cent of the total population, respectively. Over a half of this French community were in the armed forces. But there was also a significant community (around 20 per cent of the total) of *fonctionnaires* (civil servants), most of whom resided in Indochina's two main cities of Hanoi (Tonkin) and Saigon (Cochinchina), and smaller groups (each of around 10 per cent) of *professions libérales* (teachers, scholars and missionaries), and colonists engaged in the banking and the commercial sector, agriculture, forestry, manufacturing and mining.[67]

Sarraut's promotion of *l'idée coloniale* and *la plus grande France* (a Greater France) came at a time of mounting public scrutiny of empire within France and its colonies. Hundreds of thousands of Indochinese draftees and volunteers had fought in the trenches, or (more commonly) worked in French factories, during World War I.[68] Gandhi invokes "the vexed migration of soldiers from colonial outposts to the battlefields of Europe and back again as the prompt for an intercivilisational ethics devoted to rescuing the endangered spirit of Europe from the clutches of imperial-fascism, at the cost of immediate, anticolonial nationalist aims." She emphasises "the experience of reciprocity and hospitality" that stemmed from the "perverse proximity between colonizers and colonized at the battlefront" and in factories and towns.[69] On the other hand, those fortunate enough to return home saw few of the promises made to them in return for their imperial-wartime sacrifice, and anger and resentment soon mounted. In an article on "La question indigène" that appeared in the French socialist newspaper *L'Humanité* in 1919, Ho Chi Minh pondered what empire had done for his people and asserted that it had wrought only poverty, exploitation, death and disillusionment.[70] During the interwar years empire became an increasingly important barometer of public opinion about the ethos of the Third Republic, and underground radical – socialist, communist internationalist, anti-colonial and anti-racist – networks were forged not only in crucibles of war and colonial rebellion, but also in a plethora of media forms and networks.[71] French colonial and world exhibitions spawned their own counter-exhibitions – what Katelyn Knox describes as both oppositional and plural "counter-gazes" – where the exploitative truth about empire was exposed, even if such initiatives were ensnared by the very exoticism and dichotomies they sought to question (and attracted far fewer visitors than the official ones).[72]

These colonial and anti-colonial forces, and the divisions and anxieties that shaped and sustained them, were relayed through media as eclectic as advertising, art, cartography, fiction, film, glass houses, guide books, magazines, museums, newspapers, photography, theatre, travel narratives and world exhibitions. There is much to Raoul Girardet's argument that these diverse forms of representation, information and spectacle worked collectively to preserve the exotic as such – to keep it outside (*ex*) and with that 'it' frequently cast in racial terms.[73] During the first half of the twentieth century, images and commodities of empire were everywhere, Girardet continues, and it was partly on account of this ubiquity, but also due to the growing impress of anti-colonial sentiment and resistance, that between the inception of the Third Republic in 1871 and World War II France's colonial problems increasingly became matters of public debate rather than just elite concern.

Gregory Shaya suggests that during this period a new "mass press nurtured and exploited a new, mass public, one that was defined by sensations, passions, and curiosity . . . [and that] came together outside the spell of class and politics."[74] The "street crowd" that gathered to behold human misery and catastrophe (accidents and crime scenes) is his model of this public. But this was also a public that revelled in spectacles of empire and the exotic.[75] Museums and exhibitions became cornerstones of a republican "cult of the public": they were designed to instil mass patriotism and an enthusiasm for empire.[76] Historians have pondered how fully the public imbibed this imperial drive and rhetoric. However, as Alice Conklin shows, politicians and administrators believed firmly in it and invested in new disciplines (of ethnology and geography, for example) and institutions (museums, centres of study) to expedite it.[77]

The Liberation of France heightened the importance of the press as a vehicle for soliciting and shaping public opinion, particularly with regard to foreign and colonial affairs. A French public that had been starved of information during the war years, and the heavy censorship imposed by the Vichy regime, took to the national and provincial press like never before. Newspaper circulation figures soared, and by the late 1940s they were the highest they had ever been, with 370 copies sold in France per 1,000 inhabitants.[78] Then, as before, the accessibility of Indochina to the French public both fuelled and helped to dampen criticism of French colonial policies.[79] Indochina was one of the most pictured and prized yet little understood corners of France's colonial empire. Even so, by 1949 it was firmly situated in a public space of empire that was the work of the imagination as well as of a grounded colonial encounter. The *France Illustration* special issue appeared in this luminous and troublesome age and space of empire, and during the first half of the twentieth century this age and space was marked profoundly by war. It was a public age of war and empire.

Sarraut grudgingly acknowledged that it had long been difficult for France to read the mood of its colonial subjects, and that "those who prophesied a mutation in Vietnamese expectations were not listened to, and hence the errors [on our part] that were exploited by Ho Chi Minh and other extremists."[80] While Sarraut had helped to create a working colonial state, his colonial humanism of 'association' assigned Indochina's indigenous elite titular and subordinate roles within colonial hierarchies, and the bulk of the population lived from hand to mouth, often on the brink of starvation, and within the exploitative clutches of French tax collectors and plantation owners, and unscrupulous Indochinese landowners, moneylenders and administrators. It was not difficult for the rural peasant population, especially, to see what nationalists and communists proclaimed: that *mise en valeur* was skin deep, backed by violence, and as much a class enterprise dividing rich from poor as a racial one separating Europeans from non-Europeans.

In his article, Pignon retorted that "the presence of French troops in Vietnam is less our fault and more a matter of circumstance."[81] In his view, it was Vietnamese resistance to Japanese occupation, rather than French colonialism (or the Vichy regime's collusion in Japan's wartime rout of Southeast Asia) that had fuelled revolutionary sentiment. And he added: "as [Mahatma] Gandhi said, any attempt to dissociate Oriental and Occidental culture would be spiritual suicide." Pignon concurred with Sarraut that "French collaboration is still indispensable if Vietnam is to make strides as a young democracy," and his remarks might be read as a knee-jerk reaction to the exit of the French Communist Party from the government of the Fourth Republic two years before, and the ensuing 'panic' (as Alexander Keese sees it) that French communists would infiltrate anti-colonial movements across the French empire and boost their demands for independence.[82]

By the late 1940s, however, Pignon's outlook seemed out of place, if not delusional. Collaboration was used as both a strategy and as subterfuge on both sides of the colonial divide. French administrators put enormous effort into disguising the way colonial encounters pivoted on dichotomies of 'us' and 'them' rather than some *métissage* or interdependence. As with electrification, colonial health planning was put in place chiefly to serve the French, and did not start to reach farther until the 1920s (and then with the native population used to trial new medical treatments, especially against leprosy and tuberculosis).[83] Goscha argues that for the Indochinese, collaboration was often more a matter of practical necessity rather than an indication of symbiosis, and that in the long run it was as significant to the making of modern Vietnam as anti-colonial resistance. David Marr likewise reflects that "reformist efforts could serve as two edges of one sword."[84]

In key respects, then, the French colonial world in Indochina was not symbiotic but antagonistic. France found there what the British discovered in India: that they were "not at home in empire," as Ranajit Guha describes this scene, because empire "rules by a state which does not arise out of the society of the subject population but is imposed on it by an alien force. This irreducible and historically necessary otherness was what made imperialism so uncanny for its protagonists in South Asia."[85] France's colonisation of Indochina was peppered and scarred by rebellions – the most significant of them in 1908, 1917 and 1931 – that exposed the vulnerability of the French and stoked anti-French feeling among the educated Indochinese. In 1949, Pignon saw Chinese and Soviet influence as a further threat to regional stability and France's interests. French forces were now needed in Indochina, he declared, to safeguard against "the immense Asiatic Kominform scheming going on," which, he judged, would be "the ruin of South-East Asia . . . [and put] the fate of civilisation at risk."[86]

Leaving the orient and entering the tropics

The special issue included a hagiographical article on the "democrat and sportsman" Bao Dai by his private secretary, S. E. Pham Van Binh, but all of the other pieces in this publication were by French observers, and their collective message was that Vietnam would pay a heavy price for fighting its way out of the Union Française.[87] Some wrote in an aesthetic vein about the peoples, customs and splendours of Cambodia, Laos and Vietnam, and in order to underscore the contemptibility of war. Others wrote about the diplomatic and military sides of conflict, and with a view (like Pignon) to justifying France's presence. On both counts, and as Pignon intimated, Indochina was referred to as part of the Orient, and as if this locational marker and label required no explanation. France's Chief Inspector of Museums and folklorist, Pierre-Louis Duchartre, noted this in his article on "Imagerie populaire et Indochine."[88] He traced how the region had been represented as a distant and mystical realm carved out by French Orientalists. The Hanoi-based École française d'Extrême-Orient (EFEO, founded in 1900), was a fount of French Orientalism in Indochina, and there was a strong association in the French public imagination between colonial conquest and the production of knowledge by legions of French adventurers, collectors, missionaries, military personnel, scholars and colonial administrators associated with the EFEO.

The EFEO brought together various scientific traditions, professions, objects of study, and know-how, and at the outset rested largely on the enthusiasm and exploits of amateurs. However, Pierre Singaravélou suggests that it marks a "revolutionary break" in the relationship between colonial knowledge and power in Indochina in that professional Orientalists in the fields of archaeology, ethnology and philology began to take an interest, and strove, within the framework of a state institution (albeit one on the margins of university life), to examine Asiatic societies through the prism of the new human and social sciences.[89] The Orientalists of the EFEO "objectified cultures, religions and languages that the Indologists and Sinologists had left to the caprices of amateurs . . . and worked to transform these knowledges into disciplines."[90] Partisans of the colonial policy of association, in particular, privileged Orientalism (the desire for a mystical East), for this worldview pushed the French administration not only to know (and therefore

Figure 2.2 Orientalist imagery in Pierre-Louis Duchartre's "Imaginaire populaire et Indochine" *France Illustration* (1949)

Figure 2.2 Continued

effectively govern) Indochina, but also to think it was being respectful of its civilisations. Yet as Singaravélou avers, "as scholars with claims to autonomy who often found themselves serving as cultural intermediaries between the French and the natives, the members of the EFEO found themselves in an unstable situation; they were controversial and exposed to various polemics."[91] From the 1920s onwards a number of them began to question Western ethnocentrism, and some, such as Mus, became anti-colonial intellectuals.

Nguyễn Phuong Ngoc has recently described the EFEO as epitomising a new "spirit of amicableness" between French and Vietnamese intellectuals and civil servants, and as spawning a new "consciousness of the country of Vietnam" through the embrace of Western scientific tenets of fact, evidence and objectivity.[92] A younger generation of educated Indochinese did not automatically deem French canons of progress and liberty as duplicitous, bankrupt or antithetical to their hopes for independence just because they were French. To some, republican reformism through education and state modernisation seemed a viable and acceptable step in that direction, and Ho Chi Minh later recruited some of his ministers from the EFEO. In the interwar years Meiji Japan also served as an alluring educational destination for Indochinese (and especially Vietnamese) students and activists, and with the boon that Japan offered a distinctly Eastern path to modernisation.[93]

Pierre Brocheux and Daniel Hémery, on the other hand, write of the "scientific possession" of Indochina and creation of an ordered French territory "out of a geographically unknown area."[94] The EFEO and a larger number of government agencies (*services* and *ecoles*) set up around 1900 by Governor-General Paul Doumer – and not least, we shall see, the *Service Géographique de l'Indochine* – were key in this respect. The distinguished French Orientalist Louis Finot implored that this knowledge of Indochina and its peoples was "for the public good."[95] But he also grasped that the kind of work he did as an archaeologist was elementary to how knowledge was translated into colonial power.[96] The disclosure of Indochina to France by scientific and professional means was a way of placing the region at the service of the state and French colonial vision, and the Indochinese staff of these outfits and institutions were accorded lesser administrative roles compared with their French counterparts (much more so than in British colonies). The DRV's long-standing Education Minister, Nguyễn Văn Huyên was the only full Vietnamese member of the EFEO, and the first Vietnamese scholar to successfully defend his PhD at the Sorbonne, in 1934.[97]

Duchartre's way of broaching such matters was by drawing on his expert knowledge of *imagerie d'Epinal*, the cheap and brightly coloured prints and woodcuts in the style of French folk art produced by the famous Pellerin publishing house of Épinal in Lorraine. The company had been a mainstay of French imperial image-making since the days of Napoleon Bonaparte, and its work took on renewed importance after World War I. The selection of images from the Pellerin stable that Duchartre used in his article evoke the jumbled and ambivalent nature of Sarraut's 'symbiosis' (and Brocheux and Hémery's 'possession').[98] Collectively these images implied that Indochina was an Oriental 'other' that was variously separate from, subservient to, and problematic to France: conquered yet inscrutable, mystical yet mastered, in ruins yet in the process of being restored by France, knowable yet ungovernable, and venerable yet seditious. The value of Duchartre's pictorial article lies in its appeal to what Jay Winter sees as the significance of myth over reality in the way war and empire were imagined and propaganda about them was produced. *Imagerie d'Epinal* mythologised war and heroism, and France's martial virtue, and thus offered an escape from reality at moments of crisis and hardship. It "came to form an important part of the unofficial propaganda of France," Winter observes.[99]

The year 1931 was at once the apogee and nadir of France's colonial mission in Indochina: the year of the Exposition Coloniale, which was staged as an epic and triumphant celebration of empire, but also the year of the Yên Bái Uprising of Vietnamese troops in the French Army in Tonkin, and the year that the Great Depression hit the colony, decimating business. Duchartre recalled this fraught moment with his choice of illustrations. Seven of them project ambivalent messages about the dignity and violence of the colonial encounter: panoramas and woodcuts of military campaigning; lithographs of Vietnamese military 'types'; and he included a Pellerin print of the ancient Khmer temple of Angkor Wat, 'discovered' by Henri Mouhot in 1860, that came from the Orientalist Ludovic Jamme's 1892 *Supplement au progres de Saigon*. The Angkor Wat pavilion at the Exposition epitomised France's *mission civilisatrice* and served as a potent emblem of a Greater France, Marco Deyasi notes, "by combining a reconstruction of portions of [the temple] . . . with showcases for colonial educational reforms."[100] It was bathed in blue, white and red light (the colours of the French flag). Finot had edited Pierre Dieulefil's *Ruins of Angkor, Cambodia in 1909*, and *L'Illustration* used its coverage of the monument at the Exposition Coloniale to declare "we are – we French of Asia, we Western pacifiers of the Far-East – the legitimate inheritors of the ancient Khmer civilisation."[101]

At the 1900 Exposition Universelle in Paris, French colonial authorities sought to stress the sophistication of Indochinese culture, project an image of its colonial subjects as gentle and deferential, and create an exotic image of the region as part of a generic Far Eastern race that was portrayed as being naturally resistant to progress but with a strong capacity for work. Five ornate pavilions comprising the Indochinese section of the exhibition displayed a plethora of natural and human phenomena 'typical' of the region: trees, flowers, plants, gardens, birds; lanterns, costumes, porcelain, silk, mythical dragons; and a village inhabited by 100 natives. The Government of Indochina spent nearly 2 million francs creating these exotic scenes, and in so doing, Dana Hale suggests, tried to communicate "serenity in common situations and natural settings" in order to solicit the interest and approval of a sceptical public.[102] This form of Orientalism continued with and beyond the perimeters of the world exhibitions, but as time went by it became increasingly difficult to disguise anti-colonial sentiment and rebellion. Accordingly, the 1931 Colonial Exposition did not simply display French prowess; it also symbolised imperial anxiety: France's exhibitionist need (in Fanon's sense of the term) to demonstrate how Indochina had benefitted from its presence, and with figures such as Mouhot, Jammes, and the founder of the lÉcole des Beaux-Arts du

Cambodge, George Groslier, represented as patriotic figures restoring history and the possibility of progress to a fallen civilisation.[103]

The *imagerie d'Epinal* in Duchartre's article was drawn from the copybook of Orientalism and in combination with the images on its next page staged a dialectic of serenity and commotion. Duchartre picked two images of young Vietnamese women reposed on the backs of water buffalo, commented that the traditional Vietnamese festival of Tet was symbolised by an animal, noted that in 1949 the festival was "placed under the sign of the buffalo," and surmised that Vietnamese artists were now using the buffalo "to materialise the dreams they wish to see realised : Unity, Independence and Peace."[104] The water buffalo is considered a national symbol of bravery, happiness and prosperity in Vietnam, and Duchartre described the women in these images as "*gems*" (or guardians) of Indochina. The American scholar Douglas Pike later reinvested this image with revolutionary meaning, observing that "Vietnamese women were far harder workers than Vietnamese men. . . . She grew the vegetables, raised the chickens, and poled the sampans to deliver food to guerrilla bands; she ran the market struggle movement, unmasked the spies, and led the village indoctrination sessions; she made the spiked foot traps, carried the ammunition, and dug the crosshatch roadblocks. The woman was in truth the water buffalo of the Revolution."[105] In the French popular imagination of the mid-twentieth century, however, the women riding the water buffalo in these images still evoked an ancient, sexually seductive and passive Oriental realm (albeit with Cambodia and Laos often depicted as more docile and obliging than Vietnam).[106] Such imagery contributed to the wider portrayal of Indochina as "the beautiful but problematic concubine of the French republic," as Kimberly Healey puts it.[107]

Duchartre's juxtaposition (however unwittingly) of a menacing/militarised and romanticised/sexualised Indochina – his dialectic of serenity and commotion – is noteworthy. It speaks to Healey's observation that French representations of Indochina were frequently "shrouded in extremes of either romance or terror," and to Nicola Cooper's argument that the "use of analogies between the female body and the colony . . . reinvented colonial conquest in an age where allusions to violent intervention abroad had been superseded by an emphasis on civilising and protective guidance. Similarly, where there is no longer a place in fiction for the heroic conqueror, a textual equivalent is sought in the sexual conqueror."[108] Indochina was at once submissive and menacing, and picturesque and opaque, in an Orientalist capacity, and both French and Indochinese women were shrouded in these fantasies and contradictions.

Monsoon colonialism

Yet contributors to the special issue also situated Indochina in the tropics – in what the geographer Jules Sion, in 1929, termed "monsoon Asia."[109] Situated between the Tropic of Cancer and the latitude 4°N, Sion identified Indochina as marked by strong monsoon influences, with high annual average temperatures (of between 22°C and 27°C, and highs in the mid-30s), seasonal fluctuations in humidity and especially precipitation (of between 1,500 and 2,500 millimetres per year), and with a pronounced dry season (especially in the south) between November and March and very intense rain, with typhoons and flooding, during a summer wet season from April to October. Rainfall also fluctuated from year to year, he continued, making irrigation doubly important to the 80 per cent of the population that lived off the land, and particularly in the north, which has a cooler dry season than the south and warmer and very humid summers. And across monsoon Asia as a whole, Sion drew stark distinctions between the inhabitants of the mountains and plains, with the peoples of the former generally more "vigorous" and "agile" than those of the latter.

In his article, Xavier de Christen, who was newly appointed in 1949 to Pignon's Cabinet, and before that had been the head of the French Press and Information Service in New York, saw Indochina as a place of tropical as well as Oriental charm. "In spite of the war in its midst," Indochina was still "a country with exotic flowers and sun," and also spectacular rain, he wrote. However, he underscored the crucial difference between experiencing the country first-hand and looking at it from afar.[110] "When the colors that burst on to the surface of things are not fresh due to the habit of being looked at, the viewer is left with sparkling collections of images, such as the *imagerie d'Epinal*." But, as he continued, looks could be deceptive. While French adventurers, settlers and soldiers were captivated by Indochina's beauty, many of them had struggled to live in the region comfortably. The diaries of French soldiers involved in the late nineteenth-century pacification of Indochina did not follow colonial stereotypes of easy and inevitable victory over a supine race, and the region's natural – and ostensibly tropical – environment soon attained a forked place in the French imperial imagination, as remote and dangerous on the one hand, and as a theatre of adventures and dreams on the other.[111]

Mouhot was one of the first to emphasise "the sanguinary attacks of mosquitoes and leeches."[112] When the novelist André Malraux returned to the ruins of Angkor Wat in 1923, he did not find the mystical place of splendour and spiritual rebirth promised by Mouhot's otherwise romantic depiction of the place, but riotous jungle and ruins. The novels that Malraux went on to write from this experience are infused with the dizzying sights, sounds, smells, colours, glare and darkness of the tropics. Charlotte Rogers suggests that for him, "the health dangers endemic to the tropical wilderness come to represent the obstacles every human faces in his or her ill-fated attempt to control destiny."[113] This experience was not restricted to the French. The same twin experience of tropical splendour and struggle suffuses the English writer Somerset Maughan's 1923 travelogue of his journey through Burma, Siam and Indochina, *Gentleman in the Parlour*.[114] He contracted malaria, and opium compounded the effects of the heat. In a wider frame, Alison Bashford observes that "Just how different humans fared in the torrid zone was an enduring question at both expert and lays levels. . . . Could white people work and reproduce normally and effectively in the tropics?"[115]

The problematic of what Bashford dubs "human-climate interaction" is an important, if little investigated, leitmotif of France's colonial presence in Indochina. French colonists found Indochina seductive but draining, and as Eric Jennings notes, whether one lived the "Indochinese dream," or deemed it a fantasy shrouding an altogether harsher reality, hinged on one's position in a colonial society that was sharply divided by class, race and gender.[116] While the French grasped the significance of rice cultivation "both to sustain livelihoods as well as lay the foundation of viable export production," Geoffrey Gunn observes, they "had to confront the vulnerabilities of irrigated rice production in the two delta regions, alongside rising population densities and natural calamities, including long droughts and raging floods, as well as concomitant cycles of famine and poverty."[117] French settlers soon became disdainful of the spectre of environmental calamity and the peasant unrest it fomented. Van Nguyen-Marshall demonstrates that instead of being able to assume they had a right to subsistence under the French, and contrary to their claim that the schemes of agricultural modernisation introduced by the colonial administration had added to their woes, Indochina's rice-growing peasant population had to prove to the French authorities that they were deserving of colonial aid.[118] "Vestimental [and gastronomical] difference accentuated the chasm between colonizer and colonized. And petty differences constantly divided *colon* and administrative circles."[119]

The monsoon environment did not cause these difficulties and divisions, but impacted on them and the anti-republican sympathies held by the majority of the French settler community. The French chose clothes and food that distinguished them from their Vietnamese and Chinese labourers and servants, and struggled when imports of French victuals were curtailed during the two world wars.

The biggest infrastructural challenge presented to the French was the control of the Red River. This heavily silted river was prone to overflowing its banks during the flood season (June to October) and inundating vast low-lying areas. There were serious floods in the delta in 1894, 1904, 1915–1916 and 1926–1927, and, as Gunn notes, "a consensus emerged among French administrators that hydraulic control helped to tame nature . . . [and] display[ed] colonial mastery."[120] From the twelfth century, North Vietnam's imperial court oversaw the construction and maintenance of a vast dike system, with village *corvée* labour. The French only took on this challenge belatedly, but between 1918 and the early 1930s expanded the dike system threefold, and with a major dike reinforcement programme for the entire Red River Delta during the 1920s creating a new two-tier classification, with high dikes 4–8 metres above sea level, and lower ones 2.5–4 metres above sea level.[121]

At the end of the 1920s, Robequain deemed this "vast irrigation network," and the French road system connecting Tonkin, Annam and Cochinchina, a "magnificent symbol of unification."[122] High dikes and sluices protected Hanoi from the might of the Red River. Philippe Roques and Marguerite Donnadieu expanded on this upbeat assessment in their 1940 survey *L'Empire français*, a propaganda piece, produced at the behest of the French colonial information service, that was published on the eve of the fall of France, where they represent France's colonial empire as a "living organism" comprised of "common ideals," and Indochina as a "crossroads of peoples" whose "political and historical configuration is in perfect harmony with its geographical configuration."[123] This harmony, they continued, revolved around the Mekong and Tonkin Deltas and was perhaps the most earthly of harmonies anywhere in the empire, and certainly the most tropical. "The [Mekong] river carries the earth and outstanding materials in its waters, bringing fertility to the land. . . . It is the same in Tonkin. The Red River, which takes its name from the colour of the silt-laden waters, is torrential . . . [with] a wonderful network of dams and canals, a feat of technical organization . . . [shaping] an area of intensive cultivation area that caters to a very dense population."[124] This optimistic outlook goes hand in hand with the authors' view of the natives, who are considered superior to those of Africa but nevertheless compliant, with "a very pronounced taste for sedentary life," and "always thrifty, hard-working."[125] All of this created an excellent spectacle, they concluded, and made Indochina "the country of tourism *par excellence*."[126]

But by 1949, indeed by 1940, darkness and trouble had descended over this bright tropical scene, and a different genealogy of empire and environment had come more fully into view. In a 1931 report on the Tonkin Pavillion at the Exposition Coloniale, for instance, we learn that "floods, typhoons, which since 1926 have periodically ravaged the Tonkin Delta, have so exhausted its resources that is has become physically impossible to provide the protectorate with the finances it needs in relation to its degree of evolution and place within the Indochinese Union."[127] In 1933, the French colonial entrepreneur René Bouvier, evaluating the wealth and poverty of Tonkin, noted: "one of our best colonial painters, Joseph Inguimberty, has made himself a poet [of the region] who delights in the thick foliage of these overcrowded towns, and with the flickers of dappled sunlight that enchant his tropical palette, which redress even the stifling heat of the enervating summer months."[128] But Bouvier paints a different picture, of hunger, debt and overpopulation, and sings the praises of the other delta, the Mekong, which he deems more conducive to economic development.

Such cracks in the exotic spectacle of colonialism in Indochina are also exposed in the 1950 novel *Un barrage contre le Pacifique* (translated as *The Sea Wall*) by Marguerite Duras (née Donnadieu). The daughter of French settlers, Duras led an impoverished farming life in Cambodia as a child, and then had a spell in the French civil service before turning her back on Vichy France and empire, joining the French communist party, and becoming an avant-garde writer and filmmaker (and lifelong friend of François Mitterrand). In this novel she conjured with imperial dystopia. In fact, she tried to pretend that the book she had written with Roques 10 years before never existed.[129] In the novel, the narrator looks ironically at a poster stuck to a wall in Hanoi: "In the shade of a banana tree weighed down with fruit, colonial couples, all dressed in white, swayed while the natives were busy smiling around them."[130] The novel denounces "all of the great vampire colony, rice, rubber, banking, usury," and is accompanied by a dystopian vision of the Mekong Delta. At the centre of the novel is the futile struggle of a mother against the annual floods. Her misfortune is to acquire 'unused' (and largely unusable) land and to have to struggle against a corrupt colonial administration.[131]

Duras was writing about what we might call monsoon colonialism: staples of imperial fantasy and colonial power (racism, violence, division, arrogance, anxiety, degeneration) that both shaped and were shaped by the tropical environment of Indochina. This configuration of the tropics as both a metaphor and reality of dislocation and trauma was by no means restricted to the French. This experience of feeling overwhelmed can readily be found in the work of writers from other Western countries. For example, Ulrich Bach notes that in the Austrian avant-garde writer Robert Müller's 1915 novel *Tropen: Der Myhos der Reise* (translated as *Tropics: The Myth of the Journey*), published at the height of German imperial expansion and as a literary guide for potential travellers, "the Blue Danube is transposed into the muddy Amazon" and the story creates "a murky zone between fictional reality and imagination, and produces a moment of uncanny dislocation. . . . The nation-state and its fixed borders give way to a boundless swamp. Müller's exotic jungle is foremost a site of metaphysical alteration and a threat to the European intruders"[132] Anne McClintock coins the term "porno-tropics" to capture how, in more general terms, the West's 'penetration' of the tropics from the late nineteenth century was trailed by fantasy, paranoia and megalomania, how this came in gendered, racialised and sexualised forms, and how it spawned "a recurrent doubling in male imperial discourse . . . [through] the simultaneous dread of catastrophic boundary *loss* (implosion), associated with fears of impotence and infantilization, and attended by an *excess* of boundary order and fantasies of imperial power."[133]

Sion wrote at length about the inadequacy of geographical determinism alone as a means explaining the fortunes of the French in Indochina. However, he did not deny that the experience of debilitation, and fear of physical and moral degradation due to the humidity and disease, fuelled French disdain towards the native population, engendered fears of rebellion, and thickened the atmosphere of violence that historians see as pervading colonialism in French Indochina. At the start of their colonising efforts in the region, the French knew little about tropical diseases and were on a steep learning curve. The best that the first colonial physicians could do was to establish a *cordon sanitaire* to protect the French population. Not all of the serious epidemic and endemic diseases that afflicted the French – beriberi, cholera, dysentery, intestinal parasites, malaria, tuberculosis, typhoid and venereal disease – were uniquely tropical diseases, but this suite of afflictions was often attributed to the tropical environment.[134] Hanoi and Saigon were both situated at the edges of regions of low malaria incidence, but French rubber plantations, mines and forestry stations in the jungle and highlands to the north and west were stricken by outbreaks of malaria, spurring, from 1905, diverse health research initiatives and sanitation measures.[135]

For the majority of French settlers, Cooper concludes, the region's "defining features were disease, debilitation, threat and isolation," and all of this detracted from Sarraut's *mise en valeur*. The French became obsessed with health and hygiene, establishing hill stations and sanatoria (the most important of them at Dalat) away from the humidity of the jungle and deltas.[136] Indeed, it was partly on account of the enervating wet season, which was felt first by French legionnaires who were dispatched to the region in the late 1880s with heavy woollen apparel and 90 pound packs, that the colonial administration sought to discourge French nationals from emigrating to Indochina. During the late nineteenth century, the mortality rate of French troops and settlers in Indochina rarely crept below 2 per cent per annum – much higher than elsewhere in France's colonial empire. These deaths were largely from tropical diseases (especially dysentery and cholera), and as Eric Jennings notes, "colonial administrators chose some of the grimmest points of reference in tropical health to describe Indochina."[137] Doctors proselytised about "the negative influence exerted over time by tropical climates on the health of Europeans," and, in lieu of preventative measures, advocated that settlers return to France.[138] From 1927 mineral and thermal springs were sought for colonial spas that might serve as prophylactic means of "hydrotherapy" to allay what were deemed to be the 'degenerative' influence of the climate. The early twentieth century was the age of the French "doctor-hygienist," Laura Victoir suggests, and tropical hygiene manuals that had initially been written for military (and chiefly male) personnel were soon adapted for settlers and morphed into manuals of household and domestic advice.[139]

Yet these problems of insalubrity, and the feelings of death and struggle they made omnipresent, could also be used as a smokescreen for the more basic colonial desire for mastery and racial and class distinction. The greatest obstacle to colonial success, Marie-Paule Ha suggests, "was the imperative of physical and social separation of Europeans from non-Europeans." She alights on a passage from Robequain that is pertinent here because of what it says about this tropical screen: "Even if it is granted that, with certain precautions, the climate need not hinder European settlement in the tropics, the contact between Europeans and natives raises acute problems. It is generally recognised today that a European cannot engage in the same kind of agricultural tasks as the Indochinese without being degraded to the status of the 'poor white' . . . [and eventually] lowering his standard of living to the point of destitution."[140]

Cooper insists that anxiety about Indochina's monsoon environment undermined the "exotic charm of a life overseas" and promoted a fear of engulfment by nature. The breakdown of colonial humanism was connected to what she sees as this "dystopic reality of exile and disillusionment." This reality prompted many French colonists to buck the republican aspirations of political figureheads such as Sarraut and Pignon. In colonial memoirs and government reports, and in the writing of travellers and residents such as Malraux and Duras, "the organic omnipotence of Indochina's vegetation at once threatens and dethrones French control."[141] In its late nineteenth-century heyday, colonial exoticism was fuelled by the quest to escape an artificial and banal world of modern consumerism and naked individualism, and reach a distant, authentic and primeval world still based on community and characterised by closeness to nature. By the 1930s, however, this yearning and quest had faltered, and not least in Indochina, where many colonists viewed the tropical environment as annoying rather than enchanting, and had little interest in the colony's peasant landscapes and cultures.[142] Advances in science, technology and medicine had not, and perhaps never would, fully shield the French mind and body from Indochina's tropicality, and as Michael Vann surmises, *anémie* (deficiency) was the catch-all phrase the French used for the malaise of white settlers in the tropics.[143]

Pierre Gourou and his wife (Barrion Hélène-Georges) and daughter (Gilberte) lived in a well-apportioned apartment in the French quarter of Hanoi, and Gourou imported a piano from France for Gilberte in 1928.[144] But Barrion struggled with the climate and spent the summer months in cooler climes – at colonial resorts in the district of Tam Dao, 90 km north of Hanoi, on *Colonial Route 2* – where, her husband recalled, she played tennis and "lived a life of the wife of a colonial civil servant."[145] According to Gilberte, Gourou himself was not bothered by the monsoon climate and stayed in Hanoi and the delta during the summer to do his research.[146]

Between tropical spectacle and militant tropicality

In the lengthiest piece in the special issue, the Director of French *troupes coloniales*, Général Jean Valluy, who had commanded the Corps Expéditionnaire Français en Extrême-Orient (CEFEO) – the French force in Indochina – in 1947–1948, offered a frank military appraisal of the conflict. He argued that France's aim of defeating the Viet Minh militarily, and the assumption that with military defeat Vietnamese support for the revolution would melt away, had fallen flat. The more the French threw at Giáp's fighting force, he averred, the stronger popular support for the revolution seemed to become, especially in the countryside. One of the main reasons the French military was stumbling, he continued, was that Vietnam's climate, terrain, vegetation and human geography of villages – in short, its tropicality – was having an adverse effect on the CEFEO's attempt to "pacify and control the country." Valluy remarked:

> Indochina does not conform to the 'Western' style of military thinking and strategy that worked against determined opponents before: the Germans. In Indochina the opponent is altogether different: he often refuses open combat and is a master in the guerrilla art in all areas, including propaganda. The terrain is not equivalent to Europe's. Indochina has forests, thick bush, steep-sided mountains, and flooded rice fields. Only fleeting and fluid detachments of troops will work. . . . A soldier is first and foremost a soldier. But the soldier of Indochina is of a very special kind.[147]

He had been a Brigadier-General in the French First Army which spearheaded the Liberation France in 1944 and pushed on into Germany with the Allies. He was struck by the difference that the Vietnamese landscape made to the nature of the conflict. Giáp's guerrillas and regular army units exploited jungle, swamp, mountains, and a maze of rice paddies, dikes and secluded villages, to outflank, disperse and pick-off a militarily superior French fighting force. French soldiers and artillery got bogged down in the Tonkin and Mekong mud. "In view of the rugged Indochinese climate," the war reporter Bernard Fall observed, "older men without tropical experience constituted more of a liability than an asset."[148] The French were also acutely aware that the aerial bombardment of delta regions, which had the highest population densities, was counter-productive. During the summer of 1945 French intelligence became alarmed about the way U.S. Air Force sorties over the Red River Delta, ostensibly targeting Japanese supply lines (bridges, roads and munitions dumps) threatened the dike system (containing 85 million cubic metres of water) and portended catastrophic flooding which would further devastate a people already afflicted by famine and disease.[149]

'Special' has particular connotations in military lore, and not just within the West. The term connotes small and elite groups of military personnel recruited and trained to undertake 'specialist' ('secret,' 'strategic,' 'precision') operations of reconnaissance, intelligence, counter-insurgency, sabotage and so forth, outside of traditional combat roles and formal military jurisdiction, usually 'behind enemy lines,' and with a dispensation to bypass laws of war regarding the use of terror and violence, and the treatment of civilians.[150] During World War II each of the Allied and Axis powers had its own 'special' airborne, commando and marine military units and services, and Valluy knew that they had played vital (if publicly opaque) wartime roles.[151] He was also cognisant of the fact that beyond Europe, and especially in the Far East, the enemy that Allied 'special forces' encountered came in an environmental as well as human form, and that the Vichy regime's complicity in the Japanese occupation of Indochina had blurred the distinction between friend and enemy. It was not simply the fact that the vulnerable hydraulic system of the Tonkin Delta was off limits, so to speak, to aerial assault. Valluy was also alluding to the fact that specialist skills, training and fortitude were needed to combat the tropical environment of Southeast Asia and the Pacific, and that these requirements and attributes were by no means restricted to specialists. They applied to the military as a whole. Valluy was intimating that each and every fighter in the CEFEO needed to be a 'special' kind of soldier.

World War II hastened the production of specialist army field and training manuals on 'jungle' and 'unconventional' warfare that were aimed at all military personnel, and geographical knowledge (especially maps of climate, terrain, vegetation and human habitation) and medical knowledge were central to this effort.[152] In Burma and Malaya, British commanders urged that the difficulties of fighting in 'unconventional' jungle environments could be exaggerated, and that the British were logistically as – if not more – adept at it than their Japanese foe. In Indochina, however, Valluy was not so sure; and no one sloughed off the threat of disease. "Just as the French army had learned during the conquest of Vietnam in the nineteenth century and British forces more recently fighting the Japanese in the Southeast Asian crescent," Goscha notes, "disease and sickness were deadly adversaries for any army deployed in the tropics."[153] Malaria and typhus, and to a lesser extent dysentery and skin diseases "caused more losses than gunfire" for the British in Burma and Malaya, Max Hastings adds.[154] The French in Indochina were much better able to control malaria, principally due to pioneering experimental work undertaken at the *Institut Pasteur*'s three research stations in Indochina. Considerably more CEFEO troops were killed or incapacitated in battle than by disease. Nevertheless, over the duration of the Indochina War an average of 42 in 1,000 French troops suffered from skin diseases, 26 from digestive disorders, 28 from respiratory disorders (tuberculosis), and 21 from diseases of the sense organs (especially trachoma). Only 19 in 1,000 were stricken by malaria, and more than that contracted venereal disease.[155] "The

Vietnamese soldier's body was no more immune to tropical diseases than the colonial one," Goscha reminds us, and until China and the Soviet Union began to step in, from 1950, with supplies of antibiotics and anti-cholera and anti-malaria medication, the Viet Minh commanded fewer medical supplies and services than their French adversaries.[156] Vietnamese soldiers and civilians suffered from a number of diseases that were uncommon among the French – especially trachoma, leprosy and tuberculosis – and the problem for all was that while not all of these diseases killed immediately, they could put scores of troops out of action for long periods and spread like wildfire. Soldiers, like civilians, were bombarded with advice and literature on the importance of personal hygiene and of securing clean drinking water.

The semantics of the exceptional and extraordinary had been shifted on to the tropical environment of Indochina and the corpus of Valluy's fighting force, and the wider incorporation of 'the tropics' into Western military doctrine, and fields of combat and guerrilla insurgency, militarised tropical nature and turned it into a battlefield.[157] Giáp forged a people's army that was adept at living off, spreading across, and melting into it the soggy, jungle and mountainous landscapes of Indochina. The Viet Minh exploited the tropical environment for military ends, and thus forged what we might call a militant tropicality. "There was no clearly defined front in this war," Giáp recounted: "It was there where the enemy was. The front was nowhere, it was everywhere. Our new strategy created serious difficulties for the enemy's plan to feed war with war"; and between 1947 and 1949 "we launched small campaigns which inflicted considerable losses on our adversary."[158] French army maps of northern Vietnam from these years depict a chessboard of French and Viet-Minh controlled areas, and with civilians in the densely populated Mekong and Tonkin Deltas ensnared and terrified by the violence. Goscha recounts: "The Expeditionary Corps [which comprised over 200,000 troops by 1950, including some battle-hardened World War II veterans] rolled into villages in jeeps, armored trucks, and tanks. Troops touted machine guns, lobbed grenades, and carried flamethrowers as barking German shepherd dogs pulled at the chains during search and destroy missions. Officers radioed in mortar barrages, while the air force dropped paratroopers, bombs, and, from 1950s onwards, American supplied napalm."[159]

This was a brutal war. Vietnamese villagers struggled to keep themselves safe. The French onslaught left hundreds of thousands dead and maimed. Many in the two delta regions in the north and south abandoned their homes and fled to the comparative safety of surrounding hills and jungle. Many more joined the revolutionary cause, often as much to defend their homes and homeland, and in a spirit of revenge, as out of any strong ideological commitment to communism. Yet as Goscha continues, for all their firepower, "French forces were spread too thinly to control all of Indochina all of the time . . . [and] the remote hills of the north and the deepest jungles of the Mekong, and a vast stretch of central Vietnam were still left in the hands of the Democratic Republic of Vietnam."[160]

In 1950 Giáp noted that this "war of liberation" had brought the diverse peoples of Vietnam, Cambodia and Laos into a single "Indochinese battlefield."[161] Twenty years later he reflected that Valluy's 'special kind' of soldier, and later Americans troops and commanders, were strategically and psychologically ill-equipped to fight on Vietnam's "tropical battlefield."[162] Valluy made only piecemeal military gains in Vietnam and struggled to consolidate the limited success he had in northern Tonkin, delivering some heavy blows on Viet Minh strongholds, by securing *Route Coloniale 4* – a long and winding road through jungle, high mountain passes and canyons connecting French garrisons at Lang Son and Cao Bang. During his command, French flotillas were ambushed more than 30 times.[163] As military historian Adrian Gilbert relates with the testimony of French troops, the Viet Minh attacked *Route coloniale 4* in orderly waves, and

> vanished into the jungle in perfect order, unit by unit. Special formations of coolies carried off their killed and wounded, as well as all the loot they had taken. . . . The road was a graveyard, a charnel-house. Nothing was left of the convoy but a heap of ripped-opened bodies and blackened engines. It was already beginning to stink. The survivors gathered on the roadway . . . and picked up the corpses and the wounded. And what was left of the convoy set off again.[164]

French troops, Gilbert continues, equated the smell of the tropics with the smell of death, and, latterly, napalm.

The tropical qualities of the Tonkin Delta likewise afforded the Viet Minh a strategic advantage. Yet they were also advantageous to the French, at least for a while. Valluy had sensed that while the CEFEO might secure the area, they could not win it outright because to do so would necessitate the use of firepower that might wreck the dike system, causing catastrophic flooding, and undermining any strategic advantage they had attained by hunkering down there. And what would not work for the French would not work for the Viet Minh either. Viet Minh guerrillas worked clandestinely (and usually under the cover of darkness), harassing French checkpoints and artillery positions, but could not take the Tonkin Delta. It was at once a dynamic and dead zone from the point of view of military defiance and destruction.

The flamboyant and highly decorated Jean Marie de Lattre de Tassigny (the youngest French Général at the start of World War II), who took charge of the CEFEO in 1950 and whom Giáp regarded as a worthy adversary, turned this stalemate to his advantage by creating a defensive line (which became known as the De Lattre Line – see Figure 2.1) comprised of an arch of 1,200 strongpoints (in clusters of five or six) encircling the Tonkin Delta, cutting off communications between Hanoi and Haiphong and supply lines between the north and south.[165] The French sought to control the rice crop and discouraged the flight of Vietnamese villagers as a further means of 'supervising' the revolution.

The 'line,' which was peopled by over 100,000 French and allied troops, was by no means impregnable. By 1952 the French had lost over 30,000 troops there, and the Viet Minh held over a third of the delta.[166] The defensive shield was costly and logistically challenging to construct (it took over a year to complete) and maintain. However, it confounded Giáp's idea of a "war without fronts." In January 1951 he threw 20,000 troops against 9,000 French troops defending the French stronghold at Vĩnh Yên 20 miles northwest of Hanoi, at the western end of the line, as the first leg of his general counter-offensive. But he was soon repelled with nearly 700 of his fighters dead and a further 1,700 wounded (the French gave a much higher figure) from machine gun and mortar fire. He tried to break through again two months later, this time near the coast, at Mao Khé 20 miles north of Haiphong, but on this occasion his 11,000-strong force was met by napalm and rockets launched from navy vessels, and Giáp was again forced to retreat, albeit this time with fewer casualties.[167]

Giáp was one of the privileged few Vietnamese who gained a higher education. He had been helped by one of Sarraut's former police commissioners, André Louis Marty, who sanctioned the young revolutionary's release from detention for seditious activity, perhaps with a view to trying to co-opt him to the French administration.[168] Gourou taught Giáp geography and history for two years at the elite Lycée Albert Sarraut in Hanoi in the early 1930s and deemed him a very inquisitive and hard-working student. They spent a good deal of time together, and Gourou remembered well his pupil's fascination with his classes on the history of European warfare, and particularly Napoleon's strategy at Austerlitz.[169] Gourou wrote of the "tropical nonchalance" of his "hardworking Annamite pupils" at the school, and subsequently employed Giáp to collect data for him on Annam villages. The experience proved invaluable to the formulation of Giáp's idea of a "people's war," which he adapted from Mao's model of revolutionary warfare.[170]

Gourou contributed an article to the special issue on the École des beaux-arts de Hanoi (established in 1925), extolling the work of the School's figurehead Victor Tardieu, and particularly one of its professors, the artist Joseph Inguimberty, to whom he dedicated a copy of his 1936 *Les paysans du delta tonkinois*, a text lavishly illustrated with Inguimberty's sketches, "in memory of our deltaic wanderings and monsoon mensurations."[171] Gourou's concern in his article was with aesthetics rather than military or political questions, at least on the surface. It is with this short piece, and his connection to Giáp in the background, we will now move to a discussion of the aesthetics of tropicality and suggest that in post-war Paris and Indochina art and militarism were articulated in complex ways. At a moment when the French (and subsequently American) popular imagination was increasingly picturing – through art and photography – Indochina as a war zone, Gourou underscored the significance of painted landscapes as guides to what was being invaded and barraged.[172]

"One thing that all French artists attracted to Indochina had in common was a taste for the exotic," Nadine André-Pallois observes. "Whether or not they made the trip there all seemed motivated by the discovery of distant and different civilisations."[173] Indochina was not immune from what Edward Said identifies as the Orient's destiny to be watched, and the reverie of looking led the likes of Inguimberty and Tardieu to paint an eternal and quixotic Indochina.[174] Their artistic quest was also suited to the poetics of imperial display. Indeed, for imperial Japan, as much as imperial France, individual valour and national prowess could be seen and consumed, and sometimes touched, at exhibitions, and in museums and department stores, with panoramas, dioramas, photographs, paintings and mannequins.[175]

The groundwork for the display of French colonial art was laid at the 1925 Paris Exposition International des Arts Décoratifs et Industriels Modernes, staged around the Eiffel Tower. This exposition was advertised as a celebration of modernist living and break with tradition and the past, including France's imperial heritage. If this presented a challenge to imperial propagandists, then as Robert Rydell observes, they rose to it. French colonial authorities succeeded in commissioning pavilions pertaining to France's African and Asian possessions, and one for colonial art in which an array of 'productions' in ivory, lacquer, metal, shell, silk, wood and yarn were displayed.[176] And if visitors were not convinced about the value of such crafts and objects to France and a *civilisation coloniale*, then the hope was that they would be left in no doubt once they entered the African and Asian buildings reconstructed at the venue and encountered native artisans shipped in from the colonies working with their materials. "French imperial designers did their best to make the modernist sensibility on view at the fair unthinkable apart from imperialism."[177] As one of these designers remarked, the colonial art pavilion invited "the prospect of an artistic collaboration of the metropolis with the colonies."[178] The same idea was imagined and realised on a more grandiose scale at the 1931 Exposition Coloniale. Colonial art was solicited there as a means of imperial defence.

Tardieu wrote that in Hanoi he had sought to "create a school, which, while respecting local traditions, will be adapted to modern needs."[179] And those needs were imperial as much as – and for some more than – they were artistic. The display of exotic and colonised realms at world exhibitions, and in realistic terms (and often life-sized proportions), was integral to what Timothy Mitchell characterises as the impulse to create detached and privileged platforms from which "one could see and not be seen," and in so doing grasp the world as if it was a picture.[180] "The photographer, invisible beneath his black cloth as he eyed the world through his camera's gaze, in this respect typified the kind of presence desired by the European in the Middle East, whether as tourist, writer or indeed colonial power," Mitchell writes.[181] So too did the painter, albeit with one difference: Inguimberty's Orientalist manner of watching and picturing also involved a desire to immerse himself in an alien reality in order to capture its essence, and it was a desire that he shared with many other painters in the tropics. Natasha Easton thus reflects that during this era modernity and colonialism were "driven by the paradoxical desire to render the world as an object of representation and to lose oneself in this object world – experience it directly."[182]

This paradox came in a number of forms. Angela Meyer, for example, shows that French artists produced "a painted vision of the tropics as a space desirable and strange, redolent of fear and longing, bewilderment and comfort," and that they expressed this vision in a range of ways. Henri Matisse, for example, "concentrated on the formal element of its physical appearance" and how the tropics might serve his artistic technique, whereas Paul Gauguin "immersed himself in an experiential lived involvement," and Henri Rousseau "dreamed only of the tropics via his imagination."[183] A good deal of critical attention has also been bestowed on how painted visions of the tropics facilitated colonial control. Suzie Protschky, for instance, notes that during the interwar years "a veritable swarm" of Western travellers, painters and illustrators "descended on the tropics," and carrying this paradox. The art of the tropics dovetailed Orientalism and its deep complicity in imperialism through their conjoint fixation with mystery and realism – a desire to capture the tropical unknown with an eye for detail, and with the effect of entangling fantasy and realism.[184] Jean-François Staszak shows with respect to Paul Gauguin's sojourn in Tahiti that while the reality and immediacy of tropical otherness was valued, it was often realised through stereotype and as a means of comforting the European Self (which Gauguin did in a markedly sexualised way) and reassuring that Self of its superiority.[185]

Nora Taylor notes that Inguimberty was "remembered by his students to have been sincerely interested in Vietnam – its people and its life"; one of those students, Nguyễn Quang Phong, drew stark comparisons between his teacher's concern with "objective realities" and the "colonialist" work of other French artists, which he saw as mired by an uninspired and derogatory "curiosity."[186] However, there was still a colonial pecking order. The School awarded diplomas in "Indochinese" art and architecture to its Indochinese students, but they were deemed subordinate to corresponding "French" qualifications.[187]

If this kept tropical art within the clutches of Orientalism, then Gourou's point was that the School succeeded in breaking through one Orientalist barrier: of deeming only Western art canons as worthy, and only pre-colonial Asian art as 'authentic' and worthy of Western scrutiny.[188] Inguimberty was known as 'the painter of the delta' and Gourou exalted the way he captured "the harmony of colours, and the emotions, that these scenes aroused in him." Imguimberty tried to "soak up the environment," Gourou continued, by painting scenes of daily life and picturing French and Vietnam figures interacting without giving the European a superior gait. Between 1926 and 1945 Inguimberty explored the delta and mountain countryside around Hanoi both by himself and with his students, and Gourou implied that he surmounted colonial power relations, or had at least created a form of hybridity: the School did not herald the artistic possession of the delta; rather it had helped to overcome Vietnamese suspicion towards the French.

Gourou was close to Inguimberty. The two of them visited the Tonkin Delta together on many occasions and remained friends until the artist's death in 1971. Upon his return to France in 1945, Inguimberty gave his friend a number of his paintings, including Figure 2.3, which hung on the wall of Gourou's Brussels apartment. Inguimberty signed his oil paintings, but did not date or specify the scenes he composed, heightening the sense of the immutable. Gourou used Inguimberty's work to represent the Tonkin Delta as a cherished location outside war and Western politics.[189] He passed over the damage that French and Viet Minh forces, and before that the Japanese, had inflicted on Inguimberty's idyllic painted landscapes. The French intelligence team that had been so alarmed about American bombing of the Tonkin Delta probably had Gourou's 1936 monograph to hand. Gourou's 1:250,000 map of the dike system was more comprehensive than

Figure 2.3 Tropicalist imagery in Pierre Gourou "lÉcole des beaux-arts de Hanoi" *France Illustration* (1949)

anything the French colonial authorities had produced, and its attention to village patterns and population density revealed how vulnerable this human-hydraulic mosaic was to collapse.

By 1949 Gourou and Inguimberty were back in France, and both of them thought that the Japanese occupation "sounded the death knell of colonisation."[190] Inguimberty had returned to France in 1945, avoiding Marseilles, his family recall, "so as not to be thrown into the water by Communist dockers."[191] However, Gourou's message was not devoid of politics and has some affinities with the military sensibilities glossed by Valluy. For while Gourou and Inguimberty rendered the delta as a tropical spectacle – a landscape to be looked at and coveted within the frame of a painting – that spectacle was not simply about harmony, but also, and poignantly, about struggle. The 'objective reality' of which one of Inguimberty's students spoke was rooted in an age-old struggle with the land, with nature, and against foreign domination. Adaptation and struggle were two sides of Braudel's *structure* – in the Tonkin Delta a thousand-year struggle to adapt a challenging monsoon environment to human use and occupancy.

Was this message out of sync with news of the war that Gourou would have imbibed from the radio and newspapers? He wrote his piece at a time of intense Viet Minh propaganda castigating all things French. At the same time, he was well aware that French art historians tended to look down their noses at the hybrid works of colonial artists like Inguimberty and continued to venerate the 'purer' products of French and Orientalist traditions, and Western teaching.[192] Gourou perhaps deemed it important at this moment to use art, aesthetics and the *longue durée* to have a political conversation about the military conflict.[193] The School, he mused, was not content simply to "resurrect" the past, or "reconnect" with it through fiction or Orientalist (archaeological, philological or ethnographic) methods. More importantly, it sought to "create something new," produce a "new interpretation of the physical and human environment of Tonkin." Inguimberty sought to do this by mastering Vietnamese drawing and painting techniques (he had a special interest in lacquer), combining them with French methods, and passing this joint wisdom on to young Vietnamese artists, enabling them to "discover that their own country was able to inspire the strongest aesthetic emotions."[194]

Gourou still represented this as a gift bestowed by French doyens, and while he named some of these Vietnamese artists, there are no illustrations of their work in his article. One of the four examples from Inguimberty's oeuvre that Gourou used to illustrate the work of the School – *Young Indigenous Women in Front of a Lotus Pond* – bore tinctures of Orientalism (seduction and submissiveness). But in the other three – *Scene of Annamite Life Around Hanoi, Nap in a Hanoi Garden*, and especially *Water Carriers in the Tonkin Delta* (the painting Gourou had) – something else was going on. "If it was necessary to make an effort at renewal in order to feel in all its original aroma the beauty of a young paddy field, the rhythm of a peasant woman carrying the double load across her shoulders," he observed, "it had nothing to do with an exotic or superficial picturesqueness; nothing is more mysterious than this Tonkin nature, and it only delivers up its beauties to someone who has studied it for a long time."[195] This nature was tropical, and Inguimberty symbolised it with palm trees and a palette of green, brown and orange.

Gourou and Inguimberty were hankering after reality rather than spectacle, and their way of accessing the sights and scenes of the Tonkin Delta was through immersion: fieldwork and close observation. This was a particular kind of watching. Gourou called it "ethnographic," and in his view, the triptych of Inguimberty's paintings that were displayed in the Tonkin Pavilion at the 1931 Exposition Coloniale – *A Thai village, The rice field*, and *The Red River* – along with those of some of his pupils "gave proof to the extraordinary success of the School."[196] Gourou himself wrote a detailed guidebook, entitled *Tonkin*, for the Exposition. It was handsomely illustrated with photographs and maps, and fellow geographer Robequain eulogised it as an "imaginative" and "conscientious" overview of the region that talked up the benefits of France's colonial presence and superseded prior "hasty and insipid compilations."[197] While display was the exhibition's *modus operandi*, absorption was the key to the School's art and Gourou's text. "The collection of its pupils' works was probably the most captivating thing in the Exhibition. It was the sole example of a new art school being created on French territory outside Paris," Gourou observed.[198]

Gourou's and Inguimberty's aim was to faithfully capture an alien reality and scene of cultural and geographical difference. "On entering the vast [Tonkin] pavilion," Gourou noted, "one got the immediate impression of light. No-one is better than Inguimberty at reconstructing this tropical light, so different from our own. And the poetry expressed in his paintings could only emerge thanks to the sympathy he had for his subjects and the pleasure he found in looking at them."[199] But this can hardly be deemed innocent, and others saw a connection with France's civilising mission. One commentator, writing in another prominent Parisian magazine, *La Dépêche coloniale illustrée*, praised Inguimberty thus: "It seems to me that the luminous back room [in the Pavilion], at the end of the [reconstructed] village which sums up Tonkin without the French, is, thanks to the paintings of Inguimberty, a corner of heaven, a corner of France, of Asiatic France, and one faithful to its gentle genius. The allegorical triptych [furnished by Inguimberty] expresses the allegory of France's presence in Tonkin, of our contribution of serenity to an otherwise patient, mischievous and febrile race, a contribution that had to wait our coming, the artist's coming, to raise its head and look around."[200] In other words, and as Nicholas Mirzoeff observes, during this period the visual imagery of colonialism was drawn as much from photography and painting, and exhibits such as the "back room" Tonkin village, as it was from literary works, and the visual quality (realism and accuracy) of such colonial representations resided in their ability to both transport the observer to a particular scene and transmit a Western sense of place (both as location on a map and as home) to that scene.[201]

French military minds and bodies were not so far away from all of this as one might imagine – and nor, we have latterly discovered, were Viet Minh ones. There was an intimacy of sorts to French military assessments of the Tonkin environment. Much of what Valluy related was anticipated in a French army report of 4 June 1947, on the obstacles presented by the mountainous area of northern Tonkin: "The Haut Tonkin presents undeniable difficulties for military personnel: considerable inconvenience in the use of motorized vehicles, difficulties of communication and supply, difficult conditions of life for the troops." A January 1951 report assessing Valluy's decision to regroup in the delta in order to shelter French troops from both Viet Minh and Chinese attack confided that while "the guiding idea was to create, on the northern front of the Delta, a fortified line [the De Lattre Line] of such defensive value that its attack by an adversary deprived of very powerful means is unthinkable," the French were nevertheless

> stretched and dispersed in sectors that are too vast for them, the units of the delta zone are content to hold solid and well-designed positions on the whole. But worn out, they can only venture a very short distance from the posts, supported from time to time by the reserves. . . . The rebel has thus become the undisputed master of vast densely populated areas where, according to Giáp's instructions, he organizes himself, strengthens himself, and, sheltered within heavily defended and mined villages, and every day more numerous, toys with the weak elements that come to him. Our relatively recent work of pacification is likely to diminish and give way to a climate of permanent insecurity.[202]

A large part of the problem was that neither side operating inside the De Lattre Line could risk large-scale or sustained mortar fire, again because it would threaten to breach the intricate network of dikes.

Tropicalising Indochina 71

Figure 2.4 Soldats de la boue. © Service historique de la Défense, CHA Vincennes, 10 H 2351

Echoing Valluy, French intelligence introduced the terrain to newly arrived soldiers in the following way:

> When the characteristics of the terrain take on a predominant importance – in relation to the usual characteristics [of military doctrine and strategy] – the whole tactic is influenced by it and gives combat a special character. This is the case in Indochina, where the terrain is presented in a completely different aspect to that of the classical theatres of operation in Europe, and more particularly of large undulating plains, or those medium-sized mountain ranges in which the French Army is accustomed to waging war.[203]

Another report continued: "Overpopulated, the Tonkinese Delta is dotted with villages separated by very narrow lanes. They are surrounded by a fence of bamboo which is almost impenetrable, and preceded by a zone of ponds."[204] The delta had "disproportionate" significance in the north a later report added: it was eight times smaller but five times more populous than surrounding areas, produced nearly all of the rice in Tonkin, and its inhabitant were "undoubtedly far superior to the mountain dwellers" and thus "possess[ed] cultural and political influence."[205] And as French commanders reflected, the area was "checkered by streams and dikes, and large [bamboo] picketed villages, which often constitute the objective of the attack, and appear from afar as big copses." The flooded rice field "is a swamp or thin film of water, but more often a lake from which the villages emerge. Leeches and their effects [on the body] make walking difficult." In the rainy season troops become "soldiers of the mud" (*soldats de la boue*).[206]

What was a form of escape to the artist was a hassle and form of internment to the military, although the landscape details captured by Inguimberty were not lost on French commanders and military advisers. While "a French soldier, whatever his rank, his origin, his training, must never forget that his presence in Viet Nam is of an essentially noble character," Colonel Fernand Gambiez, in command of the French units in the region of Ninh Binh in the heart of the delta, reflected, "he must respect the human person and the property of others. The Vietnamese individual is as attached to his paddy field [*sa rizière*], his hut, his buffaloes, as the peasant of our country is to his field [*son champ*], his house, his flocks."[207]

The jungle areas around the Tonkin and Mekong deltas were later evoked by the French writer, film director, and veteran of Điện Biên Phủ (where he was captured), Pierre Schoendoerffer in his well-known novel *La 317ème Section*:

> For fifteen minutes, the column has struggled slowly in the gloomy atmosphere of the jungle. . . . [W]ounded bearers wade through the mud, bump against the roots, and thrash at the thorns. Each moment they stop, accumulate a little energy, and continue, they advance with clumsy steps, head down, clash with their comrades and stop again, panting, to give themselves a little respite. Blood, perspiration and rain soak their fatigues, sticking them to their skin. They leave, crushed under their loads, slide on the vicious rot of the soil, spend all their strength at once to keep their equilibrium, fall on their knees, cling to the sticky bark of the trees, get up trembling like groggy boxers and try to make it a few more metres.[208]

The Viet Minh sustained itself in this shadowy, sticky, watery world, and Gourou found an inner and alternative beauty in it. "Indochina exercises on the memory of those who knew it a particular attraction," he wrote in a 1952 article written for France's leading colonial military magazine, *Tropiques*. "Few countries solicit such charming changes of scenery," he observed, and "a great part of the seduction of Indochina is connected to the human factor she owes it to," by which he meant human guile in finding an accord with tropical nature. He went on to contrive, "it is again down to a human fact, to the intervention of France, that Indochina exists as a coherent territorial entity. The past reveals no political construction that announces the birth of Indochina, and it is not the product of a convergence of physical determinisms."[209] But his defence of the Union Française on geographical grounds must have fallen flat on the many thousands of *soldats de la boue* who had fallen on their knees in the mud.

Gourou and Giáp briefly met again in the spring of 1946, at Dalat, where they sat on opposite sides of a negotiating table, discussing cultural and educational aspects of a possible settlement with France, but on this occasion exchanged just pleasantries, the official record of proceedings suggests.[210] We shall say more about this meeting in Chapter 5. Suffice it to note here that in his reflections on this meeting, Giáp hinted that Gourou's work and association with him had helped him to forge a militant tropicality which turned the tables on French exoticism and afforded its own panorama and military aesthetic. Giáp mused:

> The French delegates were completely out of step with the radical change in the situation in the Indochinese peninsula; they were steeped in an outdated colonialist ideology inherited from a decadent imperialism. . . . That night I stayed up very late. The [Dalat] night was like ink. Through the window I could hardly make out the lake from the mountains in the background. The pine forests of the hills and the wild forests of the Lang Biang Plateau were bathed in darkness. The struggle of our compatriots, of our guerrillas on the heights and in the distant jungle continued, and would continue.[211]

Such imagery fuelled what Goscha characterises as the "heroic myth of war" cultivated by the DRV's communist leadership, with the Viet Minh waging a "sacred resistance" against the French, and over a longer history, stretching to the final withdrawal of the United States from Vietnam in 1975, it camouflaged an "unprecedented assault on the Vietnamese body . . . [and] picture of human suffering, fear and corporal destruction" that issued from the French and American wars.[212] Giáp also likened the new Vietnamese nation to the shape of the yoke used by Vietnamese peasants to carry rice baskets. His militant tropicality was about devising a counter-symbolism of tropical land and life for nationalist and revolutionary ends, and his connection to Gourou points to a much larger arena of struggle in post-1945 Vietnam over the cultural and intellectual legacies of French rule.[213]

The part that this militant tropicality played in the defeat of the French – and later the Americans – should not be romanticised, for "the noble war to liberate the Fatherland was oftentimes horrific for the soldiers who fought in it," Viet Than Nguyen stresses, and the tropicality of Indochina's long war created loops and scars of dread, violence and sorrow.[214] In 2005 American journalist Sherry Buchanan helped to bring to publication the war diary and artwork of a young Vietnamese soldier, Pham Thanh Tam, from Điện Biên Phủ. If one can find in a number of Tam's watercolours and sketches – especially, perhaps, *On the Road to Điện Biên Phủ* – echoes of Inguimberty's art and style, then it is because Tam was taught by Luong Xuan Nhi, who, in turn, had been trained by Inguimberty at the École des beaux-arts in Hanoi.[215] Tam enrolled on one of the revolutionary art courses set up by a number the school's graduates in the Viet Minh resistance zone deep in the jungle of North Vietnam in 1946–1948. Colonial art was given a renewed revolutionary purpose, and with young Vietnamese resistance fighters, many of whom had been separated from their families by the Japanese occupation and then French military incursion, designing banners, flags, posters and theatre sets. But Tam's remarkable work – one of the only Vietnamese visual records of Điện Biên Phủ known to exist – went a step further by juxtaposing the beauties of the landscape and the sorrow of war.

A subsequent painting by Tam, entitled *Barbed-wire Fence After the War* (1963), with bright tropical foliage wrapped around two lines of barbed wire, at once captures the militarisation of the landscape and the hope that its points to a brighter rather than twisted future. But things at the time looked grim, and the tropics played their part in the desolation. "[N]o imposing monument, either French or Viet-Minh, honours the 10,000 men who died here and who may have done more to shape the fate of the world than the soldiers at Agincourt, Waterloo, or Stalingrad," Bernard Fall recounted in his iconic account of Điện Biên Phủ, *Hell in a Very Small Place*: just 500 Vietnamese bodies were placed in a small cemetery at the foot of a banyan tree, and the French cemetery soon "disappeared altogether under a succession of monsoon floods and tropical underbrush."[216]

Mus, who was one of Gourou's friends during his Indochina days, and later at the Collège de France, also conjured with militant tropicality, and at two levels. He described the Tonkin Delta as a chessboard of compartmentalised rice fields, and used the chessboard as a metaphor for a revolutionary game that hinged on stability, discipline, rhythm and guile, and with the villages of the delta constituting a dissident zone out of the reach of the French.[217] And on a larger scale, he questioned the interminability of the Orientalist tradition to which he belonged, and the revolutionary one that supplanted it, using a metaphor of seasonality:

> Immutable Asia, static Asia, as we used to hastily judge it, turns out to be, when you look more closely, steeped in the scientifically proven principle of mutation, in other words, of discontinuity. Every regime becomes worn out, every State changes. The world, as these peoples have seen, is seasonal in its most varied manifestations. That is where their politics resides. Revolutions are the seasons of history.[218]

We might suggest that the seasonal – tropical – environment of Indochina was not incidental to either the revolution and war that defeated the French, or the process by which Indochina was inched out of the Orient and into tropics.

Gourou in the Tropical-Orient

Indochina did not start out in the Orient and end up in the tropics. These two locational markers and motifs of otherness were present at the birth of France's colonial odyssey in the Far East. They were both part of what Panivong Norindr has described as a "phantasmatic Indochina," and they worked their way into economics and militarism as much as they did culture and art. "At issue is not simply the fictional or real existence of Indochina," Norindr writes, "but, perhaps even more important, the question of how factual truths have been used and manipulated to construct an identity for Indochina."[219] From the late nineteenth century, the monsoon character of colonialism and conflict in Indochina became more pronounced, and, like French Orientalism, it had both positive and negative (alluring and disturbing) connotations. Norindr goes on to explore how Indochina became a subject of architectural, literary and cinematic reflection within a broadly Orientalist tradition. We have explored how the region also became a tropical scene of settler and military activity and angst. Between 1900 and 1950 Indochina became tropicalised through the experiences of settlers, soldiers and scholars such as Gourou, who studied the Tonkin Delta, and fellow geographer Robequain, who worked on the Annamite region of Thanh Hoá immediately to the south. The leitmotivs of settlers' and soldiers' tropicality were displacement, disorientation, disease and deficiency. Indochina's tropicality threatened French power and identity, and was mixed in with the aggression, fear, loathing and inequality that marked France's colonial mission. But as we shall now show more fully, for Gourou, this tropicality was about beauty, the reverie of difference, and freedom of inquiry. Gourou captured the latter sentiment in a 1964 obituary to Robequain, noting that it was with the support of the EFEO and its Orientalist apparatus that the two of them were able to "undertake without impediment" their geographical studies in a "tropical territory."[220] And they had done much to engineer this reconfiguration of Indochina as tropical.

While Gourou and Giáp came from entirely different worlds, they both conjured with ideas of place and belonging, and had more in common than one might suppose. As we shall now show, for Gourou there was more to the human condition than money, power, and success – all of them core attributes of empire – and he was deeply sceptical about the idea of a benevolent colonialism. Underlying his aesthetic appreciation of Inguimberty's art and respect for Giáp was a concern with home – or, more precisely, the art of furnishing oneself a place in the world – and the tropical hue of that place. As the French and Americans learned at considerable cost to themselves as well as Vietnam and its people, this was a core human value that would be defended to the death. It was in Braudel's time of *structure* that the Indochina War made sense to Gourou. And it was within the revolutionary frame of guerrilla warfare, and a militant tropicality, that Giáp planned Vietnam's future. Gourou saw Inguimberty's paddy field as a thing of beauty hewn from an age-old peasant struggle with nature, rather than as a scene of violence. For Giáp, on the other hand, this thing of beauty was also a key resource and ally in the fight against Indochina's imperialist aggressors.

In a recent and insightful retrospective on the ambiguous place that the Indochina War has in post-war French memory, Kathryn Edwards traverses myriad "disagreements over whether the war was primarily a war of colonial reconquest or a struggle against communism," how such disagreements are tied to varying interpretations of French colonialism in Indochina, and how in the bulk of evaluations that came from war veterans, one can find either a refusal "to engage with the colonial dimension of the war," or the argument "that military objectives had nothing to do with colonialism."[221] Following the leads of figures such as Gourou, Valluy and Giáp, our suggestion in this chapter has been that the Indochina War and how it is remembered becomes hollowed out if it is lifted out of its tropical mooring in the jungles and swamps, heat and rain, and water and mud of Indochina, and if this mooring is not seen as intrinsically bound up with the violence of war and colonialism.

Lastly, Jacques Rancière helps us to see that the politics and aesthetics of representation at work in the wide range of contributions to the 1949 special issue of *France Illustration*, which we have used here to develop a broader story, find common ground in their concern with what was conceivable or incredible, imagined or real, visible or invisible, and appreciable or insignificant about French Indochina.[222] For Duchartre, Christen, Gourou and Valluy – that is, for diverse figures from governmental, scholarly and military spheres alike – 'mystique' is one lodestar of this concern. Mystique appears throughout the special issue as a byword for the struggle that France had in delineating itself in Indochina – in saying and knowing what it was doing there. After more than 70 years of colonising, settling, and getting to know Indochina, the French were still talking about the region's secrets and opacity. Nevertheless, by 1949 a good deal of delineation had gone on, and not least by Gourou, who was an expert on questions of land and population, and whose work on the peasant culture of the Tonkin Delta had placed him in the thick of a complex set of colonial relationships linking fieldwork, scientific institutions, and imperial administration, which we shall now investigate in more detail.

Notes

1 *France Illustration* (1945–1957) was a rebooted version of the weekly *L'Illustration* (1843–1944), which was shut down upon the Liberation of France because of its links with the Vichy propaganda machine. *L'Illustration* was the one of the first newspapers to use coloured artwork, and Indochina featured regularly. From 1907 the magazine also promoted the process of colour photography pioneered by the Lumière brothers, and thus hastened the shift in the relations between art, science and colour (*bariolage*), which had shaped French impressionist art, towards new – print and film – processes of mechanical reproduction. See Laura Anne Kalba, *Color in the Age of Impressionism: Commerce, Technology, and Art* (Philadelphia: University of Pennsylvania Press, 2017); and Jean-Noël Marchandiau, *L'Illustration (1843–1944). Vie et mort d'un journal* (Toulouse: Privat, 1987).
2 Even if, as France's leading chronicler of the conflict, Yves Gras, stressed, until 1953 the fighting was almost entirely in Vietnam. For a wider and more nuanced view on the political and military rhetoric of the DRV, see Christopher Goscha, "Une guerre pour l'Iindochine? Le Laos et le Cambodge dans le conflit franco-vietnamien (1948–1954)," *Guerres mondiales et conflits contemporains* 211, no. 3 (2003): 29–58.
3 Alexander C.T. Geppert, *Fleeting Cities: Imperial Expositions in Fin-de-Siècle Europe* (New York: Palgrave Macmillan, 2010).
4 Alan MacFarlane, "Fernand Braudel and global history," Unpublished lecture, Institute of Historical Research 1996, 1–3, www.alanmacfarlane.com/TEXTS/BRAUDEL_revised.pdf.
5 Fernand Braudel, "Personal testimony," in Stuart Clark ed., *The Annales School: Critical Assessments, Volume II: The Annales School and Historical Studies* (London and New York: Routledge, 1999), 491–408, at 496.
6 Mary Louise Pratt, *Imperial Eyes: Travel Writing and Transculturation* (London and New York: Routledge, 1992), 7 and *passim*.
7 Nancy Stepan, *Picturing Tropical Nature* (London: Reaktion Books, 2001), 18.
8 The iconic postcolonial formulation of this position can be found in Kelpana Sheshardi-Crooks, "At the margins of postcolonial studies," *Ariel: A Review of International English Literature* 26, no. 3 (1995): 47–71.
9 They were elected to chairs in 1947 and 1949, respectively; and both were proposed by Lucien Febvre (see Chapter Six).
10 Paul Claval, Interview with the authors.
11 In *La Méditerranée* Braudel commented "a geographer once wrote jokingly to me [of the study] 'Not enough meat and too many bones.'" Letter from Gourou 27 June 1949, Fernand Braudel, *The Mediterranean and the Mediterranean World in the Age of Philip II* vol. 1 trans. Siân Reynolds (London: Fontana/Collins, 1972), 240, fn, 42.

12 Claude Lévi-Strauss, correspondence with the authors, 19 September 2008.
13 Ricardo Padrón, "'The Indies of the West' or, the tale of how an imaginary geography circumnavigated the globe," in Christina H. Lee ed., *Western Visions of the Far East in a Transpacific Age, 1522–1657* (Abingdon: Routledge, 2016), 19–42, at 30.
14 Corey Ross, *Ecology and Power in the Age of Empire: Europe and the Transformation of the Tropical World* (Oxford: Oxford University Press, 2017), 23.
15 Ross, *Ecology and Power in the Age of Empire*, 9.
16 Christophe Bonneuil, "Crafting and disciplining the tropics: Plant science in the French colonies," in John Krige and Dominique Pestre eds., *Science in the Twentieth Century* (London and New York: Routledge, 2003), 77–96, at 77, 93.
17 Ross, *Ecology and Power in the Age of Empire*, 9–10.
18 Trumbull White, *Our New Possessions: A Graphic Account, Descriptive and Historical, of the Tropical Islands of the Sea Which Have Fallen Under Our Sway . . . Four Books in One* (St Louis, MO: J.H. Chambers & Co., 1898), 22–23.
19 White, *Our New Possessions*, 163.
20 Frederick Upham Adams, *Conquest of the Tropics: The Story of the Creative Enterprises by the United Fruit Company* (New York: Doubleday & Co., 1914); Richard H. Grove, *Green Imperialism: Colonial Expansion, Tropical Island Edens and the Origins of Environmentalism, 1600–1860* (Cambridge: Cambridge University Press, 1995).
21 Peter Chapman, *Jungle Capitalists: A Story of Globalisation, Greed and Revolution* (New York: Canongate, 2008).
22 See, for example, Jyotirmoy Pal Chaudhuri, *Whitehall and the Black Republic: A Study of Colonial Britain's Attitude Towards Liberia, 1914–1939* (London: Palgrave Macmillan, 2018); Frédéric Thomas, "Protection des forêts et environnementalisme colonial: Indochine, 1860–1945," *Revue d'histoire moderne et contemporaine* 56, no. 4 (2009): 104–136; Michael A. Havinden and David Meredith, *Colonialism and Development: Britain and Its Tropical Colonies, 1850–1960* (London and New York: Routledge, 1995); Pierre Brocheux and Daniel Hémery, *Indochina: An Ambiguous Colonization, 1858–1954* trans. Ly Lan Dill-Klein (Berkeley: University of California Press, 2009), Ch. 3.
23 Grove, *Green Imperialism*, 486.
24 Richard P. Tucker, *Insatiable Appetite: The United States and the Ecological Degradation of the Tropical World* (Berkeley: University of California Press, 2000).
25 Jean Suret-Canale, *French Colonialism in Tropical Africa, 1900–1945* trans. Till Gottheiner (London: C. Hurst & Co., 1971), 300.
26 Frederick Cooper, *Citizenship Between Empire and Nation: Remaking France and French Africa, 1945–1960* (Princeton: Princeton University Press, 2014), 180.
27 W. Arthur Lewis, "The export stimulus," in William Arthur Lewis ed., *Tropical Development, 1880-1913: Studies in Economic Progress* (Evanston, IL: Northwestern University Press, 1970), 13–45, at 42–44.
28 Joseph Morgan Hodge, *Triumph of the Expert: Agrarian Doctrines of Development and the Legacies of British Colonialism* (Athens, OH: Ohio University Press, 2007).
29 Paul Bernard, "L'avenir économique de l'Indochine," *France Illustration. Le monde illustré*. Numéro spécial sur l'Indochine no. 190 (juin 1949): n.p. And see Andrew Hardy, "The economics of French Rule in Indochina: A biography of Paul Bernard (1892–1960)," *Modern Asian Studies* 32, no. 4 (1998): 807–848; and Martin J. Murray, "'White gold' or 'white blood'? The rubber plantations of colonial Indochina, 1910–1940," in E. Valentine Daniel, Henry Bernstein and Tom Brass eds., *Plantations, Proletarians and Peasants in Colonial Asia* (London: Frank Cass, 1992), 41–67.
30 See P.D. Curtin, "'The white man's grave': Image and reality, 1780–1850," *Journal of British Studies* 1, no. 1 (1961): 94–110; Alexandra Minna Ster, "Yellow fever crusade: US colonialism, tropical medicine, and the international politics of mosquito control, 1900–1920," in Alison Bashford ed., *Medicine at the Border: Disease, Globalization and Security, 1850 to the Present* (New York: Palgrave Macmillan, 2006), 41–59.

31 Daniel Stephen, *The Empire of Progress: Africans, Indians, and Britons at the British Empire Exhibition, 1924–1925* (New York and London: Palgrave Macmillan, 2013), 82.
32 Frantz Fanon, *The Wretched of the Earth* trans. Richard Philcox; orig. pub. Eng. 1963 (New York: Grove Press, 2004), 17.
33 Fanon, *Wretched of the Earth*, 17.
34 Michael A. Osborne, "European visions: Science, the tropics, and the war on nature," in De Roland Waast ed., *Les sciences hors d'Occident au xx'siécle* (Paris: Orstom, 1995), 20–32, at 30.
35 Stepan, *Picturing Tropical Nature*, 18.
36 Emily S. Rosenberg, "Transnational currents in a shrinking world," in Emily S. Rosenberg ed., *A World Connecting 1870–1945* (Cambridge, MA: Harvard University Press, 2012), 813–996, at 887.
37 This story is discussed in Lauren Janes, *Colonial Food in Interwar Paris: The Taste of Empire* (London: Bloomsbury, 2016), 85–92.
38 See Hugues Tertrais, "France and the associated states of Indochina, 1945–1955," in Marc Frey, Ronald W. Pruessen and Tan Tai Yong eds., *The Transformation of Southeast Asia: International Perspectives on Decolonization* (London and New York: Routledge, 2003); Stein Tønnesson, *Vietnam 1946: How the War Began* (Berkeley: University of California Press, 2010), 116. Peasants were promised greater access to communal land and credit, and rents were capped. See Martin Thomas, *Fight or Flight: Britain, France and Their Roads from Empire* (Oxford: Oxford University Press, 2014), 141–142.
39 Christopher Goscha, "A 'total war' of decolonization? Social mobilization and state-building in Communist Vietnam (1949–1954)," *War & Society* 31, no. 2 (2012): 136–162.
40 The causes of the famine were complex and remain contested. It was by no means simply a 'natural' disaster precipitated by flood and drought. DRV propagandists blamed Japan's wartime occupation, and especially the seizure of rice and other foodstuffs. The Vichy regime in Indochina was complicit in this, and long-term French mismanagement of the economy also played a role, as did Allied bombing of the north during the war. See Geoffrey Gunn, "The great Vietnamese famine of 1944–1945 revisited," *The Asia-Pacific Journal* 9, no. 4 (2011): 1–20.
41 Cited in Frank Cain, *America's Vietnam War and Its French Connection* (New York and London: Routledge, 2017), 74; Panagiotis Dimitrakis, *Secrets and Lies in Vietnam: Spies, Intelligence and Covert Operations in the Vietnam Wars* (London: I. B. Tauris, 2016), Ch. 2.
42 Võ Nguyên Giáp, *Military Art of People's War: Selected Writings of General Vo Nguyen Giap* ed. Russell Stetler (New York: Monthly Review Press, 1970), 83.
43 To the tune of 15 per cent of total expenditure by the end of 1948, and 33 per cent by 1952. See Laura M. Calkins, *China and the First Indochina War 1947–1954* (London and New York: Routledge, 2013), 8–36.
44 Official figures released soon after the war put the figure at 92,800 dead and 76,400 wounded. Michel Bodin revised this figure to 112,032. The French reckoned that around 500,000 Vietnamese people lost their lives (a high proportion of them civilians), although around 300,000 is likely a more accurate figure. For a discussion, see Christopher Goscha, *Historical Dictionary of the Indochina War (1945–1954): An International and Interdisciplinary Approach* (Copenhagen: NIAS Press, 2011), 88–89, where it is estimated that French losses accounted for around 17 per cent of the total.
45 See Charles R. Shrader, *A War of Logistics: Parachutes and Porters in Indochina, 1945–1954* (Lexington: University of Kentucky Press, 2015), 91–103. Xiaobing Li, *A History of the Modern Chinese Army* (Lexington: University of Kentucky Press, 2007), 207–208. Giáp did not heedlessly follow Chinese military advice. At crucial times in the conflict, and usually successfully, he bucked against it.
46 "The crisis in Indochina" (secret) 10 February 1950, 5. United States, Central Intelligence Agency, www.cia.gov/library/readingroom/docs/DOC_0000258850.pdf.

47 Cited in David A. Anderson, *The Joint Chiefs of Staff and the First Indochina War, 1947–1954* (Washington, DC: U.S. Government Printing Office, 2004), 36. For an overview of these developments, see Brocheux and Hémery, *Indochina*, 365–370. Also see Edward Rice-Maximin, *Accommodation and Resistance: The French Left, Indochina, and the Cold War* (New York: Greenwood Press, 1986), 84.
48 Cited in Robert J. McMahon, *Dean Acheson and the Creation of an American World Order* (Washington, DC: Potomac Books, 2009), 123.
49 William J. Duiker, *Ho Cho Minh: A Life* (New York: Theia Books, 2000), 415–419.
50 Fredrik Logevall, *The Embers of War: The Fall of an Empire and the Making of America's Vietnam* (New York: Random House, 2012), 221–224. Mao soon granted the DRV formal diplomatic recognition and sought to bolster Chinese economic and military assistance to the Viet Minh.
51 By the end of 1949 the French force required to successfully mount operations unaided had swelled to battalion size (800 troops), and that figure continued to grow. John Prados, "Assessing Dien Bien Phu," in Mark Atwood Lawrence and Fredrik Lovegall eds., *The First Vietnam War: Colonial Conflict and Cold War Crisis* (Cambridge, MA: Harvard University Press, 2007), 215–239, at 216.
52 A dilemma, we shall see, grasped by French officers at the time, and reiterated in Yves Gras, *Histoire de la guerre d'Indochine* 2nd edition (Paris: Denoël, 1992), 471–474; and Christopher Goscha, "Une guerre pour l'Iindochine? Le Laos et le Cambodge dans le conflit franco-vietnamien (1948–1954)," *Guerres mondiales et conflits contemporains* no. 211 (2003): 29–58.
53 Giáp, *Military Art of People's War*, 91.
54 The Union Française was created in 1946, under the constitution of the Fourth Republic, to replace the Empire Français (of colonies, protectorates and mandated territories), and brought an official end to the 'indigenous' (*indigène*) status of French subjects in colonial regions. The Union Indochinoise was renamed la Fédération indochinoise in 1947 and lasted until 1954. For short biographies of Pignon and Sarraut, see Goscha, *Historical Dictionary of the Indochina War*, 382–384, 417–418.
55 Nguyen Thi Dieu, *The Mekong and the Struggle for Indochina: Water, War, and Peace* (Westport, CT: Praeger, 1999), 219.
56 Président Albert Sarraut, "L'Indochine française," *France Illustration. Le monde illustré*. Numéro spécial sur l'Indochine no. 190 (juin 1949): n.p.
57 See Nicola Cooper, "Colonial humanism in the 1930s: The case of Andrée Viollis," *French Cultural Studies* 17, no. 2 (2006): 189–205.
58 See Frederick Cooper, *Colonialism in Question: Theory, Knowledge, History* (Berkeley: University of California Press, 2005), 197.
59 Martin Thomas, "Albert Sarraut, French colonial development, and the communist threat, 1919–1930," *The Journal of Modern History* 77, no. 4 (December 2005): 917–955. Also see Pascale Bezançon, *Une colonisation éducatrice? L'expérience indochinoise, 1860–1945* (Paris: L'Harmattan, 2002).
60 See Christopher Goscha, *The Penguin History of Modern Vietnam* (London: Penguin, 2017), 123–126.
61 Paul Mus, "The role of the village in Vietnamese politics," *Pacific Affairs* 22, no. 3 (1949): 265–272.
62 In Indochina and elsewhere in the French colonial world, Goscha notes, the two policies were "never mutually exclusive" and "never worked out so easily on the ground." Goscha, *Penguin History of Modern Vietnam*, 131.
63 Goscha, *Penguin History of Modern Vietnam*, 131.
64 Leela Ghandi, *The Common Cause: Postcolonial Ethics and the Practice of Democracy, 1900–1955* (Chicago: University of Chicago Press, 2014), 93 and *passim*.
65 Hugues Tertrais, "L'électrification de l'Indochine," *Outre-mers* 89, no. 334–335 (2002): 589–600.
66 Goscha, *Penguin History of Modern Vietnam*, 131.

67 Charles Robequain, *L'évolution économique de l'Indochine Française* (Paris: Centre d'Études de Politique Étrangère, 1939). Also see Brocheux and Hémery, *Indochina*, 183 who give a slightly lower figure of 34,000 for the French population in 1940.
68 See Geoffrey Gunn, "Coercion and co-optation of Indochinese worker-soldiers in World War I: Mort pour la France," *Asia-Pacific Journal* 14, no. 1 (2016), online http://apjjf.org/2016/01/5-Gunn.html
69 Gandhi, *The Common Cause*, 96–97.
70 Cited in Duiker, *Ho Chi Minh*, 66–69.
71 The political imprisonment of Vietnamese journalists for being critical of the French became a *cause célèbre* of Vietnamese anti-colonialism during the 1920s. The arraignment of free speech was deemed an exemplar of the violence of the colonial state. See Peter Zinoman, *The Colonial Bastille: A History of Imprisonment in Vietnam, 1862–1940* (Berkeley: University of California Press, 2001), 247.
72 Katelyn E. Knox, *Race on Display in 20th- and 21st Century France* (Liverpool: Liverpool University Press, 2016), 6 and *passim*.
73 V. Raoul Girardet, *L'idée coloniale en France de 1871 à 1962* (Paris: La Table Ronde, 1972), 15. Racism was not simply transacted across metropolitan-colonial divides. During the interwar years France became home to more immigrants (7 per cent of the total population) than any other European country, chiefly out of France's need for unskilled workers, and with the influx from Russia, Central Europe and the Iberian Peninsula fuelling fears on the political right about invasion and racial mixing. Clifford Rosenberg, "Albert Sarraut and republican racial thought," *French Politics, Culture & Society* 20, no. 3 (2002): 97–114.
74 Gregory Shaya, "The flâneur, the badaud, and the making of a mass public in France, circa 1860–1910," *American Historical Review* 109, no. 1 (2004): 41–77, at 42.
75 See Alice Conklin, *In the Museum of Man: Anthropology, Race, and Empire in France, 1850–1950* (Ithaca, NY: Cornell University Press, 2013); and Aline Demay, *Tourism and Colonization in Indochina, 1898–1939* (Cambridge: Cambridge Scholars Publishing, 2014).
76 Alice L. Conklin, Sarah Fishman and Robert Zaretsky, *France and Its Empire Since 1870* 2nd edition (Oxford: Oxford University Press, 2015), 66.
77 Conklin, *In the Museum of Man*.
78 Raymond Kuhn, *The Media in France* (London and New York: Routledge, 1995), 22.
79 Nélia Dias, "From French Indochina to Paris and back again: The circulation of objects, people, and information, 1900–1932," *Museums and Society* 13, no. 1 (2015): 7–21.
80 Sarraut, "L'Indochine française," n.p. Also see Albert Sarraut, *La mise en valeur des colonies françaises* (Paris: Payot, 1923), 96–98; Veronique Dimier, "Politiques indigènes en France et en Grande-Bretagne dans les années 1930: aux origines coloniales des politiques de développement," *Politique et Sociétés* 24, no. 1 (2005): 74; and Walter Schicho, "'Keystone of progress' and *mise en valeur d'ensemble*: British and French colonial discourses on education for development in the interwar period," in Joseph M. Hodge, Gerald Hödl and Martina Kopf eds., *Developing Africa: Concepts and Practices in Twentieth-Century Colonialism* (Cambridge: Cambridge University Press, 2015), 222–249.
81 L. Pignon, "Et maintenant . . .," *France Illustration: Le monde illustré*. Numéro spécial sur l'Indochine no. 190 (juin 1949): n.p.
82 Alexander Keese, "*A culture of panic*: 'Communist' scapegoats and decolonization in French West Africa and French Polynesia (1945–1947)," *French Colonial History* 9, no. 1 (2008): 131–146.
83 Nicola Cooper, *France in Indochina: Colonial Encounters* (Oxford: Berg, 2001); Laurence Monnais-Rousselot, *Médecine et colonisation: L'aventure indochinoise, 1860–1939* (Paris: CNRS, 1999), 84–90; Soghieng Au, *Mixed Medicines: Health and Culture in French Colonial Cambodia* (Chicago: University of Chicago Press, 2011), Ch. 6.
84 David Marr, *Vietnamese Anticolonialism, 1885–1925* (Berkeley: University of California Press, 1971), 182.

85 Ranajit Guha, "Not at home in empire," *Critical Inquiry* 23, no. 3 (1997): 482–493, at 482.
86 Pignon, "Et maintenant...," n.p.
87 S. E. Pham Van Binh, "Bao Dai, empereur démocrate et sportif accompli," *France Illustration: Le monde illustré*. Numéro spécial sur l'Indochine no. 190 (juin 1949), n.p.
88 Pierre-Louis Duchartre, "Imaginaire populaire et Indochine," *France Illustration: Le monde illustré*. Numéro spécial sur l'Indochine no. 190 (juin 1949): n.p.
89 Pierre Singaravélou, *L'École Française d'Extrême-Orient ou l'institution des marges (1898–1956)* (Paris: L'Harmattan, 1999), 58.
90 Singaravélou, *L'École Française d'Extrême-Orient*, 162.
91 Singaravélou, *L'École Française d'Extrême-Orient*, 185.
92 Nguyen Phuong Ngoc, "Adopting western methods to understand one's own culture: Social and cultural studies by Vietnamese scholars of the French colonial period," trans. Helene Tammik, in Regna Darnell and Frederic W. Gleach eds., *Historicizing Theories, Identities, and Nations. Histories of Anthropology, 11* (Lincoln and London: Nebraska University Press, 2017), 199–218, at 207.
93 Goscha, *The Making of Modern Vietnam*, 97–109.
94 Brocheux and Hémery, *Indochina*, 32.
95 Louis Finot, "Paul Doumer (1857–1932)," *Bulletin de l'Ecole française d'Extrême-Orient* 33 (1933): 549–552, at 552.
96 See, recently, David Biggs, "Arial photography and colonial discourse on the agricultural crisis in late-colonial Indochina, 1930–1945," in Christina Folke Ax, Niels Brimnes, Niklas Thode Jensen and Karen Oslund eds., *Cultivating the Colonies: Colonial States and Their Environmental Legacies* (Athens, OH: Ohio University Press, 2011), 109–132.
97 See Susan Bayly, *Asian Voices in a Post-Colonial Age: Vietnam, India and Beyond* (Cambridge: Cambridge University Press, 2007), 33–36.
98 See Martin Jay, *Sites of Memory, Sites of Mounring: The Great War in European Cultural History* (Cambridge: Cambridge University Press, 1995), 122–131; Patricia Mainardi, *Another World: Nineteenth-Century Illustrated Print Culture* (New Haven, CT: Yale University Press, 2017).
99 Winter, *Sites of Memory*, 127.
100 Marco R. Deyasi, "Indochina, 'Greater France' and the 1931 Colonial Exhibition in Paris: Angkor Wat in blue, white and red," *History Workshop Journal* 80, no. 1 (2015): 123–141, at 123.
101 Cited in Deyasi, "Indochina, 'Greater France' and the 1931 Colonial Exhibition," 123.
102 Dana S. Hale, *Races on Display: French Representations of Colonized Peoples, 1886–1940* (Bloomington, IN: Indiana University Press, 2008), 69–74.
103 Regarding the latter imagery, Girardet argues that what distinguishes this French rhetoric, or myth of progress, which was a staple of colonialism more generally, was France's self-referential invocation of the myth – i.e., the suggestion that France was trying to make peace with its own past. Girardet, *L'idée coloniale*, 15.
104 Duchartre, "Imaginaire populaire et Indochine," n.p.
105 Douglas Pike, *Vietcong: The Organization and Techniques of the National Liberation Front of South Vietnam* (Cambridge, MA: MIT Press, 1966), 178.
106 See Kathryn Robson and Jennifer Yee, "Introduction," in Kathryn Robson and Jennifer Yee ed., *France and 'Indochina': Cultural Representations* (Lanham, MD: Lexington Books, 2005), 7.
107 Kimberley J. Healey, "Andrée Viollis in Indochina: The objective and picturesque truth about French colonialism," *A.J.S.S* 31, no. 1 (2003): 19–35, at 35.
108 Healey, "Andrée Viollis in Indochina," 35; Nikki Cooper, "(En)gendering Indochina: Feminisation and female figurings in French colonial discourses," *Women's Studies International Forum* 23, no. 6 (2001): 751.
109 Jules Sion, *Asie des Moussons. Inde – Indochine – Insulinde*. Tome IX *Géographie Universelle* (Paris: Librairie Armand Colin, 1929). And Sion cited in Bernard Debarbieux and Gilles Rudaz, *The Mountain: A Political History From the Enlightenment*

to the Present trans. Jane Todd (Chicago: University of Chicago Press, 2015), 154. Sion was Professor of Geography at Montpellier, and had never visited Asia.
110 Xavier de Christen, "Indochine," *France Illustration: Le monde illustré*. Numéro spécial sur l'Indochine. No. 190; juin 1949, n.p.
111 Thi Tuyet Trinh Nguyen, "L'imaginaire colonial français de l'indohine 1890–1935," unpublished PhD dissertation L'université François – Rabelais de Tours 2014.
112 Henri Mouhot, *Travels in the Central Parts of Indo-China (Siam), Cambodia, and Laos During the Years 1858, 1859 and 1860* (London: John Murray, 1864), 293. 293.
113 Charlotte Rogers, *Jungle Fever: Exploring Madness and Medicine in Twentieth-Century Tropical Narratives* (Nashville, TN: Vanderbilt University Press, 2012), 64. She is commenting on Malraux's 1930 *La Voie royale* (*The Way of the Kings*).
114 Christine Doran, "Popular orientalism: Somerset Maugham in mainland Southeast Asia," *Humanities* 5, no. 13 (2016), https://pdfs.semanticscholar.org/2321/d09a7b8e35dc11741775e160826519ebe81f.pdf.
115 Alison Bashford, *Global Population: History, Geopolitics, and Life on Earth* (New York: Columbia University Press, 2014), 147.
116 Eric T. Jennings, *Vichy in the Tropics: Petain's National Revolution in Madagascar, Guadeloupe and Indochina, 1940–1944* (Stanford, CA: Stanford University Press, 2001), 137.
117 Geoffrey C. Gunn, *Rice Wars in Colonial Vietnam: The Great Famine and the Viet Minh Road to Power* (New York: Rowan & Littlefield, 2014), 25.
118 Van Nguyen-Marshall, *In Search of Moral Authority: The Discourse on Poverty, Poor Relief and Charity in French Colonial Vietnam* (Vancouver: UBC Press, 2008), 34–41.
119 Jennings, *Vichy in the Tropics*, 137 and *passim*. Also see Erica Peters, *Appetites and Aspirations in Vietnam: Food and Drink in the Long Nineteenth Century* (Lanham, MD: AltaMira Press, 2012).
120 Gunn, *Rice Wars*, 27.
121 Arthur J. Dommen, *The Indochinese Experience of the French and the Americans: Nationalism and Communism Cambodia, Laos and Vietnam* (Indianapolis: Indiana University Press, 2001), 27; Tuan Pham Anh and Kelly Shannon, "Indigenous knowledge and international practices: The case of the Red River Delta," *N-Aerus* XI (2010), http://n-aerus.net/web/sat/workshops/2010/pdf/PAPER_pham_t.pdf.
122 Charles Robequain, *Le Thanh Hoá, étude géographique d'une province annamite*, 2 vols. (Paris and Brussels: Van Qest, 1929), II, 613.
123 Philippe Roques et Marguerite Donnadieu, *L'Empire français* (Paris: Gallimard, 1940), 9–10, 104–108.
124 Roques et Donnadieu, *L'Empire français*, 104–108.
125 Roques et Donnadieu, *L'Empire français*, 110–111, 116–117.
126 Roques et Donnadieu, *L'Empire français*, 113.
127 ANOM FOM 532/D54.
128 René Bouvier, *La crise en Indochine: Sémaphore du 23 Mai 1933* (Marseilles: Le Sémaphore, 1933), 12–16.
129 Laure Adler, *Marguerite Duras* (Paris: Gallimard, 1998), 139.
130 Marguerite Duras, *Un barrage contre le Pacifique* (Paris: Gallimard, "Collection Folio," 1950), 23.
131 Duras, *Un barrage contre le Pacifique*, 209, 159.
132 Ulrich E. Bach, *Tropics of Vienna: Colonial Utopias of the Habsburg Empire* (Oxford: Berghahn, 2016), 116.
133 Anne McClintock, *Imperial Leather: Race, Gender and Sexuality in the Colonial Contest* (London and New York: Routledge, 1995), 26 (emphases in original).
134 Mathieu Rayssac, *Les médicins de l'assistance médicake en Indochine (1905–1939)* (Paris: L'Harmattan, 2015).
135 Michitake Aso, "Patriotic hygiene: Tracing new places of knowledge production about malaria in Vietnam, 1919–1975," *Journal of Southeast Asian Studies* 44, no. 3 (2013): 423–443.

136 Nicola Cooper, "Disturbing the colonial order: Dystopia and disillusionment in Indochina," in Kathryn Robson and Jennifer Yee eds., *France and 'Indochina': Cultural Representations* (Lanham, MD: Lexington Books, 2005), 79–94, at 81, 87, 88, 91; Cooper, *France in Indochina*, 149–150.
137 Eric T. Jennings, *Imperial Heights: Dalat and the Making and Undoing of French Indochina* (Berkeley: University of California Press, 2011), 8.
138 Cited in Aline Demay, *Tourism and Colonization in Indochina (1898–1939)* (Newcastle Upon Tyne: Cambridge Scholars Publishing, 2014), 23.
139 Laura Victoir, "Hygienic colonial residences in Hanoi," in Laura Victoir and Victor Zatsepine eds., *Harbin to Hanoi: The Colonial Built Environment in Asia, 1860–1940* (Hong Kong: Hong Kong University Press, 2013), 207–230, at 232. Also see Olivier Le Cour Grandmaison, *L'Empire des hygienists: Vivre aux colonies* (Paris: Fayard, 2014).
140 Marie-Paule Ha, *French Women and the Empire: The Case of Indochina* (Oxford: Oxford University Press, 2014), 116–117.
141 Cooper, "Disturbing the colonial order," 81, 91.
142 Our line of reasoning and history here follows Christopher Bongie, *Exotic Memories: Literature, Colonialism, and the Fin de Siècle* (Stanford, CA: Stanford University Press, 1991).
143 Michael Vann, "Of le Cafard and other tropical threats: Disease and white colonial culture in Indochina," in Kathryn Robson and Jennifer Yee eds., *France and 'Indochina': Cultural Representations* (Lanham, MD: Lexington Books, 2005), 95–106, at 95. Also see Milton Osborne, *Fear and Fascination in the Tropics: A Reader's Guide to French Fiction on Indochina* (Madison: University of Wisconsin Press, 1986), 96.
144 ANOM: FM EE/11/4453/8/GOU – (Gilberte), 12 September 1928.
145 John Kleinen, Interview with Pierre Gourou, 24 August 1994, n.p., unpublished typescript, cited with the permission of the author.
146 Gilberte Bray. Interview with the authors.
147 Général Valluy, "Les corps expéditionnaire," *France-Illustration: Le monde illustré*. Numéro spécial sur l'Indochine no. 190 (juin 1949), n.p.
148 Bernard B. Fall, *Street Without Joy: The French Debacle in Indochina* (Mechanicsburg, PA: Stackpole books, 1961), 280.
149 See Gunn, *Rice Wars*, 248–249.
150 See Sibylle Scheipers, *Unlawful Combatants: A Genealogy of the Irregular Fighter* (Oxford: Oxford University Press, 2015), Ch. 4 and 5.
151 Of a vast literature, see recently, Michael E. Haskew, *Encyclopaedia of Elite Forces in the Second World War* (London: Casemate, 2007); Max Hastings, *The Secret War: Spies, Codes and Guerillas 1939–1945* (London: Harper Collins, 2015).
152 See Judith A. Bennett, *Natives and Exotics: World War II and Environment in the Southern Pacific* (Honolulu: University of Hawaii Press, 2009); Daniel Clayton, "Militant tropicality: War, revolution and the reconfiguration of 'the tropics' c.1940 – c.1975," *Transactions of the Institute of British Geographers* 38, no. 1 (2013): 180–192.
153 Christopher E. Goscha, "'Hell in a very small place': Cold War and decolonisation in the assault on the Vietnamese body at Dien Bien Phu," *European Journal of East Asian Studies* 6, no. 2 (2010): 201–223, at 208.
154 Max Hastings, *Nemesis: The Battle for Japan, 1944–1945* (London: William Collins, 2007), 83–84. Also see Bennett, *Natives and Exotics*.
155 Brigadier General Andre J. Ognibene and Colonel O'Neill Barrett, *Internal Medicine in Vietnam*, Vol. II: *General Medicine and Infectious Diseases* (Washington, DC: Center of Military History, U.S. Army, 1982), Ch. 1.
156 Goscha, "'Hell in a very small place'," 208.
157 For a wider and longer twentieth-century history, see Razmig Keucheyan, *Nature Is a Battlefield* trans. David Broder (Cambridge: Polity Press, 2016).
158 Võ Nguyên Giáp, *People's War, People's Army* (Hanoi: Foreign Languages Publishing House, 1961), 8.

84 Tropicalising Indochina

159 Goscha, *Penguin History of Modern Vietnam*, 241.
160 Goscha, *Penguin History of Modern Vietnam*, 241. Also see Brocheux and Hémery, *Indochina*, 369.
161 Cited in Goscha, "Une guerre pour l'Iindochine?" 30–31; and on the broader revolutionary semantics of this idea of a unitary battlefield, see Christopher Goscha, *Vietnam or Indochina? Contesting Concepts of Space in Vietnamese Nationalism (1887–1954)* (Copenhagen: NIAS Press, 1995), 67–81.
162 Giáp, *Military Art of People's War*, 70.
163 Logevall, *Embers of War*, 156–170.
164 Adrian Gilbert, *Voices of the French Foreign Legion* (New York: Skyhorse, 2010), 209.
165 Stanley Karnow, *Vietnam: A History* (New York: Viking Press, 1983), 185; Martin Windrow, *The Last Valley: Dien Bien Phu and the French Defeat in Vietnam* (London: Weidenfield & Nicolson, 2004), 81–96.
166 Anthony James Joes, *Modern Guerrilla Insurgency* (Westport, CT: Praeger, 1992), 104–105.
167 See Fall, *Street With Joy*, 35–48, which remains one of the most graphic and insightful accounts of the French and American wars in Vietnam, although he did visit Indochina himself until 1953. In this and other reporting he makes frequent reference to the impact of the tropical sun, sky, and rain, and describes many scenes as instances of "tropical warfare." Fall, *Street Without Joy*, 93 and *passim*.
168 Marty secured Giáp's release from prison for seditious activity and smoothed his educational passage to Hanoi. See Cecil B. Currey, "The mystery of André Louis Marty," *Journal of Third World Studies* 29, no. 1 (2007): 97–108, although Currey is not entirely sure about Marty's motives.
169 Pierre Gourou, *Terres de bonne espérance, le monde tropical* (Paris: Plon, 1982), 12–18 Dany Bréelle, "Interview with Pierre Gourou, Bruxelles 29 August 1995," in Dany Bréelle, "The regional discourse of French geography: The theses of Charles Robequain and Pierre Gourou in the context of Indochina," Appendix H, unpublished PhD dissertation, Geography, Flinders University, 2003, 336–337. Giáp's military thinking and strategy also incorporated the teachings of Lenin, Mao, Sun Tzu, Clausewitz and T.E. Lawrence. See Cecil B. Currey, *Victory At Any Cost: The Genius of Viet Nam's Gen. Vo Nguyen Giap* (Washington, DC: Brassey's, Inc., 1997), 152–155.
170 Gourou, *Terres de bonne espérance*, 15.
171 Pierre Gourou, "lÉcole des beaux-arts de Hanoi," *France Illustration: Le monde illustré*. Numéro spécial sur l'Indochine no. 190 (juin 1949), n.p.; Dedication, shown to the authors by Inguimberty's children, at their house in Menton.
172 The French and American sides of this formulation about this war warping of the popular imagination are developed in Nora A. Taylor, *Painters in Hanoi: An Ethnography of Vietnamese Art* (Singapore: National University of Singapore, 1999).
173 Nadine André-Pallois, *L'Indochine: un lieu d'échange culturel? Les peintres français et indochinois (fin XIX–XX siècle)* (Paris: Presses de l'Ecole française d'Extrême-Orient, 1997), 25.
174 Edward W. Said, *Orientalism* (New York: Random House, 1978). Our commentary here follows Sarah Anderson, "Introduction" to Euguène Fromentin, *Between Sea and Sahara: An Orientalist Adventure* orig. pub. 1859 (London: I. B. Tauris, 2004), xxi–xxii.
175 For this comparison, see Kenneth J. Ruoff, *Imperial Japan at Its Zenith: The Wartime Celebration of the Empire's 600th Anniversary* (Ithaca and London: Cornell University Press, 2014), 104.
176 Robert Rydell, *World of Fairs: The Century-of-Progress Expositions* (Chicago and London: University of Chicago Press, 1993), 67–68.
177 Rydell, *World of Fairs*, 68.
178 Cited in Rydell, *World of Fairs*, 68.

179 Cited in Danielle Labbé, Caroline Herbelin and Quang-Vinh Dao, "Domesticating the suburbs: Architectural production and exchanges in Hanoi during the late French colonial era," in Laura Victoir and Victor Zatespine eds., *Harbin to Hanoi: The Colonial Built Environment in Asia, 1840–1940* (Hong Kong: Hong Kong University Press, 2013), 251–272, at 253.
180 Timothy Mitchell, *Colonising Egypt* (Cambridge: Cambridge University Press, 1988), 24.
181 Mitchell, *Colonising Egypt*, 24.
182 Natasha Easton, "Art," in Philippa Levine and John Marriott eds., *The Ashgate Research Companion to Modern Imperial Histories* (London: Ashgate, 2012), 543.
183 Angela Mary Meyer, "Painting in the tropics," Unpublished PhD dissertation, James Cook University, Australia 2015, vi. She refers to Matisse's "The Parakeet and the Mermaid," (1952); Gaugin's "Tahitian Women on the Beach" (1891), and Rousseau's "Tiger in a Tropical Storm" (1891).
184 Suzie Protschky, *Images of the Tropics: Environment and Visual Culture in Colonial Indonesia* (Leiden: KITLV Press, 2011), 95.
185 Jean-François Staszak, *Géographies de Gauguin* (Paris: Bréal, 2003).
186 Taylor, *Painters in Hanoi*, 31.
187 Labbé, Herbelin and Dao, "Domesticating the suburbs," 255.
188 Gourou, "lÉcole des beaux-arts de Hanoi," n.p., and a point developed in Taylor, *Painters in Hanoi*, 31–33.
189 For more on this position, and its longer history in French thought and radicalism, see Lisa Lowe, "*Des Chinoises*: Orientalism, psychoanalysis and feminine writing," in Kelly Oliver ed., *Ethics, Politics and Difference in Julia Kristeva's Writing* (New York and London: Routledge, 1993), 150–163.
190 Pierre Gourou, "La géographie comme 'divertissement'? Entretiens de Pierre Gourou avec Jean Malaurie, Paul Pélissier, Gilles Sautter, Yves Lacoste," *Hérodote* 33, no. 1 (1984): 50–72, at 58.
191 Michel and Dominique Inguimberty, Interview with the authors. The French Communist Party was excluded from France's coalition government in May 1947 and took an increasingly anti-colonial line thereafter. Inguimberty returned in ill-health, with chronic colitis. ANOM: FM EE/11/4998/9/ING Inguimberty personal dossier.
192 See Nora Taylor, "Orientalism/Occidentalism: The founding of the École des Beaux-Arts d' Indochine and the politics of painting in colonial Việt Nam, 1925–1945," *Crossroads: An Interdisciplinary Journal of Southeast Asian Studies* 11, no. 2 (1997): 1–33; Phoebe Scott, "Imagining 'Asian' aesthetics in colonial Hanoi: The École des Beaux-Arts de l'Indochine (1925–1945)," in Fuyubi Nakamura, Morgan Perkins and Oliver Kirscher eds., *Asia Through Art and Anthropology, Cultural Translation Across Borders* (London: Bloomsbury, 2013), 47–61.
193 See Lisa Bixenstine, "Art at the crossroads: Lacquer painting in French Vietnam," *Transcultural Studies* 1, no. 1 (2015): 126–170.
194 Gourou, "lÉcole des beaux-arts de Hanoi," n.p.
195 Gourou, "lÉcole des beaux-arts de Hanoi," n.p.
196 Gourou, "lÉcole des beaux-arts de Hanoi," n.p.
197 Charles Robequain, "Le Tonkin d'après Mr P. Gourou," *Annales de Géographie* 41, no. 232 (1932): 430–434, at 430–431.
198 Gourou, "lÉcole des beaux-arts de Hanoi," n.p.
199 Gourou, "lÉcole des beaux-arts de Hanoi," n.p.
200 ANOM: FOM C 249/ D. 370: *La Dépêche coloniale*, 3 September 1931.
201 Nicholas Mirzoeff, "Photography at the heart of darkness: Herbert Lang's Congo photographs (1909–1915)," in Tim Barringer and Tom Flynn eds., *Colonialism and the Object: Empire, Material Culture and the Museum* (London and New York: Routledge, 1998), 167–187, at 170.
202 SHD: 10 H 900, Colonel Gambiez, "Note relative aux rapports entre l'armée et la population," 25 January 1951.

203 SHD: 10 H 2351, Fascicule à l'usage des cadres français nouvellement arrivés au Nord-Vietnam, Bureau de Guerre psychologique, December 1953.
204 SHD: 10 H 2351, Documentation sur l'Indochine du Nord, 4 June 1947.
205 SHD: 10 H 2351, Documentation sur l'Indochine du Nord, November 1954.
206 SHD: 10 H 2351, Fascicule à l'usage.
207 SHD: 10 H 900, Colonel Gambiez, "Note."
208 Pierre Schoendoerffer, *La 317ème Section* (Paris: Robert Laffont, 2004), 45.
209 Pierre Gourou, "Panorama de l'Indochine," *Tropiques: Revue des troupes colonials* no. 342 (1952): n.p.
210 ANOM: FP 568PA/6 Papiers Monguillot, annex 5, "note sur le conférence de Dalat, 18 avril – 11 mai 1846; "Instructions politiques pour la délégation française."
211 Võ Nguyên Giáp, *Des journées inoubliables* (Hanoi: Editions en langues étrangères, 1975), 266–267.
212 Goscha, "'Hell in a very small place'," 202–203.
213 Giáp, *People's War, People's Army*, 11. Also see Robert Guillain, *Orient extrême: Une vie en Asie* (Paris: Arléa, 1986).
214 Viet Thanh Nguyen, *Nothing Ever Dies: Vietnam and the Memory of War* (Cambridge, MA: Harvard University Press, 2016), 30.
215 Pham Thanh Tam, *Drawing Under Fire: War Diary of a Young Vietnamese Artist* ed. Sherry Buchanan (London: Asia Ink, 2005).
216 Bernard Fall, *Hell in a Very Small Place: The Siege of Dien Bien Phu* (New York: Da Capo Press, 1966), 448.
217 John T. McAlister, Jr. and Paul Mus, *The Vietnamese and Their Revolution* (New York: Harper & Row, 1970), 46.
218 Paul Mus, *Le Viet Nam chez lui* (Paris: Centre d'études de politique étrangère, 1946), 33.
219 Panivong Norindr, *Phantasmatic Indochina: French Colonial Ideology in Architecture, Film and Literature* (Durham, NC: Duke University Press, 1996), 2.
220 Pierre Gourou, "Charles Robequain (1897–1963)," *Annales de Géographie* 73, no. 395 (1964): 1–7, at 2, 4.
221 M. Kathryn Edwards, "Traître au colonialisme? The Georges Boudarel affair and the memory of the Indochina war," *French Colonial History* 11 (2010): 193–209, at 194, 204.
222 Jacques Rancière, *The Politics of Aesthetics,* trans. Gabriel Rockhill (London: Bloomsbury, 2013).

3 Romancing the tropics

Technics of tropicality in *Les paysans du delta tonkinois*, 1936

By 1949 Pierre Gourou had spent more than 25 years writing about the Far East. Too young to fight in the World War I, he had lived for nine years in Indochina, which was riven with colonial violence (a mixture of racism and repression); and back in Europe (Belgium and France) from 1936 he had witnessed the rise of fascism, survived German occupation, and served in the French Resistance. His exposure to the ugliness and sorrow of war perhaps helps to explain why he was drawn to Inguimberty's art and saw it as an emblem of peace. Gourou maybe wanted readers of *France Illustration* to see something other than malevolence and destruction, and one wonders how the magazine's readers viewed his contribution: as a meditation on the virtues of order and harmony, and by implication French order, perhaps; or as a useful distraction from the military difficulties that France was facing at the time. Either way, the art of war – by painting, photograph or map – is conspicuous by its absence from this 1949 publication, and, for his part, Gourou spurned violence, from wherever it came, and sought to look past conflict and find peace and order in the world; and so did his friend Paul Mus, with whom he shared an office at the Collège de France.[1] This did not necessarily mean that either Gourou or Mus operated in a serene space of inquiry that was devoid of power. Among other things, we have already shown how, in Indochina, power came in different forms, and with the French claiming they were a civilising force for good, and the Viet Minh waging what it believed was a 'sacred resistance' against the French. Gourou was uncomfortable with both of these visions of power. Nonetheless, our suggestion in this chapter is that a certain type of metropolitan power and influence accrued from the way he and other French scholars lodged their ostensibly humane objection to both colonial violence and the idea (later enshrined in the thought of Frantz Fanon) that decolonisation would only truly occur if the West's glorification of itself and denigration of the colonised (what Fanon called "the cultural problem") was also extirpated with violence.[2]

We discussed how and why Aimé Césaire dubbed Gourou's 1947 primer *Les pays tropicaux* an "impure and worldly geography" that purveyed a specious *tropicalité*: a picture of tropical lands as backward compared with temperate ones, and one that downplayed the role that colonial exploitation and rule had had in creating this backwardness.³ The Marxist historian Jean Suret-Canale added: "The general misery and the ruin of the soil were taboo subjects until the Second War," and Gourou's antidote was to advance "the view that the tropical countries of Africa are doomed to a retarded civilization and a low level of life, owing to natural conditions unfavourable to human life and activity."⁴ Gourou's tropicality was not all doom and gloom, however; and nor was it power-laden in the same way over time and in every place. Suret-Canale, for one, knew that Gourou had a more favourable view of the delta regions – rice bowls – of tropical Asia and had been critical of French colonial policy in Indochina. In his 1936 regional monograph *Les paysans du delta tonkinois* (hereafter *Les paysans*), upon which this chapter pirouettes, Gourou forged what we see as an affirmative – admiring rather than admonishing – tropicality. He romanced the tropics. Moreover, *Les paysans* is still hailed as one of the 40 most influential scholarly works on Southeast Asia and deserves detailed critical examination as an example of how tropicality is made in particular texts, projects and circumstances.⁵

Like Césaire, we are interested in the discursive nature of tropicality: the machinations of truth and power bound up with its making and influence, and particularly in how it privileges the Western observer as the fount of knowledge. We track how Gourou fashioned a form of tropicality that differed from the one Césaire and Suret-Canale chastised, and we do so by deploying the expression 'technics of tropicality,' which we derive from the American thinker Lewis Mumford's 1934 *Technics and Civilization* and use to flag our concern with Gourou's techniques of representations and the interpretative choices he made in his study.

In a talk given at a meeting of Institute of Pacific Relations in Stratford-Upon-Avon, England, in 1947, Gourou regretted the disastrous effects of the Indochina War and argued that peace might only be restored through the resumption of France's colonial leadership. He argued: "Out of various mutually alien and hostile elements, France has molded a peaceful whole from which domestic wars were excluded. Irrespective of France's right to intervene in Indo-China, the fact is that she brought about a state of affairs which, viewed in terms of the peaceful relations established among the peoples of the Federation, was certainly not disabling. . . . [Furthermore], leaders in the various Indo-Chinese countries, having become accustomed to French patterns of thought, would experience confusion if they had now to adopt a new orientation."⁶ Such arguments provided the geopolitical scaffolding for his cultural musings from around this time (discussed in the last chapter) about Indochina's "varied landscapes," "mosaic of peoples," "disorienting charm," and how the region offered researchers "such a wide choice of problems."⁷ Geo-politics and exoticism were linked, in this instance with Vietnamese "confusion" pitched in binary opposition to an orderly and reason-bearing French influence.

The idea of Indochina only made sense to Gourou as a French geo-political construct – "a reasoned creation by France," as he put it in a 1934 school textbook co-authored with Jean Loubet – an entity based on economic, political and intellectual ties, investments and infrastructure that made the country into a coherent whole surmounting its pre-colonial diversity and disunity.⁸ This text belonged to a colonialist system of representation which countered anti-colonial and revolutionary discourses by emphasising the divisions between indigenous regions and factions.[9] The pedagogic tactic deployed, of identifying and justifying French colonialism on geo-political grounds, can be found in a plethora of educational works and moral primers published at the behest of the French colonial state during the interwar years. They purveyed the ideological fiction that French scholars like Gourou possessed a superior – detached and analytical – wisdom, and that it was Western sagacity that the Vietnamese needed if they were to become modern subjects capable of self-government, even if, as in Gourou's case, the French admitted that their colonial presence was neither unblemished nor entirely efficacious.

In *Les paysans*, however, which was the chief product of Gourou's nine-year sojourn in Indochina, he tried to keep the geo-politics and "disorienting charm" of French Indochina apart. How he sought to do so with a distinct scientific outlook and literary sensibility is our quarry here. During his entire career he only ever used the word *politique* in the restricted and neutral sense of 'policy.'[10] Furthermore, "France's arrival in Indochina was based on a geographical error," he observed in 1984, and early French maps of Indochina were "completely useless": "We were carried away by the mad hope of reaching the heart of China via the Mekong and then the Red River. . . . But let us get us get back to geography" – something, in his mind, more focused and less dreamy.[11] "It was impossible," he confided elsewhere, "for the colonial administration of Indochina to keep for long its grip on a Vietnamese population which was rapidly assimilating our ways. That said, how sad it is, for someone who thinks nostalgically of Indochinese horizons, that the inevitable effacement of the colonial system gave way to terrible disorders and oppression."[12]

In *Les paysans*, his 666-page study of the peasant landscape and culture of the Tonkin Delta, he wrote sparingly about French colonialism. As Jeanne Haffner neatly observes, "the Tonkin Delta was neither exploited for its natural resources nor settled by French emigrants. Instead, the region was created by the French solely for administrative purposes, especially tax collection. As a result, villagers had very little contact with the Western world."[13] The French never attained the same level of command over the Tonkin Delta that it had over the Mekong Delta in the south, and there have been intense debates since the 1920s about whether French colonial policies benefitted the peasant population or stoked rebellion and paved the way for the Viet Minh's ascent to power.[14] However, there was a contrived quality to the way Gourou connected with the Tonkin Delta: he wrote about this colonial region as if the French were not there at all. For the French were there, not just taxing the peasant population but also building dikes, roads and railways, conducting censuses, undertaking a land use survey (1929–1931, which was overseen by Indochina's Inspector-General of Agriculture and Forests, Yves Henry), repressing rural rebellion, regulating trade and migration, recruiting (often coercing) Tonkin peasants to work on agricultural land (especially French rubber plantations) in Cochinchina, and seeking to know the mind-set of the Tonkin peasants in order to bring them within the orbit of France's colonial mission.[15]

Gourou said next to nothing about these dynamics, and his removal of the French from the peasant landscape he beheld in his study enhanced its allure. *Les paysans* has an academic focus and no direct political purpose. Gourou had no formal attachment to the French colonial state and presented his work as that of a lone scholar, and one that was driven by his own imagination and inspiration. But this was to some extent a ruse, and herein lays an important aspect of the tropicality his work exudes. Felix Driver and Brenda Yeoh argue that tropicality is best viewed as a "transactional" process of exchange between 'temperate' and 'tropical' peoples, places and experiences rather than a one-way projection of Western categories over the exotic.[16] But this was an unequal process, they continue, because the 'temperate' remained the predominant frame of reference, and the 'colonial' became an *aide-de-camp*. While Gourou did not talk openly about the matter, important elements of his research were made possible by a colonial apparatus of knowledge and power, and a disciplinary – geographical – outlook that was not immune from the drives and tensions of empire.

In this and the next chapter we attend in detail to how this tropicality worked as a technology of seeing that rested on a mixture of image and technique, and with the "amalgam of imperial vagueness and precise detail" that Edward Said associates with Orientalism.[17] By technology of seeing we mean, broadly following Mumford, Martin Heidegger, and other thinkers from this era, both the "complex network of machines and activities" by which the world is represented, and "the essence of technology," or 'enframing,' meaning "the way reality discloses itself to us" (which is not identical with this 'complex'), as Slavoj Žižek glosses Heidegger.[18] By the 1930s, Žižek, continues,

> the paradox of technology as the concluding moment for Western metaphysics is that it is a mode of enframing which poses a danger to enframing itself: the human being reduced to an object of technological manipulation is no longer properly human; it loses the very possibility of being ecstatically open to reality. However, this danger also contains the potential for salvation: the moment we become aware and fully assume the fact technology is, in its essence, a mode of enframing, we overcome it.[19]

Gourou's Tonkin tropicality might be seen as both a mode of technological manipulation and a means of salvage. We will see that it did not amount to a straightforwardly Eurocentric framing hinging on a diffusionist understanding of technology as advancing from the imperial centre outwards to the colonial margins, for a start because Gourou's focus was on rice, which, Steven Topik and Allen Wells observe, was "contrarian" from an imperial-technological point of view.[20] Ninety per cent of the rice produced in the world is lowland rice, they continue, "which thrives in the hot lowland tropics with abundant rainfall and enough water to either naturally or artificially flood fields . . . [and] the great majority of rice farmers had very plausible reasons for eschewing new technologies."[21] Gourou's approach to the peasants of the Tonkin Delta exemplified a larger concern within

Western culture at this time: that by studying 'their' technologies one might also cast light on 'our' (Europe's) technological history, superiority or woes.[22]

The comparative study of what Gourou later termed *techniques d'encadrement* – the production and landscape moulding techniques of peasant populations – became a key facet of his post-war research agenda, and it had its origins in the Tonkin Delta. *Encadrement* literally means framing, and Gourou understood the term broadly in the sense that Heidegger conceived it: he sought to study how the Tonkin peasants disclosed reality to themselves through the way they worked a challenging monsoon environment. This primary research objective opened out on to a complex – fourfold – engagement with the paradox Heidegger identified.

First, and as he later finessed the view, Gourou maintained that "People, these makers of landscapes, only exist because they are members of a group which is itself a tissue of techniques," and that civilisations only exist if their members are *encadrés*. Accordingly, he saw the condition (integrity and vulnerability) of *encadrements* as a measure of the vitality of landscapes and civilisations. But second, he was not simply an "indigenist" (as Arnold describes the outlook that comes into view here), venerating indigenous traditions and seeing Europe simply "as an external, intrusive and coercive force."[23] We shall see that he saw the Tonkin peasants as "slaves" to their 'tissue of techniques,' and, by implication, as unamenable to change. Yet third, for Gourou as for many other Western scholars working in the non-Western world at this time, an 'indigenist' sensitivity to peasant and native traditions came with a 'salvage' mentality – a quest both to study and record 'traditional' societies before they were overrun and corrupted by the West, and to use them to diagnose the pathologies of modernity. And fourth, Gourou wielded a technology of representation that sought to at once see through and protect the "frame of a fundamental fantasy, which, as a transparent background, structures the way we relate to reality," as Žižek describes this fraught endeavour to explain, and thus impose oneself upon, the very thing (structure) that one does not want to hurt.[24] We shall go on to show that Gourou was an ambivalent tropicalist. Our point here is that such ambivalence – his avowal and disavowal of tropical difference – rested on paradoxical indigenist sensibilities.

In this chapter, then, we explore how Gourou constructed the peasant society of the Tonkin Delta as an object of study – as a traditional civilisation in a monsoon environment – and are interested in his "sites of representations": both the landscape on which he worked, and the intellectual location from which he observed it.[25]

Gourou divided his study into three main sections, on: (1) the physical landscape; (2) population structure and distribution; and (3) peasant life and culture. He worked with cadastral maps, aerial photographs and census data; distributed his own questionnaires; conducted interviews with local headmen in a quarter of the delta's 8,000 villages; compiled his own census; and produced definitive inventories and maps of terrain, house designs, village names, administrative structures and population density. He paid great attention to questions of habitat, and his discussion of Tonkin houses alone fills over 150 pages. The book contains nine maps, 80 aerial and ground-level photographs (the bulk of the latter taken by him),

125 tables and figures, and an inventory of village names extending to 43 pages. *Les paysans* was a prodigious feat of research. At a theoretical level, it might be viewed as a prime example of the 'power of landscape': the important place that real and imagined landscapes have in accounting for social realities and power relations, and with landscape seen and questioned as "an object to be seen or text to be read . . . [and] as a process by which social and subjective identities are formed."[26]

An important way into the question of how Gourou made *Les paysans*, and the consequences of this making, is with Mumford's idea of 'technics.' In *Technics and Civilisation*, Mumford assesses the advent, nature and effects of the modern capitalist-industrial-machine age. He was fascinated by how Western civilisation had shaped and venerated the machine, and the "radical transformation of the environment and the routine of life" it had wrought, involving a transformation in concepts of time and space away from a heavenly supernatural world and towards a natural-secular one.[27] But writing in the light of industrialised warfare and the utter devastation it had wrought during World War I, he was acutely sensitive to the paradoxical nature of "the conquest of the environment and the perfection of human nature."[28] He thought he was witnessing something unique in world history: not a lapse into barbarism through an enfeebling of civilisation, but the advent of brutality and an instrumental calculus that stemmed from the very forces and interests that had originally been directed towards this conquest and perfection.[29]

For Mumford, as for Heidegger, technology was integral to human life, and they had a mutual interest in the clock and the measurement of time. To both of them, Don Ihde notes, "technologies 'take account' of nature in some way" and "point back to a primitive existential base."[30] The clock, for example, accounts for the rhythm of time, and like all technologies is "based upon the human sense of finitude."[31] Heidegger and Mumford both saw time as foundational to the question technology: what does and can one do in a given time (hour, season, lifetime, and so on), and how does one concern oneself with that time as an integral part of one's encounter with the world.

This foundational temporality is also social and spatial: it is always directed towards and within a given area and set of social goals. Technologies are shaped by, adapted to, and characterise particular spaces and environments in intentional and socially orchestrated ways. Mumford argued that the positives surrounding such intentionality and orchestration – which involved finding, recognising and striving for space 'in-order-to' and 'in-which-to' deal with the finitude of human life – became fragmented and lost as the machines and technologies that were its handiwork became fetishised. Modernity had lost sight of the value of many inventions that had been produced without machines within a pre-modern 'container culture,' including vases, cellars, barns, dikes and reservoirs, and houses and granaries. Mumford was an avid student and admirer of machines and technical processes (from arms and munitions, to the steam engine and photography, as well as the clock), and argued that it was the *choices* that people had made in creating a mechanised society, rather than the machines they use, that had spawned the deformations of the modern age. His technics were about both choice and constraint.

Gourou never discussed either Mumford or Heidegger, but there are strong echoes of the interwar philosophical horizon of 'technology' in his work. Like Mumford, he was concerned with how people are essentially organisers of space and time who devise and tailor particular techniques and know-how to survive and thrive in particular circumstances, and give them finitude and meaning. And for Gourou, as for Mumford, the modern conquest of nature by technology entailed embryonic (often conflated with 'traditional') forms of know-how. Gourou's complex, if elusive, idea of 'civilisation' was caught between these two understandings of technology – as something generic and historical – and with the paradox of how "agriculture marks a great technical advance over the economy of hunter-gathers, yet . . . does not ensure a life of relative stability and ease" (to use Yi-Fu Tuan's gloss on scholars such as Gourou and Mumford) centrally in the picture.[32] Gourou's idea of *techniques d'encadrement* flags his twin concern with rural peasant technologies as they had been handed down, and the contemporary choices and dilemmas facing peasant communities. His eyes were implanted in both the past and present.

But by technics of tropicality we also mean his techniques of representation and interpretative choices, and should note at the outset that he did not ruminate on how he went about his work until late in life, and then barely at all. This observation forms a vital part of our story here, and as we noted in Chapter 1, matters are complicated by the fact that few private papers from his Indochina years have survived. *Les paysans* is not framed by a clear-cut discussion or justification of methodology, and Gourou says nothing about the often emotive language and imagery he uses to describe peasant life. Again, this paucity of reflection (by today's standards) on his ideas and methods worked to bolster the aura surrounding his 1936 text. However, this does not mean either that its findings are timeless, or that the making of his study lies beyond the bounds of critical scrutiny. Rather, this dearth of methodological reflection presents us with an interpretative licence and responsibility, which we shall cultivate through an inquiry into Gourou's romancing of the Tonkin Delta. To borrow Paul Carter's formulation about 'spatial history,' the Tonkin Delta was "a text that had to be written before it could be interpreted."[33] We are interested in the nature of this writing.

Finally by way of introduction, we seek to expand on what we said in the last chapter about *structure* (in the Braudellian sense), immersion and place, which were central to how Gourou looked at Inguimberty's art and the Tonkin landscape. These were core elements of his Tonkin research. First, Gourou's concern with the *longue durée* was realised as a quest to find and understand what he called an "original geography" – a geography of difference; a geography that was unique in terms of the problems it posed; and a geography that could be sealed off from its colonial surroundings and dissected.[34] Second, the significance he accorded to immersion via fieldwork and sustained observation was presaged by, and never fully left behind, an attraction to the Far East as a daydream. And third, the Tonkin

94 *Romancing the tropics*

Delta would give up its 'secrets' through a suite of geographical ideas and techniques that were brought from France and adapted to Tonkin circumstances.

Buffalo and books

In one of the few interviews that Gourou gave, late in life, he recalled the water buffaloes that pervaded the Tonkin landscape and were portrayed on the Inguimberty painting that hung on the wall of his Brussels apartment. While one could get a certain way into the delta by car or bicycle, he noted, most of the area was only accessible by foot. Numerous remote villages that he visited were connected by steep and narrow mud tracks, and the throng of humanity in the delta shared them with their draft animals. Water buffalo were adept not only at making large holes in the roads, he continued, but also at sticking to the same holes time and again, making them progressively deeper and more treacherous.[35] He recalled a photograph he had taken in 1932 (which is reproduced in *Terres de bonnes espérance*) of Tonkin women riding buffalo like horses across sodden rice paddies.[36] There was no scent of the Orientalism that wafted across the water buffalo images used by Duchartre (see Chapter 2) – only the smell of mud and manure. Gourou placed the animals, perhaps with the Inguimberty painting jogging his memory, in an earthy tropical scene of work that he had beheld at close quarters.

This materiality was vital to the making and influence of *Les paysans*, and the way Gourou romanced the tropics. Readers were convinced that he had visited every corner of his study area. However, his work and memories of the region were laden with a more numinous sense of the exotic, and in our view the earthy and the exotic were two sides of a geography that, for him, was both imagined and real, and involved both *jouissance* and hard work. "Our consciousness of the landscape in its historical and physical depth," Gourou wrote in a later treatise on human geography, "is always a source of joy, a school for progress, and certainty one of inexhaustible activity."[37] But one only gained such a consciousness through fastidious fieldwork, and it gave one access to a reality that could not be explained by recourse to *a priori* theories or models.

The identity of the Tonkin Delta as a tropical environment was fundamental to the 'depth' of the landscape that Gourou beheld, although, as he was at pains to point out, physical factors did not explain the human landscape. He opened his 1936 study by identifying the Tonkin Delta as a monsoon region which sloped gently (by 15m) to sea level from the northwest to the southeast, and was dominated by the Red River, which flowed through Hanoi and had served as a natural barrier protecting the city from northern enemies, and especially the Chinese. The river originated in the mountains to the northwest and its hydraulics were shaped by a mixture of natural and human-made levees, a wet season of intense rainfall, heavy alluvial deposition and summer flooding, and a diurnal tide which affected a larger portion of the delta in the dry season than in the summer months. Gourou represented the Red River as the environmental anchor of an advanced civilisation that was of Chinese origin but had become more Vietnamese as its people had sought to protect themselves from the might of tropical nature by building dikes.

The region was characterised by three types of land with different relationships to water and tides: the clay-soiled lower reaches of the delta, which were unprotected from summer flooding and would be inundated at high tide; the middle reaches, in the coastal and eastern part of the delta, which had some protection from natural levees but would be submerged at high tide during the wet season; and the sandy-soiled upper reaches, in the western and northern parts of the delta, which were protected to a considerable degree by natural levees. Dike building was central to life in all parts of the delta, and particularly in the lower and middle reaches, where they were needed to abate the advance of brackish water inland during the dry season as well as to curb and channel flood water during the wet season. Of the region's 1.1 million hectares of rice fields, there was also a broad congruence between these land types and the three main forms of rice cultivation, with one dry season (November) rice harvest prevailing in the lowlands (covering roughly 250,000 hectares); one rainy season (June) harvest prevailing in the highlands (350,000 hectares); and two (dry and wet season) harvests prevailing in the most densely populated and intricately diked midlands (500,000 hectares), which were irrigable during the dry season.[38]

The "varied and rich" relief of the delta, and most significantly the dikes, had vital human ramifications, Gourou noted, because "just minimal differences of elevation protect one region from flooding, or result in another being submerged during most of the year. A few *décimètres* higher and here is a country which cannot cultivate rice in winter, and where the villages get wider and the houses more dispersed; a few centimetres lower and the villages are drawn in, the houses squeeze up against each other, and rice cultivation in the rainy seasons comes to a stop."[39] The region and its rice crop were prone to flooding between June and November – the peak time of agricultural activity – and rice plants inundated for more than four days soon died. Between 1884 and 1923 there were more than 20 floods in the Tonkin Delta lasting longer than this in the days and weeks leading up to the June harvest and one of Gourou's refrains was that the peasants were constantly *battling* nature.[40]

This configuration of the Tonkin Delta stemmed from a prodigious programme of field research, and one that challenged the dominant French perception that the region was a completely flat and monotonous plain. At the same time, the Tonkin Delta was a product of Gourou's imagination, and he wrote at length about how it was a coherent geographic entity and might be treated as a hermetically sealed laboratory in which to think about the reciprocal connections between society and environment.[41]

But if buffalo (as emblems of a raw and majestic materiality) were important to Gourou, then so were books. He was born in Tunisia, in 1900, to parents from Florensac in L'Héraut, southern France.[42] His father was a civil servant in the French colonial administration in Tunis, and Gourou was schooled there until he was 18, and then at the University of Lyon, where he was taught geography by Maurice Zimmerman. Gourou deemed his teacher rather uninspiring and inattentive to his pupils.[43] Zimmerman was in fact a more lively and complex character than Gourou recalled. In the early twentieth century much of the momentum of

French overseas expansion came from the Lyon Chamber of Commerce and the city's Geographical Society (established in 1873), and Zimmerman was closely involved with both.[44] For over 30 years he taught a class on colonial geography and history, initially at the École de préparation coloniale funded by the Chamber of Commerce, and then at the university. He was deeply interested in questions of exploration and colonisation, and became embroiled in debates with fellow geographers Marcel Dubois and Lucien Gallois about the nature of 'colonial geography': whether there was anything intrinsically colonial about the geography of France's overseas territories, or anything patriotically or practically French about their study.[45]

Gourou became embroiled in a similar debate about 'tropical geography': did it amount to a distinct research field, with its own objects, or was it a means of applying mainstream geographical ideas and methods to a particular part of the world? Gourou claimed to be a human geographer interested in the study of tropical lands but was mostly seen as the doyen of a bespoke research area – tropical geography.[46] However, the relations between geography and empire that swirled around Lyon and Zimmerman either passed Gourou by or he ignored them. When asked about Zimmerman, he did not mention his teacher's travels in French North Africa and the Middle East, or his colonial advocacy work. Rather, he described his teacher as a "Scandanavian specialist" with an applied – and to Gourou's mind staid – vision of the discipline.[47]

Gourou passed his *Licencé ès lettres (histoire géo)* in 1919, and his *Agrégation d'histoire et de géographie* in 1923, whereupon he returned to teach at his old school in Tunis, the Lycée Carnot.[48] Disaffected by Zimmerman, he had written a thesis on the commercial policies of Louis XI. But he never lost his love of geography, and his association of the subject with adventure and pleasurable distraction (*divertissement*) never waned. He reminisced that he had the "curiosity of a child charmed by the exotic," and "loved to trace journeys across maps."[49] The Far East interested him from an early age. He was entranced by Marshal Hubert Lyautey's 1920 *Lettres du Tonkin et de Madagascar* (1894–1899), which painted a picture of Tonkin as a charming and enticing, yet remote and challenging, theatre of colonisation, and by Victor Segalen's 1922 semi-autobiographical novel *René Leys*, which is set in China on the eve of the Revolution of 1911.[50] He had also read the colonial novels of British writers such as Joseph Conrad, E.M Forster and Rudyard Kipling, and late in life noted that this reading furnished "both a smattering of culture and a just measure of my ignorance."[51]

He was particularly drawn to Segalen's *René Leys*, which is a meditation on the art of storytelling, and like Segalen's better known and posthumously published *Essai sur l'exotisme*, drew distinctions between gaudy (salacious, dreamy, demonising and covetous) exotic desire, and a 'true' exoticism the pull of which was "nothing other than the ability to conceive otherwise."[52] Gourou remembered well the first part of the novel, in which a French resident in Beijing, named Segalen, engages the services of the eponymous René Leys, a young and precocious Belgian, to teach him Chinese. René draws Segalen into his world of intrigue and cunning, explaining to his pupil that he was close to the fallen Emperor and is now adviser to the young Regent, lover to the Emperor Dowager, and a professor at the College of Nobles, which gives him full access to Forbidden Palace of the Qing Dynasty. He also claims to be head of the secret police, and tells Segalen that he is using his police informants, and a network of spies that he ran out the city's red light district, to foil a number of assignation attempts on the Regent's life which had been ordered by the Dowager. The story thus quickly becomes a dreamy and murky tale of divided loyalties that floats between the real and the imaginary, and probes the challenges and compromises in the way of the 'self' attempting to grasp the 'other' and the unknown.

René's loyalties lie ultimately with the Regent, and the story is important because it enacts a fantasy and fear that had a much broader significance in Western colonial and Orientalist literature and aesthetics. On the one hand, the brilliant Belgian's skills as a linguist, diplomat and lothario bring him great trust and authority, and fuel the romantic fantasy of the foreigner knowing China better than it knows itself. "You have gone further in your penetration of China than any other European known or unknown. . . . You have attained the heart of the center of the Within – nay, better than the heart: the bed!," Segalen tells his tutor.[53] But on the other hand, this comes with the fear of 'going native' and losing oneself in that alien world. We have already contemplated this fear of engulfment in connection with Inguimberty's art, and Segalen marks it with reference to René's 'Oriental' appearance, his love affair with the Dowager, and with reminders about his European heritage. Julia Kuehn notes that while Segalen objected to both "Gauguin's tropicalism and Loti and Kipling's imperialist exoticism because they moved away from the original, shifting, relatively value-neutral overtones that exoticism, to his mind, had once possessed . . . [he] was aware of the difficulty of producing value-neutral representations of otherness beyond cliché and ideology."[54]

For Gourou, "familiarity killed curiosity," and his reading of Segalen and other factual and fictional works on the Far East at an impressionable age fostered an expectation of difference, which he brought to Indochina in 1926.[55] He worked initially at the Collège Chasseloup-Laubat in Saigon, but recalled that the relatively Westernised landscape of Cochinchina did not satisfy his desire to desire to study an agrarian landscape that had not been radically altered by colonialism.[56] The following year, then, he obtained a transfer to the Lycée Albert Sarraut in Hanoi (1 of 69 schools in the French empire named after Indochina's former Governor-General), catering to 11–18-year-old French, Vietnamese and Chinese pupils. He obtained his *Diplôme d'études supérieures (d'histoire et de géographie)* 1929, and in 1932 became professor at the city's University of Indochina (established in 1906).[57] From Hanoi he spent the next nine years teaching and studying the delta to the southeast of the city, mainly during weekends and school summer holidays. "It was exactly the kind of land I wanted to examine. It was small, fifteen thousand square kilometres, that is, the equivalent of two or three French departments, populated in an extraordinarily dense way, with about five hundred people per square kilometre on average, with peasants and rice-growing; in other words, it had everything I had wanted to see."[58]

This was Gourou's "original geography," and he had made an interpretative choice. In his preface to the second (1965) edition of *Les paysans*, he insisted, "Born of purely scientific preoccupations, this book came to express an emotional attachment to this hard-working, ingenious and deeply civilised population."[59] It was a book that made space for plenty of Inguimberty's drawings of buildings and villages, but not for his paintings. The imagery in the study came instead from words, maps and photographs, and Gourou represented himself as a lone scholar who had followed his own inspiration.[60] He imagined the delta as an exotic object of study in suspended animation, and perhaps as a 'true' exoticism in the sense Segalen meant. His work was implicated (however unwittingly) in what Panivong Norindr sees as France's attempt to preserve an image of Indochina as "an idyllic world untouched by political cataclysms."[61]

Former Governor-Generals of Indochina, Pierre Pasquier (1928–1934) and Albert Sarraut (1911–1914, 1917–1919), contributed to the cache of guidebook material produced for the 1931 Exposition Coloniale, and emphasised Indochina's "picturesque" and "charming" qualities.[62] Dany Bréelle sees many affinities between their way of describing the landscape as a painting (with colours, forms and lines) and Gourou's attempt in *Le Tonkin*, and subsequently in *Les paysans*, to draw the eye away from the modernising and colonising activities of the French and towards a 'pure' and 'authentic' peasant culture in symbiosis with its natural surroundings, and with a keen interest in the daily toil on both men and women.[63] While Gourou (to our knowledge) never met these politicians, their work gained meaning within the same exhibitionary horizon, diverting metropolitan French eyes away from the passions and contingencies of the colonial encounter and towards an appreciation of Indochina as a picture (and from which colonial violence had been removed).

Engagement and exhibitionism

Yet Segalen's urge to supplant a lazy, if often succulent, exoticism with something more realistic and meticulous was also close to Gourou's heart and it detracted from colonialist modes of picturing Indochina. *Les paysans* signalled a departure from both colonial geography and any sense of imperial nirvana. His focus was on the "humanised landscape," as he put it, meaning the details of peasant life on the land, which colonial geography treated as either extraneous to its concerns or as a function of the economic potential, development or backwardness of particular regions.[64] "In this country, people count above all else, and it is them that the geographer must study with the greatest of care if he wishes not just to account for the human features of the country but also its physical aspects."[65]

Nor did Gourou's desire to find and capture in scholarly form an original geography spring solely from the imagination and books (listed as *études générales* in the bibliography to his study).[66] It was also piqued by the World War I. Many of his friends from Tunis and Lyon had lost their lives in the trenches, and their deaths made Gourou feel that "the European idea" – Western civilisation and reason – had come to a horrific end.[67] A stint of military service with the *4ème régiment des Zouaves* between 1920 and 1922 seemed to reaffirm the sentiment. World War I, along with a racial turn in colonial discourse from the end of the nineteenth century, and growing disenchantment with the consumerist banalities of modernity, had two main impacts on young French scholars, artists, and travellers. For some it rekindled the idea that France's overseas empire was the product of a superior civilisation and vital source of national prestige. For others, including Gourou, it rejuvenated interest in how 'the exotic' and 'the primitive' could help Europe come to terms with the pathologies of modernity.[68] "The modern entailed its antithesis, the primitive," Marc Matera and Susan Kingsley observe, and during the interwar years, "Both modernism and primitivism were the products of distinctly modern and imperial forms of interaction."[69] Kelly Enright adds that from the end of the nineteenth century, and with vigour after World War I, "Many American nature writers used wilderness as an escape from, and a metaphor for human conflict, including war."[70]

Imperial propagandists exploited this reaction to war, and the interwar period was marked by a great expansion in "the French national community's colonial consciousness," as Elizabeth Ezra puts it.[71] Awareness of empire should not be equated with either wholehearted support or full understanding. Nevertheless, *l'idée coloniale* pervaded French popular culture like never before, with events such the 1931 Exposition Coloniale, and the counter-exposition, *La Vérité sur les Colonies*, staged alongside it, putting the glories and sins of an aestheticised, racialised and bureaucratised empire on public display and negotiating distinctions between the 'domestic' and the 'foreign.'[72]

The 347-page guidebook, *Le Tonkin*, that Gourou produced for the event had been commissioned by colonial government of Indochina. Copies could be bought or consulted in the lavishly decorated Tonkin Pavilion opposite the reconstructed ruins of the temple of Angkor Wat. Gourou spent most of the book contrasting the different monsoon – mountain and delta – areas of the colony, and ended by describing France's colonial record there as "a most brilliant balance-sheet. Although much still remains to be done."[73] In *Le Tonkin*, and more systematically in *Les paysans*, Gourou used the maps and aerial photographs of the region produced by the Service Géographique de l'Indochine (SGI). But the 'balance sheet' – a common accounting metaphor used to talk about the course of empire – with which he ended the book was barely in keeping with the preceding analysis, or the research that ensued, which pushed the French out of the picture.

In fact, between 1931 and 1936 Gourou grew increasingly ambivalent about the benefits of France's presence in Indochina, and his fieldwork became more and more aligned with the paradox of commitment that was captured so vividly in the opening lines of Claude Lévi-Strauss's famous 1955 travelogue *Tristes Tropiques* (sad tropics): "I hate travel and explorers. Yet here I am proposing to tell the story of my expedition."[74] The tropics may have been 'sad,' both in their own right (backward) and by virtue of this bind, but Lévi-Strauss's point was that this judgement was a figure of the imperial imagination and a consequence of centuries of colonial plunder. If the West was to be rejuvenated through an engagement with the primitive and the exotic, he continued, then adventure and the imagination would remain important but needed to be subordinated to the more disciplined outlook of the ethnographer. Gourou and Lévi-Strauss later became colleagues at the Collège de France, and the geographer struck a similar chord in his reflections on his Indochina days. He was carried to the region by the stories of travellers, generals, scholars and pundits, he mused, but sought to place knowledge of the region on a solid scientific footing. He did so by studying rather than travelling, and his recollections about the water buffalo and mud might be treated as a metaphor for his quest to traipse the landscape in a laborious search for its routines and to overcome pot holes in understanding.

If Segalen's fantasy and fear formed part of Gourou's anticipation of the alien reality to which he ventured, then it would soon be subordinated to a methodical fieldwork project that sprang from French geography. In 1931, however, and with little fieldwork then under his belt, the exhibitionism of empire still loomed large in his work. The Exposition Coloniale brought to a head the vexed imperial question of whether France should civilise its colonial subjects through a process of 'assimilation' (in line with French revolutionary tradition), or pursue the putatively less hierarchical policy of 'association' advocated by Sarraut (and in line with a reformist republicanism) (also see Chapter 2). *Le Tonkin* projected an idyllic image of Tonkin that was geared to association, and it was complemented by Inguimberty's dioramas of 'typical' Tonkin scenes (peasant villages and paddy fields) displayed in the Tonkin Pavilion: an image of an environmentally rich yet undeveloped colonial space with a sophisticated and compliant indigenous culture. This was a highly selective truth that obscured the depth of Vietnamese resistance to the French. Indochina was whittled down to a set of typical motifs that captured the ambiguous character of the region as both a picturesque scene of French pride and a pacified place of chronic backwardness. In the Tonkin Pavilion geography and art conspired to buttress the deceptive allegory of France's colonial presence in Indochina articulated in the official guidebook for the exposition: that of an enlightened France offering to protect and nurture a refined yet resigned tropical backwater and race.[75]

Peasants and populations

The year 1931 was a turning point for Gourou for a further reason. It was in that year, in Paris, that he met one of France's leading geographers, Albert Demangeon, and secured his patronage as a long-distance supervisor for his *thèse d'etat*.[76] Gourou had returned to Paris in 1930 to advise over the layout of the Tonkin Pavilion at the Exposition Coloniale, and to attend the International Geographical Congress meeting in the city, where he presented two short papers, on village types and the mapping of population density in the Tonkin Delta (beginning a long fascination with both subjects).[77] Gourou and another of Demangeon's students, Jean Gottmann, described their "master" as "a patron who chose his disciples carefully and encouraged curiosity, honesty, precision and clarity."[78] Demangeon worked ostensibly on France and Europe, and according to Gourou and Gottmann, the originality of his work lay in his emphasis on "the human factor," and with the regions on which he worked "among the most humanised on earth."[79] "The study of rural habitat, which gained momentum (and not just in geography) during the interwar years," Omnia El Shakry adds, "was in large part as a result of the enthusiastic efforts of Demangeon."[80] Yet Demangeon was no stranger to either non-French or colonial geographies. In 1925 he published a book on the British Empire in which he suggested that the discipline as a whole, and colonial geographers especially, needed to look past how colonies served metropolitan economic and political interests and pay more attention to how colonialism had impacted on native social and territorial systems, rural conditions and peasant life.[81]

Demangeon's work on Europe and empire was permeated by a peasantist discourse that stretched back to Antiquity and became revitalised in interwar France: a discourse, dubbed *retour à la terre* and encompassing literature, philosophy and politics, which rooted social morality in the compact between earth and peasant, and which extolled the virtues of toil. Twentieth-century French republicans seized on this tradition at a moment of profound change and uncertainty, envisioning peasants as noble, hard-working and acquiescent citizens of a modernising nation that was going awry. During the 1920s and 1930s these values and interests were spurred by the Catholic Church, and from 1940 Marshal Pétain's Vichy regime used the idea of a 'return to the land' as a strategy of both national revolution (based on the slogan *Travail, Famille, Patrie*) and wartime survival (to boost the agrarian population and production).[82] Petain promoted the study of geography, creating a separate *agrégation* in the subject (it had been part of the history qualification).[83] *Retour à la terre* also worked its way into colonial ideology, fostering forms of colonial romanticism and paternalism that jarred with rationalist strains of colonial thought and policy.

Pierre Pasquier, who was Governor-General of Indochina (1928–1934) while Gourou was there, implored French colonists to befriend the Vietnamese people, respect their customs and traditions, and realise that "Learning to know each other would be the best way to love one another. Do not destroy the old Asian edifice."[84] Regardless of the extent to which Pasquier believed his own rhetoric, it remained lop-sided: it was the Vietnamese who were enjoined to learn and love French ways, and little pressure was put on French colonists to keep their end of the colonial bargain. Pasquier's plea was reduced to a colonial wish image that was bequeathed to scholars and institutions such as the EFEO to expedite, but even scholarship for the purpose of imagining and idealising a harmonious colonial relationship was less evident than an exoticism skewed towards the treatment of 'traditional societies' as antique objects of curiosity, and the incorporation of non-Western ('primitive' and 'naïve') forms of representation into Western art and letters, as contrary to the materialism and anomie of modern life, and, in the process, making the 'tribal' part of the 'modern.'

Gourou never met Pasquier and certainly had no truck with Pétain. Indeed, there are more shades of Henri Lefebvre's 1930s Marxist work on the peasant communities of southern France (the subject of his *doctorat d'état*) in Gourou's peasantist position than there are strokes of primitivism or patriotism.[85] However, these metropolitan sensibilities in French society and politics undoubtedly shaped Gourou's interests and influenced how his work was read and received. From start to finish, *Les paysans* is imbued with a curiosity about the social and moral world of the Tonkin peasants, and the "perfection" that was to be found in their rice growing techniques and the way the peasants "draw all possible advantage, thanks to the very artful adaptation they manage to achieve with these techniques, from the conditions of the environment."[86] "In all their agricultural activities," he continued, "Annamite peasants use exceedingly judicious methods, with many nuances that are adapted to diverse milieux. . . . They have inherited ancient techniques, a rich experience coming from Prehistory."[87]

This curiosity was geared to a distant past and side-lined colonial dynamics. However, there was an imperial politics to his discussion, and one that once more hinged on the tropicality of the scene he revered. "An essential element of the [Tonkin] landscape, the village plays a primordial role in the moral and social life of the peasants. The peasant is not an isolated individual, an arbitrary citizen of a community in whose life she participates only from afar, like the inhabitant of the French countryside," he wrote in this veiled political way; "quite the contrary, the religious, political and social life of the Annamite community is intense and immediate, and all the peasants take part in it with faith, with ardour, and with the ambition of playing a bigger and bigger role."[88]

This contrast was symptomatic of Gourou's tropicality, and it was an unorthodox tropicality in that his Tonkin peasants were not being seen or judged against the perceived superiority of their French counterparts. In fact, in some ways quite the reverse: the ardent and intense Tonkin peasant was being valorised as a member of a civilisation, rather than as an 'isolated' and 'arbitrary' – and by implication fallen and broken – individual. This 'indigenist' position, as Arnold calls it, was set up in geographic terms, and with a feeling for alienation that was central to Lefebvre's critique of everyday life. In his reflection on his 1930s fieldwork in southern France, Lefebvre wrote of "rough peasants, full of joviality and vitality, and fairly poor . . . [who] lived so to speak on the level of nature and natural life, in its elemental violence, its uncomplicated freshness and also its ignorance . . . and understood their basic humanity through the phenomena of nature, animals and plants, the heavens or the bowels of the earth."[89] Peasant communities had a "natural and a relatively stable balance, achieved and preserved by a peasant wisdom, by a set of techniques and a spontaneous skill," and to preserve this peasant order "man cooperated with nature; he maintained and regulated its energies, both by his real work and by the (fictitious) effectiveness of his magic."[90] But in Europe the "human fulfilment" found in ancient peasant life had all but disappeared. Organised religion had besieged the rituals and symbols that were the glue of the peasant order, and "tended to dispossess human actions of their living substance in favour of 'meanings'"; and social life had gone from being "on the level of natural life and the 'world' . . . [to] become pyramidal, with chiefs, kings, a State, ideas, abstractions, at its apex."[91] The result, Lefebvre argued,

> has been a deprivation of everyday life on a vast scale, by religion, by abstraction, by the life of the 'mind,' by distant and 'mysterious' political life. . . . Bit by bit everything which formerly contributed to the elementary splendour

of everyday life, its innocent, native grandeur, has been stripped from it and made to appear as something beyond its own self. Progress has been real, and in certain aspects immense, but it has been dearly paid for. And yet it is still there, not unchanged, but degraded rather, humiliated.[92]

This 'deprivation' made the bucolic image of France's patchwork of villages, farms, fields, paths and hedgerows deceptive. "Everything is calculated on a cut-price basis. A penny for heaven," Lefebvre concluded, in a manner not dissimilar to Gourou's assessment of the 'isolated individuals' and 'arbitrary citizens' of the French countryside, where everyday life was now accounted for with an economic language of the 'balance sheet.'

It might be argued, however, that France and the agency of the Western observer still came first: the above passage from Lefebvre can be read as a warning about what had gone wrong in France, and how it might come to pass in the Tonkin. Gourou was acutely conscious of the fragility of his tropical peasant scene. One of the chief findings that came from his tour of the villages was that while over 20 per cent of the land was communally held, the average size of communal plots, which were worked in petite sections, was a mere 360 square metres, and landless peasants accounted for around one half of the population. Furthermore, while those parts of the delta with large concentrations of communal land were coastal marshes and mud flats, and were generally ill-suited to permanent occupation and cultivation, the same peasant logic of subdivision and small plots prevailed, and considerable effort went into working them.

Gourou marvelled at the labour and organisation that went into keeping the dikes intact and the daily toil of peasant men and women in intricately subdivided paddies. He was not alone in this veneration of the Tonkin peasant. As David Biggs shows, many colonial writers contributed to an "expanding myth of the heroic Tonkinese peasant."[93] "Gourou and other colonial observers helped create a notion of Tonkinese peasants as almost superhuman in their ability to overcome extreme environmental conditions," Biggs continues, and Mekong peasants were condemned, by the same yardstick, as "lazy and unfocused."[94] Tonkin peasants lived in autonomous village communes and their lives were defined by a noble hand-to-mouth rural existence far removed from the aspirations, comforts and vanities of educated and westernised urbanites. Indeed, as Gourou observed, nationalists and revolutionaries who had left their villages for work or education in the towns found it difficult to re-connect with their countryside kinfolk and convince them of the need for, or merits of, change. He thus saw the village as a litmus test of how far both France's project of *mise en valeur* and a Vietnamese revolutionary impetus might reach.[95] "Social conditions, inherited from a distant past, dominate the village institution," he insisted, favouring the concentration of social life in the village, and presenting many obstacles to progress and change.[96] In turn, he helped anti-colonial leaders like Võ Nguyên Giáp see the village as the fulcrum of political and military mobilisation.

Gourou began his research in October 1927, but undertook the bulk of it between end of 1931 and the summer of 1935.[97] His fieldwork was self-funded, was planned and undertaken with barely any input from Demangeon, and was mostly conducted during school summer holidays, in the hottest and wettest months of the year (May–August). Gourou worked out of the regional tradition of French geographic inquiry spearheaded by Vidal de la Blache and his students and acolytes, such as Demangeon: a tradition that focused on human landscapes, regional identity and questions of rural habitat (agrarian practices, field patterns, settlement types, architectural styles and handicraft traditions), venerated peasant traditions, celebrated France's regional diversity, and was deeply suspicious of the centralising aspirations of the Third Republic. *Les paysans* was one of the first and most impressive attempts to adapt the regional paradigm in French geography to a colonial situation, and Gourou took seriously Vidal's proposition in his 1922 (posthumously published) *Principes de géographie humaine* that

> the most suggestive geographical relationships are those between the number of inhabitants and a given area, in other words, the density of population. If detailed statistics of population are compared with equally detailed maps . . . it is possible, by analysis, to find a connection between human groups and physical conditions. Here we touch upon a basic problem of human occupancy. For the existence of a dense population . . . means, if one stops to think of it, a victory which can only be won under rare and unusual circumstances.[98]

High population densities came in various forms and had diverse drivers, and maps and statistical information (survey and census data), in conjunction with extensive fieldwork, were key to their interpretation.

This 'basic problem' lay at the heart of Gourou's study, and he added: "there is nowhere else in the world a natural region better defined than this Tonkin Delta, which is clearly individualised in relation to its mountainous surroundings by physical and human characteristics, is self-supporting, and has for a long time been closed off to any foreign ethnic influence."[99] The vegetation in the delta "is nowhere natural," he continued: "over the 15,000 square kilometres, not a strip of forest, not a grove, not any large trees, which are rare . . . all vegetation is cultivated."[100] Indeed, "it is impossible to find a landscape more soaked with humanity and, sometimes, alas, more stinking of the terrible smells of the human race."[101] In the 1930s nearly 7 million people were crammed into this small area and the region had population densities as high as anywhere in the world. In other words, while he used the idea of natural region, he insisted that it was dangerous to use it in isolation from the idea of civilisation or cultural region. On this latter count, he greatly admired the *Annale* historian Lucien Febvre's 1922 *La terre et l'évolution humaine*, because it acknowledged the influence of the physical environment on cultural evolution while escaping environmental determinism and placing human toil, struggle and ingenuity at the centre of social evolution and environmental change.[102] Indeed, in significant respects *Les paysans* is more Febvrian in tone than it is Vidalian.

Gourou observed that one might begin to understand the Tonkin Delta by starting with two striking and interrelated geographic facts that could be gleaned from the maps he had produced: "that it is extremely heavily populated, and almost exclusively inhabited by peasants."[103] In making this move – a second choice in his technics of tropicality – he was not following the patriotic and instrumental vision of colonial geography as it was pursued by the likes of Zimmerman, but a putatively purer – value-neutral and professional – conception of geography as a 'science' of observation and interpretation that started with the attempt to visualise and enumerate how people were situated in the landscape.

Gourou's project was ideologically suited to a French modernist movement, which, as Jacques Marseilles has shown, enjoined France to pride itself on its science and modernity rather than empire, albeit recognising the downsides of modernity.[104] The high population densities of the region, Gourou explained in 'scientific' terms, were the result of intense rice cultivation by a huge workforce, and were supported by a range of rudimentary handicraft industries (especially pottery and footwear). Agriculture predominated everywhere in the delta, however, and spawned a meagre peasant existence that bordered on starvation in many places.

Paul Claval argues that during the interwar years French geography was largely built around the paradigm it had chosen: simplified versions of Vidal's principles and stereotyped (and often descriptive) practices of regional analysis.[105] No one put this in doubt, he argues, and the only new problems that were posed to geographers revolved around how they should respond to the widening of their field of inquiry.[106] One of the main questions that arose from the effort to bend *le régionalisme* in colonial directions was whether French and foreign landscapes displayed the same levels of cultural and historical complexity, and thus whether regional geography, as it emanated from the *pays* of France, was a universally cogent framework of analysis. To Gourou, the Tonkin landscape was as sophisticated as any rural region one might find in France. Indeed, Gourou did not use Vidal's concept of *genres de vie* (ways of life) because he thought it simplified the multiple (agricultural, trade and craft) roles and complex routines of the Tonkin peasants.

And while the Tonkin Delta was a *region naturelle par excellence* – an area delimited by physical characteristics of climate, terrain, vegetation and hydraulics – Gourou delineated it foremost in human terms. He began his study by proclaiming that the Tonkin Delta "cannot be compared to the regions of Europe where swarming populations flourish. European countries with a comparatively high population density are all industrial countries with pronounced urban development . . . whereas the prodigiously high density [of Tonkin] is exclusively agrarian." He mentioned Lancashire and the Ruhr, and as his study continues a temperate/tropical contrast is gradually spliced into industrial/agrarian and urban/rural ones. Gourou did not deem the teeming urban and industrial populations and landscapes of Europe as normal, in the sense of customary, inevitable or superior. Indeed, in some respects he saw them as disorienting blots on the landscape. Yet he did not disown either the association of rapid economic, social and geographical change with progress, or the suggestion that the tropical mud, smells and intrigues of the Tonkin village belonged to something more primordial.

So it was that the idea of tropicality – of comparing and contrasting regions using markers of 'temperate' and 'tropical' – came into Gourou's field of vision, and in both unorthodox and ambivalent ways. This tropicality was expedited through the disciplinary project of French geography and Gourou claimed that the Tonkin peasants belonged to a more ingenious and venerable civilisation than others in the French colonial world, where "the lethargic native population exploits oft-fertile lands with a minimum of effort."[107] Over the course of his career he argued that the tropical contexts in which the geographer worked inevitably influenced how questions were posed and tackled, and it can be seen from the above discussion that he did not follow Vidal slavishly. While French geographical ideas and methods might be transposed overseas, he began to argue, details of landscape organisation varied and what was appropriate to France and a Vidalian programme of inquiry there could not be projected seamlessly to the tropical world.

Rice and romance

Gourou was entranced by the rice-growing culture of the delta, which he construed as retaining a distinctively Chinese "administrative organisation" but as indubitably Vietnamese in character and entirely distinct from surrounding mountain areas.[108] For him, the region also raised important questions about human development. In a final section of his study, on "the beauties of the delta," he waxed lyrical about the splendour, harmony and cohesion of this landscape. Its beauty impressed itself on the senses, he noted, but its order, symbolism and meaning only came to him after a considerable amount of work. In his view, the delta housed "a stabilised civilisation in material and aesthetic harmony with its natural conditions," and "one of the charms of the Delta lies in this perfect accord between people and nature."[109]

In 1955, Charles Robequain, who worked around this time on the neighbouring region of Thanh Hoá, remarked that *Les paysans* was "heated" by its author's warm sympathy for Tonkin civilisation and "commiseration for the Tonkin peasant, who [Gourou says] is unaware of his poverty and finds in the framework of the village, and in family and communal feasts, a distraction from his tenacious labour."[110] The peasant can "only live thanks to frugality and resignation," Gourou himself observed, and in good measure because of the seasonal and often violent swings of an unruly tropical nature. Peasants lived in "misery but not despair" thanks to their "intense and well organized village life."[111]

Gourou described the Tonkin Delta as a "beautiful landscape" set in a "hot climate but broken by a cool season," albeit with "not a month in which some crop does not ripen," and with the "Tonkin peasant never losing a crop from excessive heat or cold."[112] There is a matter of fact quality to his writing, about which we shall say more shortly, but also a good deal of lyricism, and the latter worked to "preserve the exotic as such," to rehash Raoul Girardet's important observation: Gourou kept the Tonkin Delta outside an intrusive and, in to his mind, damaging and ugly colonial and modern world.[113]

The "drizzle" of December to February, which afforded a rice harvest in the dry season, was one way in which this romanticist move was precipitated (plainly), because it helped to shape the density of distribution of people across the Tonkin paddy field.[114] Village life played a similar role in his narrative, both as the most palpable expression of high population density and as the place to which one needed to venture (as most colonists never did) to get "a true picture of human facts."[115] The charm and harmony of the delta were also etched into its natural and human colours: "For centuries, the peasants have organised harmonious relationships with the surrounding environment. Their clothes are often ragged and dirty, but brown or greyish colours and simply cut clothes, sometimes cheered up by a brilliant green belt, illuminate rather than stain the natural environment."[116]

And then there were the dikes: "The defence against river flooding, the seawater, and [dike] drainage and irrigation dominate Tonkin peasant life," he proclaimed, underscoring the tropical identity of the region. "The dikes are essential to the character of the delta landscape . . . [and] the study of the regime of the rivers shows why people resign themselves to this immense work."[117] For over a thousand years the Tonkin peasants had constructed "a vast network of dikes, moulded the soil with their hands; determined the relief of the country as we now see it; and made productive a territory which would otherwise have been given itself over to swamp."[118] In these ways and more, "the Delta is the work of people" – the local work of peasants and their way of organising and managing the land, and averting flood and the threat of crop failure and starvation, rather than the work of the colonial state or mechanical technology.[119]

This is a perfect example of how Yi Fu Tuan characterises a "romantic geography": a twofold way of seeing that stems, first, from a profound, if always selective, understanding of place, and second from a process of immersion which spawns a desire to escape familiarity, embrace difference, and see the outlandish as both a thing of beauty in its own right and a puzzle to be solved by the observer.[120] Another way to put this is with the words of Henri Copin. In French literature on Indochina between 1900 and 1954, Copin observes, "two movements confront one another: exoticism and otherness. The first is the western gaze upon a foreign reality that surprises and makes you dream, whereas the second expresses a movement towards the Other. Exoticism organizes the contemplation of a spectacle, whereas otherness admits a difference that it tries to understand."[121] Gourou organised a spectacle (and in some ways entertained a dreamworld) and an intriguing reality that was based, another of his admirers, Georges Condominas, told us, by his deep humanity and refusal to treat the Tonkin peasant as an exotic object.[122] Yet in Gourou's work there was a fine line between what was routine and what outlandish about the Tonkin peasant.

In Indochina, and elsewhere in the tropical world, Michel Bruneau notes, Gourou's emphasis on landscape and population density led him to privilege the *longue durée* of civilisations and push shorter-term social and political dynamics (of colonialism, decolonisation and development) into the background.[123] From *Les paysans* onwards, there was a "standard Gourou line," another astute observer-cum-critic of Gourou's work, Georges Courade, told us: "The rural. The peasant. The superiority of tropical Asia over Africa. Population density. The absence of the state. The use of the map and comparison. The quality of his writing (something

108 *Romancing the tropics*

now 'lost to science')"[124] These were the elementary technics of tropicality that Gourou began to assemble in the Tonkin Delta, and it is the latter two elements – his maps and writing – that we shall now probe.

Data delta

Gourou was always more interested in the character of the tropical landscapes and populations he beheld than in the methods he used to analyse them. He produced and synthesised a range of locational, quantitative, textual and visual – archival, cartographic, census, questionnaire, photographic sketch and survey – data. However, it is his veneration of the map as an interpretative tool that shines through *Les paysans*. "I always try to reason my geography out from maps," he declared in 1972, in one of the very few papers in which he reflects on questions of method. It was maps that had allowed him to grasp that "people are first and foremost organisers," he noted elsewhere: "members of a society that is more or less capable of organising a more or less large number of human beings, in a more to less vast territory, for a more or less long period of time."[125] He said that two facts pertaining to the Tonkin Delta, namely that is was "relatively homogenous and rigorously defined, and had been precisely administered and mapped," were key to his understanding of the area.[126]

Frédéric Dufaux observes that a cartographic understanding of population density founded and oriented Gourou research in many parts of the tropics, and that the Tonkin Delta, in particular, facilitated "an abstract reading of the lines of the terrain, and the map of densities, as they arose directly from the landscape." There was an aesthetic to Gourou's fascination with the contours and patchwork of the Tonkin landscape. But it also belonged to a realm of technique (method, procedure and choice) that chimed with other elements of Gourou's approach. Jean-Pierre Raison recalls Gourou's penchant for writing about landscape features – buildings, field and village patterns, and agricultural practices – in immense detail, his meticulous attention to the design of his questionnaires, and his love of indexes, with every village precisely located and described.[127] For all the lyricism in Gourou's description of the 'beauties of the delta,' he also coded the delta meticulously, even if he did not write about this proclivity.

The diverse forms of data that Gourou used and produced were bound up with different material and figurative sites where the meaning and purchase of particular forms of knowledge were forged. In Gourou's case, these sites – or what David Livingstone has called "spaces of knowledge" – were the field, the archive, the state (census and map) office, and the aeroplane.[128] They each spawned a different 'data delta,' and they were combined in *Les paysans* in both smooth and uneven ways. Gourou's fixation with population density, and the delta's associated village geography, stemmed from the extensive research he undertook in the library of the EFEO – alone, he later ruminated, and with only mosquitos to keep him company.[129] Gourou imbibed the School's Orientalist objectification of architecture, culture, language, myth and religion – a research agenda which was designed to help the French administration govern the peoples of Indochina. He worked from two lengthy EFEO bibliographies of French publications on Indochina up to 1914, and selected Indochinese works from 1913 to 1926, and the school's own scholarly *Bulletin*. Only a small percentage of the research published in the latter was by the EFEO's so-called 'annamitisants.'[130] The EFEO was not beholden to the colonial state. Rather, and as Pierre Singaravélou shows, it played a more dissimulated and, in some respects, ambivalent role in the culture of colonialism in Indochina.[131] It attracted specialist French scholars who received a formal Orientalist schooling at the École de langues Orientales and École pratique des hautes études in Paris, academic affiliates like Gourou, and non-specialist travellers, missionaries and amateur collectors. Researchers had both autonomous pretensions and a desire to serve the imperial cause by acting as cultural intermediaries between the French and Indochinese.

It was at the EFEO, in the heart of Hanoi's French Quarter (and close to the Lycée Albert Sarraut), that Gourou befriended and recruited an *annamitisant* who was crucial both to this traditionalist vision of the delta and his ability to cultivate 'the field' as a space of knowledge: Nguyễn Văn Khoan. John Kleinen suggests that the recovery of lost voices likes Khoan's "helps us to identify a commonality of direction and focus among Indochinese and French scholars during the colonial period as well as the hierarchical relations of knowledge production which, during the 1930s, had kept Khoan's work in a subordinate position to that of his French colleagues."[132] Khoan wrote on "the Vietnamese communal house, the Dinh, and its protective spirits," Kleinen continues, but was "best known in the West for being one of the key informants of Pierre Gourou."[133] It was Khoan, Gourou later acknowledged, who secured interviews with villagers, tried to teach him Vietnamese, served as a translator in the field, and encouraged him to study the distribution of village names.[134] Mus likewise regarded Khoan as his "workmate" and a "brilliant analyst of village ceremonies and rituals."[135]

Gourou later recalled that he began his research by walking extensively in the delta, and then by working his way through the Indochina bibliographies at the EFEO; and he acknowledged that Khoan helped to position him between the field and the archive.[136] But Gourou also insisted that he tried to follow his own instincts, and did so in good measure by venturing to the map room of the Service Géographique de l'Indochine (SGI) located in the administrative district of Hanoi. Established in 1899, and taken over by the Gouvernement Général in 1926, the SGI undertook geodesic triangulation, cadastral surveying, topographical studies, and map publication, and oversaw the work of the Service Aéronautique (SA), whose aerial photography and reconnaissance assignments, first over Hanoi and Saigon, and then over the landscapes of the Mekong and Tonkin Deltas, began in 1920, as a wing of the French army.[137] Both outfits employed a mixture of civil and military personnel, and were integral to Sarraut's *mise en valeur*.

In 1930, the colonial lobbyist Pierre Cordemoy deemed the aeroplane "a marvellous colonial tool particularly adapted to Indochina," and between 1920 and 1940, Gregory Gelzer notes, "aircraft were used in Indochina for creating cadastral registers, completing public works projects, drawing up maps, assisting scientific research, aiding local businesses and helping to package the colony for public consumption."[138] The work of the SGI and SA was spurred by public pressure to remedy the devastating flood of 1926, which had inundated more than a third of the delta and caused widespread famine. The Hanoi Chamber of Commerce grasped that this and previous floods (of 1894, 1904, 1909 and 1915) had bred unrest and fettered the aim and ability of French engineers and colonial planners to bring greater areas of Tonkin land into cultivation.[139] By the late 1920s the SGI had produced scores of 1:100,000 maps of Tonkin and Annam, and 1:250,000 maps of the Tonkin Delta, and both were expedited by aerial photographs taken from vertical and oblique angles.

The view over the hedge

With the assistance the SGI Gourou developed an aerial tropicality that was geared to the view of village autonomy propounded by the EFEO and Khoan – one associated with the Vietnamese proverbs 'the emperor's writ stops at the bamboo hedge of the village,' and 'each village strikes its own drum and worships its own deity' – and that culminated, Sophie Quinn-Judge notes, "in the mystification of the peasant farmer as the incarnation of virtue and wisdom."[140] Gourou was implicated in the creation of this village fable and traditionalist reading of peasant localism, which was deeply rooted in French intellectual culture, and which he developed with an image of concentric circles. "A complex civilisation surrounds the individual with a network of family and village relations," he observed; and "At the same time as protecting against external threats, the bamboo hedge is also a sort of sacred limit of the village community, the sign of its individuality and independence."[141] *Les paysans* features a photography he had taken of a 'typical' bamboo hedge (thicket) concealing a village in order to drive his point home, and the symbolism of the bamboo hedge left a deep impression on his colleague Mus too. "The typical village of Viet Nam is enclosed within a thick wall of bamboo and thorny plants," Mus observed and continued in terms as stylistic as Gourou's: "the villagers used to live behind a kind of screen of bamboo, or perhaps it was more like living within the magic ring of a fairy tale."[142]

The two of them toured the delta together during the early 1930s, and the traditional village was central to how they fomented what we might call an essentialising erudition. Their learning was on the Orientalist spectrum described and critiqued by Edward Said: their fastidious scholarship and affectionate attachment to Vietnamese culture was framed by images of "the absolute and systematic difference" between the West and the Orient-tropics, and with the mythic knowledge and collectivism of traditional Vietnamese civilisation pitched against the rationalism and individualism of a new generation of educated and urban Indochinese.[143] Neither of them suggested that 'Asian minds' lacked the capacity for critical judgement. Indeed, Mus wrote in his copy of *Les pays tropicaux*, "the annamite peasant's taste for risk-taking leads to social upheaval and a break in the social equilibrium on which Gourou concludes."[144] As Susan Bayly says of Mus (and we think the point applies to Gourou too), their "discussion of Vietnamese civilisational ideas entails a correspondence between the individual's mental landscape and the ordered, cultivated physical landscape in which . . . the Vietnamese believe civilized human existence is framed."[145]

While Gourou's work on the village was developed by Mus into a complex and nuanced body of ideas about the "experimental" sociality of the Vietnamese, and eschewing "a picture of mute, passive 'Orientals,'" Mus's ideas were simplified in the process of being popularised by the American journalist Frances FitzGerald in her 1972 Pulitzer Prize-winning *Fire in the Lake* (excerpted in *The New Yorker*), which is a searing account of the American war in Vietnam and its impact on the Vietnamese population.[146] "These anonymous masses of bamboo hedge," she wrote, "make up a moving and continuous horizon which always encircles the spectacle of the rice fields and the manifold work it implies. . . . After going mile after mile in the Vietnamese countryside one has the impression of having stood still."[147] FitzGerald latched on to Mus's idea that the Vietnamese term for socialism, *xa hoi* (frequently used by Ho Chi Minh), was derived from the word *xa*, which signifies the spiritual unity and communal traditions of the village, and argued: "The problems of the village were, finally, national problems, and the NLF alone among the southern political groups offered a solution on a national scale."[148] Throughout the country the communal dynamics of village life were pivotal to the communists' ability to renew their revolutionary capacity and determination in the face of colossal death, injury and destruction.

Tuong Vu identifies a consistent technique in *Fire in the Lake* whereby FitzGerald "juxtaposes the two nations [United States and Vietnam] in a way that essentializes and exaggerates the contrasts between them"; a technique that "is Orientalist pure and simple."[149] That the book buttressed the anti-war movement in the United States, not least on the grounds that the United States was pulverising a land 'where time stood still' and with tactics that would have been unthinkable during World War II, give us reason to describe it as purveying a militant Orientalism.[150] At another level, however, it amounted to a troublesome tropicality which romanticised the landscape in some of the ways Gourou had.

Gourou's legacy is still keenly felt and debated within Vietnamese studies, and is not hard to find it in 1950s and 1960s American social scientists' fascination with the peasant village either.[151] "The village, the basic peasant community, shaped in the course of struggle against natural calamities and foreign invaders, was in the course of history the cell of our rural society," a 1993 Vietnamese collection maintains.[152]

To be sure, Gourou's determination to look over the bamboo hedge using interviews and questionnaires, and to work in provincial archives on local customs and rituals, was a grounded affair. But it also hinged on a vertical mode of understanding – looking at by looking down upon. Once inside the village, Gourou observed, "The streets are all similar to one another; they are narrow and terribly muddy when it rains, except in the villages where they have built a central pavement out of limestone slabs or bricks, the sides being reserved for the buffalo, who like to squelch around in the soft soil. . . . The streets are not laid out randomly and there often exists a very visible layout in the village."[153] As Paul Lévy noted, when Mus later wrote about his attempt to forge a 'visual sociology,' he modelled

112 *Romancing the tropics*

it to some extent on Gourou's aerial view, noting that he had sought to 'read' a society through its forms in the manner that Gourou had read the spatial form of the Tonkin landscape as an *ensemble* and *montage* (assemblage) of points, lines, surfaces, clusters, distributions, shadings, synchronisations and geometries.[154]

Les paysans came with nine large, immaculately produced, maps – four of them at 1:250.000 scale and five at 1:500.000 scale – folded into wallets at the back of the book. We have used two of them, the density of population in the Tonkin Delta, and Tonkin Villages (Figures 3.1 and 3.2; Gourou's Map 2 and 3; both 1:250.000), to illustrate this chapter and underscore how central maps, diagrams, sketches and

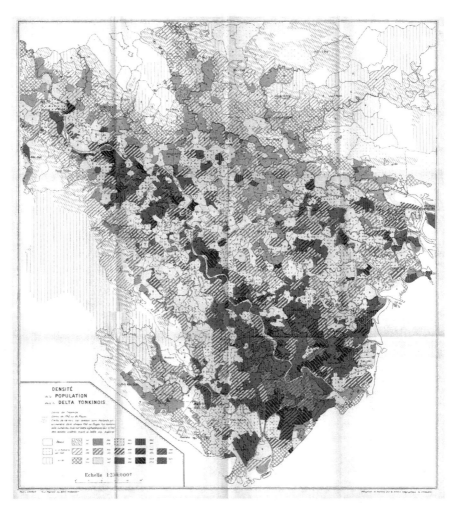

Figure 3.1 The density of population in the Tonkin Delta. From *Les paysans du delta tonkinois. Etude de géographie humaine* © EFEO

Romancing the tropics 113

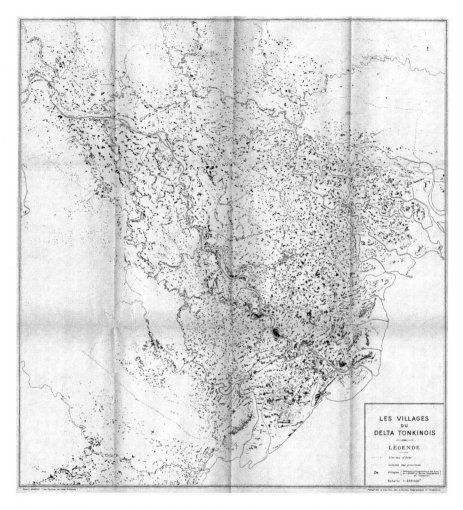

Figure 3.2 Villages in the Tonkin Delta. From *Les paysans du delta tonkinois. Etude de géographie humaine* © EFEO

photographs were to the way he interpreted the region. Maps, particularly, were at once starting points for, and the crowning glory of, his study. Large swathes of his analysis proceed from them, and both the maps themselves, and Gourou's silence surrounding their deployment, constitutes a further potent technic of tropicality. But Gourou also mobilised a plethora of other visual data – (1:100,000) maps, tables, diagrams, drawings, and aerial and ground photographs, all of which, somewhat confusingly, he lists as "Figures" at the end of the book – and over the next few sentences we shall refer to them in lower case, to distinguish them from our two map illustrations.[155] Gourou used the most up-to-date census figures from 1926 and 1931 to calculate average population densities by canton (administrative district), and rendered the density of population in cartographic form (Figure 3.1) with 17 colour-coded (red/yellow) classes from "deserted" to "1350+" inhabitants.

He used the same statistical and geographic logic to map the average number of land parcels per hectare (map 7); average (and lower and higher) population densities, with 400–600 per km2 the average (fig. 31); the average area of villages by canton, in three classes, of 110–150, 150–200 and +200 hectares (fig. 34); and village types (fig. 46). He also sought to map the relationships between population density and village agglomeration (maps 2, 4 and 6); relief and population density (map 4); and village agglomeration and the dike system (fig. 12).[156] In short, he looked at the organisation of space from every angle.[157] Such fastidiousness was exactly what the French archaeologist and post-war custodian of the Angkor Wat ruins, Bernard-Philippe Groslier, had in mind when, in 1961, he wrote of the "scrupulous science" that had descended over interwar Indochina.[158]

This cartographic apparatus allowed Gourou to mobilise Vidal's concern with "connections" in the landscape, and anchor his argument about the tropical (monsoon) symbiosis between people and nature in the delta in a seemingly precise geographic reality. Vidal's own 1908 *La France: Tableau géographique* was driven in no small measure by his analysis of 13 aerial views, and Jean Bruhnes used the image of photographing the world from a balloon as a metaphor for the discipline's objectivity, method and (literally) elevated position among the human sciences.[159] Marie-Claire Robic sees these pioneers of Geography in France as promoting a "geographical culture of the visual" in which "the ideal of an absolute seeing eye" combined the synoptic (gatherings in the same space) and panoptic (of all surrounding things being visible from a particular location).[160] She invokes "the aerial gaze as a metaphor of geographic method" – for what she terms "the *paysagisation* of geography" in France.[161]

Such inflections were different from geographers' and surveyors' use of ground and aerial photographs in other parts of the world at this time. Matt Dyce suggests that in North America and India, for instance, they were used to depict fieldwork and show "the human interface with nature," and many geographers did not seek to elide their presence in what they were doing.[162] However, around the world the advent of aerial photography, and its different oblique and perpendicular means of capturing landscape features, and stereoscopic methods of aligning photo and map, raised complex questions about how or whether this way of studying the surface of the earth mimicked and supported the logic of the map or constituted a different way of seeing.[163] *Les paysans* is lavishly illustrated with aerial photographs produced by both methods. Gourou tended to use oblique shots when discussing terrain, and vertical ones when dealing with villages.

French geographers' embrace of maps and the aerial view as the basis of a rational and rigorous description of regions brought tropicality in its train and disguised the military reach and roots of this methodology. As Robic notes, the aerial view played into the hands of metropolitan political discourses that envisaged 'civilised' landscapes as dense and ordered mosaics that were monotonous – and by implication compliant – in their human regularity, and the imagination of 'uncivilised' landscapes as savage, irregular and untamed, and with 'jungle' often a synonym for 'savage.' Moreover, this geographical lens was part and parcel of what Martin Jay and Sumathi Ramaswamy describe as a more fulsome and sinister "imperial lens" with "overt affinities... between the gun and the camera in the arsenal of empire."[164] Gourou's liberal use of photos produced by the SGI and SA pulls him into this field vision, and there are thus intimate connections between his romancing of the tropics and the ordering of this zone as a visual scene with a militarised tint.

Haffner notes that the Tonkin landscape "had much to teach its colonizer" about how to look at maps.[165] "The northwest portion of the delta appears on the [hypsometric] map as higher and more serrated than the surrounding area," Gourou observes in a passage where he thinks about the connections between the map and the aerial view:

> The rivers are bordered by natural levees which rise above the alluvial plain... although by no more than two or three metres. There is, however, an important coincidence in the landscape, in that the spots at which the rivers afford protection from regular flooding are also the spots that attract villages. This fact is confirmed by the map of villages, the examination of which [in turn] will clarify the hypsometry of the delta.[166]

Similarly, with respect to connections between water, soil, monsoon seasonality and human habitation, he observes:

> The fertility of the soil is reflected in the density of the population which reaches its highest values on the richest soils rather than in the most anciently populated regions.... The landscape is dominated by the village settlements, [and] in this country, where the Tonkin peasants have worked with their hands, and defended an immense network of dikes, they are the slave of these things.[167]

Such arguments could have been developed without maps, but they had more clout with them and the numerous photographs that Gourou took with his Leica camera.

But Gourou was once again almost completely silent about how he made such connections, and paid scant attention to Vietnamese maps and conceptions of space.[168] In only two places in *Les paysans* does he betray his commitment to the ideology of accuracy and precision surrounding maps as 'objective' tools of explanation: first, in the chapters on villages, where he mentions in passaging how he felt beholden to visit over a quarter of the habitations in the delta by foot (first in the south, then the west and north, and visiting an average of around five villages per day), there to undertake his own census and survey of land use and village names, because he did not trust some of the questions and methods used by French census-takers, who did not converse in Vietnamese; and second, in the bibliography, where he describes the SGI as a "wonderful working instrument."[169]

116 *Romancing the tropics*

This mapped landscape was integral to Gourou's tropical romanticism. "An appropriate series of *plans* [maps and diagrams], leading the eye to the horizon, adds to the infinite greatness and attraction . . . of the river landscape" and the appreciation of "its delicate and peaceful harmony."[170] With regard to this and similar passages, Yves Lacoste later remarked, in an essay entitled "What is a beautiful landscape":

> When Pierre Gourou describes . . . the 'beauties of the delta,' it is not only the uniform plain . . . with the paddyfield in the foreground to the observer who observes it from his height alone. . . . [It is also] the plain which appears, very vast and differentiated, with the green spots indicating big villages hidden in the trees, with the network of dikes and canals: it is the landscape that can be seen from the top of a tall dike; it is from these dikes that we discover the beauty of the landscape.[171]

In other words, for Gourou as for Lacoste, 'beauty' is a technical achievement as well as an aesthetic response. Lacoste was reflecting on his 1972 visit to the Tonkin dikes, when he stood atop the dikes in order to ascertain whether there was a pattern to how they were being bombed at the time by the U.S. Air Force, and he knew that Gourou had had the luxury of an even more elevated view. We say more about this forward connection to 1972 in Chapter 7.

Gourou had already made use of the maps and aerial photographs of the SGI and SA in his guidebook for the Exposition Coloniale, and both of these outfits had produced their own volumes for the event, outlining their mission.[172] The Orientalist Louis Finot viewed the SGI's 1920 *Atlas de l'Indochine* as a "sign of incontestable progress" and "point of perfection" in France's spatial grasp of its colonial possession.[173] However, he also noted that the SGI's "ground surface maps (of forests, plains, paddy fields, etc.)" cried out for specialist attention, and Gourou heard him. The work of the SGI was geared first and foremost to colonial administration, and by soliciting its data and assistance Gourou's research was implicated in France's colonial project. Indeed, he confided in interviews that such assistance was given freely and gladly. Nonetheless, care is needed in how this complicity is evaluated. Kapil Raj, for instance, notes in this vein that maps should not be seen as true or biased according to poles and measures of objectivity or subjectivity, but as "hybrid" creations "resulting from the circulation of people, incorporated skills, instruments, procedures, etc."[174] Complicity in the colonial project might be gauged in terms of how fully or insidiously this hybridity – as it applies to the 'wonderful working instrument' of the SGI, for example – was expedited or disguised.

Gourou was one of the first geographers to make extensive use of aerial photographs in a colonial setting. In Indochina aerial photographs of occupied, abandoned and subdivided land provided a new and important way of visualising agricultural crisis in the Mekong and Tonkin Deltas, and of cementing colonial views of traditional agriculture, landholding and village life, and thus of 'the native mind.' Their colonial use proliferated after World War I when aircraft and cameras equipped for the war in Europe became available, and as new flight and camera technologies, and new commercial possibilities surrounding them, developed. "Aerial photographs not only redefined the landscape in visual terms," David Biggs argues, "but in the hands of influential geographers [he has Gourou in mind] they also redefined colonial understanding of the power and potential of local people":

> the birds-eye view afforded by aerial photos of Tonkin shifted colonial knowledge of the hydraulic landscape by showing the interior spaces of villages and fields to be a densely intricate patchworks of fields, orchards, hedgerows, and homes. The blank spaces on topographic maps were replaced with vibrant grayscale mosaics of fields, dikes and villages. Engineers and scientists quickly recognized the value of such interior landscapes subdivided into so many cells of dikes and canals, and they used the term *casier* to describe the case-like landform.[175]

This aerial view played into the hands of an inveterate Orientalist imagination. The toil of legions of Tonkin peasants keeping their dikes and paddy fields intact solicited "frequent comparisons to masses of laborers constructing the pyramids in Egypt or the temples at Angkor Wat," Biggs notes.[176] During the interwar years aerial photography fostered what Martin Thomas describes as new "empires of intelligence," "revolutionizing the working practices of military staffs and civilian specialists whose decisions demanded accurate information about topography and climate, agriculture and water distribution, or sites of archaeological interest."[177]

What Gourou achieved with maps and photographs, however, was neither simply derived from metropolitan salons of Orientalism and geography, nor just about the projection of Western and French ways of seeing. We heed Felix Driver and Luciana Martins's observation that "the complex genealogy of tropicality reveals a complex pattern of relations and mutations, within and beyond Europe."[178] While Gourou sought to disclose an underlying and external reality with maps, photos and words, the mapped, photographed and written landscape he assembled shaped that reality, and his story was caught up in complex patterns of investigation, comparison and contrast.

We have seen that Gourou wrote and thought with both France and Indochina on his mind, and with a subversive twist. Haffner, for example, suggests: "Rural villagers in northern Vietnam were much 'happier' and more sociable than their counterparts in France, according to Gourou, because the diverse areas of their lives were harmoniously integrated. . . . Gourou's aerial study of the Tonkin was intended, in part, to open the eyes of French citizens to the artificiality and emptiness of life under consumer capitalism."[179] Gourou's aerial gaze may have been implicated in the colonisation of the Tonkin Delta, but he perhaps saw it as offering France a different – liberating or critical (depending on from where one looked) – way of seeing itself.

The aerial view developed by Gourou altered Orientalist fascination with what lay over the bamboo hedge by drawing a series of connections between visual

118 *Romancing the tropics*

order and tropical happiness. Orientalist fascination with the Far East came with a historic and comparative view of landscape that conceived of exteriors and interiors (of buildings, cultures, landscapes) in ostensibly ground-level terms. Prospects and perspectives did not generally take very elevated forms. The world exhibitions raised viewing platforms higher above the ground, and began to associate distance and difference with elevation, elevation with objectivity, and objectivity ultimately with superiority. It was not until the advent of aerial photograph as a military and civilian tool of reconnaissance, however, that a vertical sense of depth and height became central to the task of working out the 'inner workings' of societies and landscapes.

Gourou's way of asking 'what lay over the bamboo hedge' involved a tropical mutation. He was steeped in monsoon patterns and relationships in the landscape, and as one saw them from above, and from a much higher plane than just the top of dikes. His maps and photographs fostered a tropical rather than Orientalist way of seeing that connected aesthetic appreciation of the landscape of the delta to an abstract and vertical understanding of its patterns, regularity and variety. He muses: "At first sight, the Delta is monotonous, greyish and without charm. But its prettiness and beauty reveal themselves little by little to the eyes of whoever agrees to undertake the necessary initiation, who follows dikes and tracks in the various seasons, and who goes into the villages."[180] Field immersion is coded as revelatory, and the access he desires comes from the air as well as the ground. "The map of population density within the villages might at first seem confusing," he notes, "but it helps us to see very quickly the distribution of loose and tight-knit villages in the Delta."[181]

He reproduced numerous aerial photographs to illustrate and bolster his cartographic rendering of this crowding and confusing village scene in abstract form – on a map of 'village types' (fig. 46 in *Les paysans*), for instance, and from which he inferred, in relation to his large map of villages (Figure 3.2), further connections between population density, village concentration, relief and land use:

> The impression produced by this map [of villages] is very strong, and on first sight we are struck by the countless multitude of black spots representing the villages and by the very big surface they occupy on the map. It seems at first that these spots are thrown about in a disorderly manner, as if traced randomly by an insect that has dipped its legs in ink. But very quickly guiding lines appear; the study of the habitat will consist of bringing out these guiding lines and determining the different kinds of villages.[182]

"The villages of the Tonkin Delta solicit our attention," he continues:

> From all sides our eyes are drawn to the dark masses, with clear contours, with the hedges around them. A village does not appear in the landscape as an accumulation of houses, but as a block of greenery. It is surrounded by a barrier of bamboo whose tight links stem from a solid rampart, and whose shimmering poles give the village a luxuriant and noble frame.[183]

With reference again to his map of villages: "Trails of extended villages attract our attention and are the most striking feature of the map: these are relief villages, villages which have been established on higher land to find shelter from floods."[184]

In 1943, and with the French air force patrolling the skies of North and West Africa, the French ethnographer Marcel Giraude wrote in similar, albeit more chilling, terms about his fieldwork patch in Chad:

> Perhaps it's a quirk acquired in military aircraft, but I always resent having to explore an unknown terrain on foot. Seen high in the air, a district holds few secrets. Property is delineated as if in India ink; paths converge in critical points; interior courtyards yield themselves up; the inhabited jumble comes clear. With an aerial photograph the components of institutions fall into place as a series of things disassembled, and yielding. Man is silly: he suspects his neighbor, never the sky; inside the four walls, palisades, fences, or hedges of an enclosed space he thinks all is permitted. But all his great ambitions for growth appear on an aerial photograph.[185]

James Clifford notes that "It is not clear whether this passage should read as enthusiastic publicity for a new scientific method . . . or as a somewhat disturbing fantasy of observational power."[186]

Clifford's work is important for a wider reason. We concur with his central formulation that "with varying degrees of explicitness, ethnographies" – and geographies – "are fictions both of another cultural reality and of their own mode of production" and "expression[s] of the performed fiction of community that had made the research possible."[187] Part of this 'performed fiction' – or narrative – of real and imagined peasants, villages and rice paddies, was about peace and conflict, and stability and upheaval. Mus's annotations on his copy of *Les paysans* are strident in this regard. "The bamboo hedge shelters those who respect conventions," he scribbles, and he proceeds to both rehearse and politicise Gourou's observations on the matter:

> We experienced this in March 1945 after the Japanese occupation, in a countryside swarming with enemy patrols. "At the same time as protecting against external threats," as Pierre Gourou so aptly analyses, "the hedge is also a kind of sacred limit of the village community, the sign of its individuality and its independence." When, in times of unrest, villages take part in the agitation or give shelter to rebels, the first punishment inflicted upon them afterwards [by the occupier/aggressor] is to force them to cut down their bamboo hedge. It is a serious wound to their self-esteem, an infamous mark; the village feels as embarrassed as a human being would be left undressed and abandoned in the middle of a dressed crowd.[188]

Gourou's judgement about the bamboo hedge as a sacred limit needs to be read in this more political light, Mus avers. It is not simply a marker of exoticism and alterity.

Another part of this 'performed fiction,' we now seek to observe, revolved around what Olivier Orain characterises as a particular style of writing and stance in the world, and one whose poetics and politics flew in a different direction.

De plain-pied dans le monde

Orain uses the complex and clever expression *De plain-pied dans le monde* – which variously translates as 'level with,' 'on a plain footing with' or 'barefoot in' the world; and also as 'to walk in the real,' 'to be true to reality,' and 'to be a clear, wide-eyed witness to the world' – to capture the rhetorical and stylistic methods by which French geographical writing became persuasive and authoritative.[189] As Orain observes, "The realist illusion of being level with the world, in the immediacy of the object, rests on the removal of what might emerge in the text about the conditions of its enunciation, and its roughness [noisiness and imprecision] as mediator in the act of communication."[190] He argues that a belief in the immediacy of the world and one's ability to imbibe and transcribe it invested French geographical writing with both realism and complacency about the need to probe the epistemological grounds on which geography was trying to operate as a privileged mirror of nature. Crucially this meant that the geographer had to convince the reader that he or she (it was usually a he) was level with the world, faithfully observing and recording its facts and relations as they appeared on the ground, yet at the same time not there – above it all, in a kind of observational fantasy (to use Clifford's image), as if those facts and relations were unclouded by personal whim or methodological artifice.

Orain describes this as a metaphysical pincer movement designed to make geography appear not as a translation of the world into words and images but as the world's very expression, and one that had obvious political implications at a time when geography, a relatively new discipline, was seeking to gain a foothold in school and university curricula and trying to impress other disciplines with its analytical gravitas. The geographer's attention to terrain, and fastidious recourse to historical documents, maps and fieldwork, was what anchored this realism. But it was also delivered, Orain maintains, in a high literary style that involved both lyricism and level-headedness to create the illusion that there was no authorial intervention in the passage from world to word. Accordingly, geography's status as a performance – a doing and a construction – was obscured and what came to the fore were regions invested with clear-eyed identities, objects and relations that seemed to have arisen from the landscape itself rather than from a trail of consenting or conflicting opinion. The trail of how one worked – the sources and methods one used, the voices one solicited, and the views one accepted or discarded – were consigned to the background, or the footnotes, if they were mentioned at all, again making interpretative judgements appear as if they were warranted by the landscape itself.

Such machinations help us to gauge Gourou's writing and give a more precise disciplinary context to the standard (and often much too generalised) postcolonial tenet that colonial discourses attain their ability to manipulate meaning and exude authority by concealing how colonisers went about seeing and knowing the things they did, and often by disguising and denigrating Native influence over what was seen and known. Said powerfully argued that Orientalist learning "was premised on the silence of the native, who was to be represented by the Occidental expert speaking *ex cathedra* on the native's behalf, presenting that unfortunate creature as an undeveloped, deficient, and uncivilised being who couldn't represent himself."[191] Gourou sidestepped this issue, which had been thrashed out in the 1920s by rival anthropologists Bronislaw Malinowski and Lucien Lévy-Brühl, with the former arguing that no people "however primitive . . . were lacking in the scientific attitude or in science," and the latter replying that 'primitive' peoples were "pre-logical" and unable to fully comprehend the laws of nature and society.[192] Gourou did not represent the Tonkin peasant as primitive – in fact, quite the reverse – but nor was he open about how he situated peasant wisdom in his work. In his interview with John Kleinen, Gourou suggested that he did not name many of his valuable Vietnamese collaborators in his published work because at that time some of them, such as Võ Nguyên Giáp, were revolutionaries and former political prisoners and such acknowledgement may have detracted from the scientific spirit of his work. Science and politics were to be kept apart. We regard this as a partial, and in some way deceptive, statement which reveals as much at it conceals about the complicity of knowledge in power in Gourou's case.[193]

While Gourou's fieldwork was transactional, in the sense that he used Vietnamese scholars and guides to generate and help interpret his field data, the transactional process in which he was involved was unequal. Gourou subjected his data, and the fieldwork situations from which it arose, to a set of 'knowledgeable manipulations' (as Said describes Orientalism – see Chapter 1) that obfuscated the role his native informants played in the making of that knowledge. This process of epistemological depreciation is most evident in the intellectual sovereignty that Gourou blithely wielded over his study: from the fact that he says next to nothing about how he worked in the field or wrote his book. It is difficult to discern from *Les paysans* precisely how he worked in the two spaces of knowledge that were crucial to his endeavours – the field and the archive. In important respects, this did not come down to a set of individual interpretative choices but to the wider French geographical research culture captured by Orain.

The idea that *Les paysans* is 'on the ground level with the world' starts with a rhetorical sleight of hand: not with the immediacy of a world bristling with life but a vacuum in understanding; with the notion that, at first blush, what was beheld was a previously unseen problem of population density shaping the landscape and needing to be explained. Gourou presents his 'original geography' as a *tabula rasa* which he will go on to fill with geographical meaning. The map was pivotal to this undertaking, and Gourou made no attempt to investigate or acknowledge Vietnamese histories and practices of map-making.[194] The influence of Khoan's work on the Dinh on his understanding of the village is plain to see, and Gourou acknowledges his debt to his *annamitisant's* "majestic" work at one point in the text, and in a number of footnotes.[195] But only late in life did Gourou talk openly about how his valued assistant arranged his visits to villages, knew the dialects, and worked tirelessly as a translator. Roger Lévy perceptively remarked that Gourou's style was "a style without hyperbole," and we might add that Gourou expedited it as one without accomplices.[196]

Gourou thought that neither the French nor the Vietnamese saw him as a threat; that it was important that he was not perceived to be either a journalist or a tax collector; and that his sole aim was to study the people "in their civilisation" and fathom "roots and borrowings" going back into the mists of time.[197] He took small presents (often cigarettes) as a means of introduction, and never asked questions about incomes and prices, he said, lest villagers saw him as a tax collector in disguise. Living in a village, or staying overnight, "was not done at the time," he continued, but both French colonial officials in provincial towns and village Mandarins in the delta helped him to check census figures. Overall, he recalled that he never got into any bother, although he was reticent about dogs because rabies was rampant.

These omissions in *Les paysans* were not inherently colonialist. For neither did Gourou mention the French friends and visitors who accompanied him to the delta, and including some of the leading scholars and administrators of his day (not only Mus and Robequain, but also Paul Rivet, Lévy and Jacques Walreusse). Nor did he discuss his methodological debts to Vidal or Demangeon in any detail.[198] He was silent about a lot of things that contemporary fields of scholarship, concerned about the grounds upon which their knowledge and data are produced, would ponder. Indeed, when asked these sorts of questions late in life by Kleinen, Gourou claimed that he had worked alone, had never spoken to any other geographers about his thesis, and had never wondered about what they thought.[199]

Lastly, while the Tonkin Delta was undoubtedly less colonised than its southern Mekong counterpart, Gourou later admitted that the dike system was based "on the combination of a colonial apparatus and already existing *encadrements*," and that while "the Tonkin peasantry in 1936 lived in the framework of traditional *techniques d'encadrement*, since there was hardly any modern economic sector, a French administration existed in parallel with the traditional administration."[200] He grasped the latter by talking to French provincial and district officers, and from working in the Vietnamese archives in Ha Dong and Nam Dinh. In 1939 Robequain produced a map highlighting areas with traditional dikes and those with modern irrigation canals and dikes constructed by the French following the 1926 flood.[201]

Such details were extraneous to the type of story Gourou sought to tell, which, in spite of the way he invoked his work as an individual effort, might broadly be situated within the framework sketched by Orain. It is important that these dynamics and voices are now recovered, as Kleinen does with Khoan, and Gourou himself partly did long after he had left the scene of his study. For his part, Khoan noted in his 1930 study of the dinh that many of the villagers with whom he dealt were suspicious of Vietnamese researchers associated with the EFEO, and, with respect to the legends surrounding "village geniuses" (ritual codes and figureheads), that "the villagers hide them jealously."[202]

It is also important to acknowledge that the project and model recounted by Orain often operated more as an epistemological wish image than as a smoothly expedited procedure. The gap between knowledge and the world was not easily closed, and the knowing subject was not removed from the page without difficulty. Such fissures shine through Gourou's discussion of what he called "geomancy," and by which he meant ceremonial and architectural forms of divination in Tonkin villages. He drew this term from Renaissance and Baroque literature and *in lieu* of an appropriate Vietnamese alternative, and used it to complicate hard and fast distinctions between realism and mysticism. When seeking to explain the concentration of villages, he notes, "one should not insist too much on rational explanations which, in this country not more than in others, do not produce a true picture of human facts."[203] "A more searching examination reveals that the peasants' religious preoccupations are manifested in a large number of buildings of all sizes," he continues, and one might "illustrate this fact" by drawing up "plan[s] of what could be called the points of religious sensibility which show with certainty that the villages quite often attain their shape for geomantic reasons."[204] But he laments: "We are too ignorant of this art" to do it full justice, and "if, after having examined the architecture and the plan of the house, one wishes to get from the inhabitants an explanation for particular arrangements, one encounters much caginess."[205]

While not colonial in a political sense, *Les paysans* might still be thought of as *émigré* and paternalist in the way it assumes and claims to know the Tonkin Delta. Maps and photographs were particularly suited to claims to realism and objectivity, and to eschewing impressionism and the picturesque. Yet as we have shown, Gourou's writing abounds with the picturesque and his 'walk in the real' came with vivid colours and rhetorical flourishes, and also doubts – as with his reflections on geomancy – about whether he had got the full measure of what he had witnessed.

> Life in the village is, beneath its dreary and miserable appearance, intense and rich in emotions; it brings the peasant interest and passion. The many events of the political, religious and social life of the commune provide the peasant with the chance to experience the satisfaction of dominating, the abundance of feasts, the rancor of defeat, the bitterness of humiliation, the pleasures of intrigue, the great pomp of a beautiful celebration in which the village participates unanimously; and all of these things help the peasant to avoid meditating on his all too mediocre condition, drawing the balance sheet of his poor resources, thinking of the debts he labours under and whose repayment will absorb the best part of his meagre income.[206]

In this passage, which takes us back to Lefebvre, Gourou uses metaphors of 'debt' and 'balance sheet' as comparative means of pointing both to the degraded condition of peasant life in France, and to the threat that the Tonkin peasants' awareness of their own poverty and misery might pose to the colonial order. On the former count, his Tonkin peasants are suspended in exotic contrast to the 'dreary' calculations of their French counterparts leading less fulfilling lives. Gourou found in the literary pretensions of French geography not merely the means of 'walking in the real' but also of escaping to another reality. The delta is a conceptual and stylistic device for drawing distinctions between a bland and in some ways broken temperate-peasant France and a colourful and harmonious, if threatened, tropical-peasant Tonkin. However, in Vidal's and Bruhnes's works, as much as in Gourou's, peasants are not discussed by name, and the conversations they may have had with him and other interlouctors are not used as a form a testimony or means of questioning. French and Tonkin peasants are written about in aggregate and abstract terms, as figures on maps, members of particular types of villages, builders of exotic buildings, and as bearers of particular economic traits.

A Foucauldian coda

Gourou's Tonkin peasants and landscape were both imagined and real, and the tropicality he developed in *Les paysans* had both horizontal and vertical dimensions. Gourou launched a ground and air interpretative offensive that revolved around books, maps and field reconnaissance. The names of individual peasants, and conversations he had with them, are virtually absent from his book. *Les paysans* contains a profusion of village names, but these are used chiefly as appellations for the dots on his village maps and diagrams. His concern was with a peasant civilisation and a landscape that he marshalled from above and at a scale the peasants themselves did not see. The military metaphor of marshalling is perhaps out of place here, for there was nothing ostensibly belligerent about Gourou's arrogation of the Tonkin Delta to his way of seeing and working. Admiration and respect were the watchwords of his 'indigenist' engagement with the peasants of the region. But we use it as a reminder that just as militarism comes with fantasies of conquest, control and pacification, so might Gourou's technics of tropicality – his techniques of representation and interpretative choices – be interrogated in terms of how they captured and coded the Tonkin landscape and its inhabitants, and fostered fantasies of order, harmony and authenticity.

Gourou's project had colonising inflections in the sense that it staged and manipulated meaning. Yet we have also sought to demonstrate that care needs to be taken over how such claims are registered and positioned. The detailed moves by which tropicality works or might be arraigned as an instrument of othering need to be studied and situated. *Les paysans* can be read as an attempt to insulate – at least allegorically – the peasants of the Tonkin Delta from the inroads of colonialism and modernity. The book is shaped by this paternalist interpretative drive. But Gourou could not completely sequester either the peasants or himself from the divisive modern and colonial forces swirling around his hallowed landscape and romantic geography, and this gave his tropicality an interesting twist. For at crucial points in his narrative the normativity of a 'temperate' France formed not only a constitutive outside to the Tonkin peasant world he pictured, but also a poor outside – poorer for being bland and disenchanted – and in this regard a military metaphor that Michel Foucault used to theorise power is apt.

In famously characterising politics as "war continued by other means," Foucault was deploying a military metaphor, in part as an alternative to dominant linguistic understandings of power (as a "great model of language and signs").[207] He argued that war is senseless, and that so too is the idea that there is a totality of power relations. As Joseph Rouse explains, "meaningful actions and situations do occur within specific alignments of power, but these have only local intelligibility, which Foucault understands as tactical. That is they make sense only as responses to a particular configuration of forces within an ongoing conflict."[208] In Foucault's model, the only rules and meanings governing war – and by extension knowledge and scholarship – are those involved in the play of forces within a specific and ongoing struggle. How such rules and meanings embrace and interdict, and divide and contrast, phenomena need to be understood at this 'local' level rather than with some grand theory of power.

Foucault was elected to the Collège de France in the year before Gourou retired, and according to Gilberte Bray her father felt uneasy about the philosopher's anti-humanism and seeming disinterest in the world beyond Europe.[209] However, we think that there was a Foucauldian play of forces at work in *Les paysans*. Gourou captured this 'play' later in life by way of a response to a provocation. When asked, in 1984, why he seemed uninterested in questions of politics and the state, he responded, "I would rather see a peasant than a politician."[210] This assertion says a lot about Gourou's technics of tropicality. While French colonialism in Indochina was shot through with hierarchy, division, and disorientation for French settlers, Gourou placed a premium on "the density of the social networks that characterize the nucleated villages of northern Vietnam," as Philip Taylor put it: on a landscape and imaginative geography of "social connection," and one that in Gourou's mind needed to be protected from colonial politicians and administrators.[211]

Gourou's denouement – his local and tactical configuration in a Foucauldian sense – was that "Only this traditional civilisation can give its people the happiness they deserve; outside this [state of affairs] there is only disorder and despair."[212] His geography of 'social connection' was threatened by encompassing forces of disconnection. This was the 'play of forces' at work in *Les paysans*, we think, and it chimed with Mumford's critique of the machine age. Mumford wrote of an interwar "reawakening of the vital and the organic in every department" of economy and society in the West, of how it was weakening "the authority of the purely mechanical," and of how "we can now act directly upon the nature of the machine itself, and create another race of these creatures, more effectively adapted to the environment and to the uses of life."[213] He had in mind various "organic" orientations in Western industrial, urban and infrastructural design that were undermining the "jerrybuilt" qualities of the mechanical age – objects and buildings "hastily clapped together for the sake of immediate profit, immediate practical success, with no regard for the wider consequences and implications" – and restoring "balance and harmony between the various parts" of a "neotechnic" environment that might overhaul the principles of "power, work and regularity" defining the mechanical age.[214]

In a recent attempt to supplement geography's current fascination with war by also thinking about geographies of peace and how interwar geographers connected with them, Federico Ferretti has argued that "early geographies of peace [from this era] came from the international spread of anarchist geographies . . . and were characterised by a strong voluntarist commitment, generally taking place in extra-institutional contexts."[215] He contrasts the experience of establishment French geographers such as Albert Demangeon, Emmanuel de Martonne and Lucien Gallois, who had been involved in World War I, and in designing the peace settlement that followed, and "extramural geographers" such as Paul Dupuy and his daughter Marie-Thérèse Maurette (based in Geneva), who worked through "*engagé* geographical networks," and with anarchist leanings, to devise new forms of "peace education." If Gourou can be considered a peace geographer (and an extramural one too), then his work on Indochina shows that these early geographies of peace had colonial as well as metropolitan dimensions, involved questions of technology as well as ones of geo-politics, and did not promote amity over enmity in a straightforward way.

Akin to Mumford, Gourou used the question of technology as a means of questioning the idea that the environment could be "invoked as a self-explanatory and scientifically legitimating factor . . . marking and effective difference between technology in Europe and technology in the colonial/postcolonial world," as Arnold captures this dominant early twentieth-century outlook.[216] This was the other element of Gourou's 'play of forces,' to continue in our Foucauldian way: technology was neither a direct product of environmental opportunity and constraint, nor a clear-cut barometer of development. Rather, it was a more conditional part of a particular civilisation's 'tissue of techniques.'

Gourou viewed colonialism as 'jerrybuilt' in precisely the sense that Mumford meant: as hasty, and often counter-productive. He was concerned with the organic (simultaneously ecological, cultural, long-evolved, and holistic) orientation of Tonkin peasants and how they had 'effectively adapted' to the delta's monsoon environment. At the same time, we have shown that he relied heavily on a technology of seeing that fashioned the Tonkin peasant as an exotic object and furnished himself with an expansive view of the delta (literally a view from above). In so doing, he fixed understanding of peasants and their traditions and village life, and cemented a vision which it took a good deal of anti-colonial politicking and debate to contradict: namely, as David Hunt put it, that "peasants are incorrigible traditionalists who cannot function in the modern world."[217]

To return to the question (*pace* Césaire and Said) with which we started the book, concerning where and when putatively objective and compassionate modes of inquiry lapse into something more sly or domineering, our investigation in this chapter suggests that while Gourou did not denigrate the Tonkin peasant as an inferior 'other,' his vision of the delta was nonetheless shaped by aspirations and sensibilities that made the delta a figment of a Western – French, geographical and tropical – imagination. The Tonkin peasants were respected as pure (exculpatory) objects in contrast to 'dreary' French peasants living in a degraded modernity. But at the same time, if one looks at Gourou's maps, diagrams and photographs, and replaces the word 'countryside' with 'city,' one soon zones into Susan Sontag's famous difficulty with photography as she encountered it through Baudelaire's urban *flâneur*: "The photographer is an armed version of the solitary walker reconnoitering, stalking, cruising the urban inferno, the voyeuristic stroller who discovers the city as a landscape of voluptuous extremes. Adept of the joys of watching, connoisseur of empathy, the *flâneur* finds the world 'picturesque.'"[218] In Gourou's tropical romance the normality of the northern temperate zone is a soulless and colourless normality, and the few things he admires about it – chiefly, its architecture and institutions – seem out of place in his scenic Tonkin: at one point in *Les paysans* he opines that French schools, churches, tax offices and prisons there "are just ugly, pale imitations of western buildings."[219]

Gourou romanticised what he saw as a harmonious connection between society and environment in the Tonkin Delta and the meagreness of peasant existence. Was this innocent? Akin to how Said sees the blush and romance of Orientalism, Gourou couches his imagery of the beauties of the delta in a "language of truth, discipline, rationality, utilitarian value, and knowledge."[220] Gourou implies that the technology of seeing he wielded could not be fully or adequately exerted by Vietnamese people themselves. Colonial government publications from around this time make explicit the idea that traditional Vietnamese knowledges were less rational and systematic than the reason-bearing knowledge attainable by France.[221] There are echoes of this position in the school text by Gourou and Loubet, mentioned at the start of the chapter: that Indochina was "a reasoned creation by France." Before the arrival of the French, the country was fractured and disorderly, and the Vietnamese lacked the means to see themselves as a whole. The notion that the scene Gourou beheld was reality – the actuality of this geography – was hewn from a system of representation – from this technics of tropicality.

If the Tonkin Delta might be regarded as the 'local field' of intelligibility shaping and shaped by Gourou's study, then it laid the groundwork for wider, regional and global/zonal, fields of study and understanding. Gourou took some of the technics of tropicality he had fashioned in the Tonkin Delta – field immersion, mapping and the aerial view; a concern with population density and the *longue durée* – to other parts of the tropics. As he did, however, 'tactical arrangements' shifted. Questions of imperial crisis, development and decolonisation, and practices of comparison and expertise, came more fully into view; and as we shall now show, his tropicality became more 'networked.'

Notes

1. On war art – which is strewn between its subject and its overcoming – see, recently, Joanna Bourke ed., *War and Art: A Visual History of Modern Conflict* (London: Reaktion Books, 2017). Also see Christopher E. Goscha, "'So what did you learn from war?' Violent decolonization and Paul Mus's search for humanity," *South East Asia Research* 20 (2012): 569–593.
2. Frantz Fanon, *The Wretched of the Earth* trans. Richard Philcox; orig. pub. Eng. 1963 (New York: Grove Press, 2004), 4 and *passim*.
3. Aimé Césaire, *Discourse on Colonialism* trans. Joan Pinkham (New York: Monthly Review Press, 1972), 34, 37.
4. Jean Suret-Canale, *French Colonialism in Tropical Africa, 1900–1945* trans. Till Gottheiner (London: C. Hurst & Co., 1971), 301.

5 Editorial, "The most influential books of Southeast Asian Studies," *SOJOURN: Journal of Social Issues in Southeast Asia* 24, no. 1 (2009): 1–5.
6 Quotation from the published version of this talk: Pierre Gourou, "For a French Indo-Chinese Federation," *Pacific Affairs* 20, no. 1 (1947): 18–29, at 19.
7 Pierre Gourou, "Panorama de l'Indochine," *Tropiques* 342 (1952): 1–68, at 9.
8 Pierre Gourou and Jean Loubet, *Cours de géographie: Enseignement primaire supérieur franco-indigène, 4e année: L'Asie moins l'Asie russe, l'Indochine* (Hanoi: Imprimerie tonkinoise, 1934), 42.
9 On this historiography, see Phillip G. Altbach and Gail P. Kelly, eds., *Education and Colonialism* (London: Longman, 1978); Kate Frieson, "Sentimental education: *Les sages femmes* and colonial Cambodia," *Journal of Colonialism and Colonial History* 1, no. 1 (2000): 18–33; Gail P. Kelly, "The presentation of indigenous society in the schools of French West Africa and Indochina, 1918 to 1938," *Comparative Studies in Society and History* 26, no. 4 (1984): 523–542; and David Marr, *Vietnamese Tradition on Trial, 1920–1945* (Berkeley: University of California Press, 1981), 54–100.
10 Jean-François Dupon, "Le refus du politique?" in Henri Nicolaï, Paul Pélissier and Jean-Pierre Raison eds., *Un géographe dans son siècle. Actualité de Pierre Gourou* (Paris: Karthla, 2000), 205–215.
11 Pierre Gourou, "La géographie comme 'divertissement'? Entretiens de Pierre Gourou avec Jean Malaurie, Paul Pélissier, Gilles Sautter, Yves Lacoste," *Hérodote* 33, no. 1 (1984): 50–72, at 59. Gourou was familiar with Milton Osborne's *River Road to China: The Search for the Sources of the Mekong, 1866–1873* (London: Allen & Unwin, 1975), which tells the story of this 'geographical error,' but had made the point in his 1936 study: Pierre Gourou, *Les paysans du delta tonkinois. Etude de géographie humaine* (Paris: l'École française d'Extrême-Orient, Les Editions d'Art et d'Histoire, 1936), 126, and referencing Louis Malleret's *L'exotisme indochinois dans la littérature française depuis 1860* (Paris: Larose, 1934). Gourou's study is hereafter cited as *Les paysans*, and unless otherwise stated, we will refer to the original edition rather than the second (unabridged) 1965 edition.
12 Pierre Gourou, *Terres de bonne espérance, le monde tropical* (Paris: Plon, 1982), 403.
13 Jeanne Haffner, *A View from Above: The Science of Social Space* (Cambridge, MA: MIT Press, 2013), 19.
14 For a recent (and to our mind the best) way in see Christopher Goscha, *The Penguin History of Modern Vietnam* (London: Penguin, 2017).
15 For a range of perspectives on these colonial activities, which of course did not reach across the Tonkin Delta alone, see Goscha, *The Penguin History of Modern Vietnam*, Ch. 6; Geoffrey Gunn, *Rice Wars in Colonial Vietnam: The Great Famine and the Viet Minh Road to Power* (Lanham, MD: Rowman & Littlefield, 2014); Martin Thomas, *Violence and Colonial Order: Police, Workers and Protest in European Colonial Empires, 1918–1940* (Cambridge: Cambridge University Press, 2012), 141–176.
16 Felix Driver and Brenda Yeoh, "Constructing the tropics: Introduction," *Singapore Journal of Tropical Geography* 21, no. 1 (2000): 1–5, at 2.
17 Edward W. Said, *Orientalism* (New York: Random House, 1978), 50.
18 Slavoj Žižek, *The Event: Philosophy in Transit* (London: Penguin, 2014), 30–31.
19 Žižek, *The Event*, 31.
20 Steven C. Topik and Allen Wells, "Commodity chains in a global economy," in Emily S. Rosenberg ed., *A World Connecting 1870–1945* (Cambridge, MA: Harvard University Press, 2012), 593–812, at 730.
21 Topik and Wells, "Commodity chains in a global economy," 730.
22 See David Arnold, "Europe, technology, and colonialism in the 20th century," *History and Technology* 21, no. 1 (2006): 85–106; Lewis Mumford, "Authoritarian and democratic technics," *Technology and Culture* 5, no. 1 (1964): 1–8, at 2.
23 Arnold, "Europe, technology," 101.

24 Pierre Gourou, *Pour une géographie humaine* (Paris: Flammarion, 1973), 9; Žižek, *The Event*, 30.
25 James Duncan, "Sites of representation," in James Duncan and David Ley eds., *Place/Culture/Representation* (London and New York: Routledge, 1993), 39–56.
26 W.J.T. Mitchell, *Landscape and Power* (Chicago: University of Chicago Press, 1994), 1.
27 Lewis Mumford, *Technics and Civilization* (London: Routledge & Kehan Paul, 1934), 3, 28.
28 Mumford, *Technics and Civilisation*, 154.
29 For a summary, see Lewis Mumford, "Appraisal of Lewis Mumford's 'Technics and Civilization'," *Daedalus* 88, no. 3 (1959): 527–536.
30 Don Ihde, *Technology and the Lifeworld: From Garden to Earth* (Bloomington, IN: Indiana University Press, 1990), 59–60.
31 Ihde, *Technology and the Lifeworld*, 60.
32 Yi-Fu Tuan, *Landscapes of Fear* (New York: Pantheon Books, 1979). Gourou dwells on these questions in *Terres de bonne espérance*, 29–31. He met Mumford at the "Man's role in changing the face of the earth" symposium hosted by the Wenner-Gren Foundation at Princeton in 1955 (see Chapter 6). Both of them lamented the deleterious nature of human impact on the earth's environments, but also invoked the possibility of finding and protecting less environmentally destructive ways of life.
33 Paul Carter, *The Road to Botany Bay: An Essay in Spatial History* (London: Faber and Faber, 1987), 41.
34 Pierre Gourou, "Itinéraire" interview, with Christian Taillard, *Lettre de l'Afrase* 29 (1993): 3–11, at 7.
35 Dany Bréelle, Interview with Pierre Gorou, Bruxelles 29 August 1995, in her "The regional discourse of French geography: The theses of Charles Robequain and Pierre Gourou in the context of Indochina," Appendix H, 328–339, at 336–337, Unpublished PhD dissertation, Geography, Flinders University, 2003.
36 Bréelle, Interview with Pierre Gourou, 337; Gourou, *Terres de bonne espérance*, plate 6–7.
37 Pierre Gourou, *Pour une géographie humaine*, 13. Also see Pierre Gourou, *Riz et civilisation* (Paris: Fayard, 1984), 207.
38 Gourou, *Les paysans*, 78, 81–108. Gourou's lengthy analysis is summarised in Yves Coyaud, *Le Riz* (Saigon: Office Indochinois du riz, 1950); and Sophie Devienne, "Red River Delta: Fifty years of change," *Moussons* 9–10 (2006): 255–280.
39 Gourou, *Les paysans*, 20.
40 Gunn, *Rice Wars*, 22–23.
41 And rather than as a 'laboratory of modernity' wherein the mechanisms and limits of western rationality progress might be gauged – which is how other colonial domains were treated by metropolitan administrators and scholars at this time. See Paul Rabinow, *French Modern: Norms and Forms of the Social Environment* (Cambridge, MA: MIT Press, 1989).
42 Paul Claval, *Géographies et géographes* (Paris: L'Harmattan, 2007), 298.
43 Gourou, "Itinéraire," 6.
44 Pascal Clerc, "Des connaissances pour l'action: La géographie coloniale de Marcel Dubois et Maurice Zimmermann," *Revue germanique internationale* 20 (2014): 135–146; Pascal Clerc, "Maurice Zimmermann (1869–1950)," in Hayden Lorimer and Charles W.J. Withers eds., *Geographers: Biobibliographical Studies* 34 (London and New York: Bloomsbury Academic, 2015), 97–126; Hugh Clout, "Popularising geography in France's second city: The rôle of the Société de Géographie de Lyon, 1873–1968," *Cybergeo: European Journal of Geography* no. 449 (2009), http://journals.openedition.org/cybergeo/22214; doi:10.4000/cybergeo.22214.
45 Marcel Dubois, "Leçon d'ouverture du cours de géographie coloniale," *Annales de géographie* 10 (1894): 121–137, at 124, 135.

46 See Jean-Pierre Raison, "Tropicalism in French geography: Reality, illusion, or ideal?" *Singapore Journal of Tropical Geography* 26, no. 3 (2005): 323–338.
47 See Pierre Singaravelou, "The institutionalisation of 'colonial geography' in France, 1880–1940," *Journal of Historical Geography* 37, no. 2 (2011): 149–157; Dominique Lejeune, *Les sociétes de géographie en France et l'expansion coloniale au XIXe siècle* (Paris: Albin Michel, 1993); Zimmerman "amassed material on Scandinavia and the polar regions, which Vidal had entrusted to him in 1908 as part of his grand design for the *Géographie Universelle*": Hugh Clout, "Geographical pioneers in Lyon, 1874–1927: A biobibliographical essay," *Géocarrefour* 86, nos. 3–4 (2011): 189–199, at 192.
48 ACDF: Pierre Gourou – Education.
49 Gourou, "Itinéraire," 4; Gourou, *Terres de bonne espérance*, 11.
50 John Kleinen, Interview with Pierre Gourou, 24 August 1994, n.p. Unpublished typescript, courtesy of the author.
51 Pierre Gourou, "Etude du monde tropical," in La Decouverte ed., *La biobiothèque imaginaire du Collège de France* (Paris: Le Monde, 1990), 111–114, at 111.
52 Segalen, cited in Camille Boussut, "Une esthétique du divers," *La plume francophone* (2008), https://la-plume-francophone.com/2008/08/01/victor-segalen-essai-sur-lexotisme/.
53 Bréelle, Interview with Pierre Gourou, 335; Victor Segalen, *René Leys* trans. J.A. Underwood (New York: Random House, 2003). For recent and detailed examinations of Segalen's exoticism and the diverse influences on it, see Jean-Luc Coatalem, *Mes pas vont ailleurs* (Paris: Stock, 2017); and Weijie Song, *Mapping Modern Beijing: Space, Emotion, Literary Topography* (Oxford: Oxford University Press, 2018), Ch. 4.
54 Julia Kuehn, *A Female Poetics of Empire: From Eliot to Woolf* (New York: Routledge, 2014), 5.
55 Gourou, "La géographie comme 'divertissement'? 53.
56 Gourou, "Itinéraire," 4–5.
57 ACDF: Pierre Gourou – Education.
58 Gourou, "Itinéraire," 5.
59 Pierre Gourou, *Les paysans du delta tonkinois. Etude de géographie humaine* 2nd edition (Paris: Mouton & Co., 1965), 2.
60 This paean to this fieldwork is especially evident in the interview he gave to Bréelle.
61 Panivong Norindr, *Phantasmatic Indochina: French Colonial Ideology in Architecture, Film, and Literature* (Durham, NC: Duke University Press, 1996), 7.
62 Pierre Pasquier, "L'Indochine pittoresque," in Georges Maspero ed., *Un empire colonial français*, 2 vols. (Paris: G. Van Oest, 1930), 243–253; Albert Sarraut, *Images du monde: Indochine* (Paris: Firm-Didot et Cie, 1930), 17 and *passim*. For discussions, see Jean-François Klein, "L'histoire de l'Indochine en situation coloniale: Entre histoire et orientalisme (1858–1959)," in Oissila Saaïdia and Laurick Zerbini eds., *La construction du discours colonial: l'empire française aux XIXe et XXe siècles* (Paris: Karthala, 2009), 89–124.
63 Bréelle, "The regional discourse of French geography," 292–294.
64 Claval, *Géographies et géographes*, 298; Marion Solotareff, "Naissance et evolution de la géographie tropicale (1930–1960)," in Paul Claval and André-Louis Sanguine eds., *La géographie française à l'époque classique (1918–1968)* (Paris: L'Harmattan, 1996), 243–257.
65 Gourou, *Les paysans*, 110.
66 Gourou, *Les paysans*, 581–582.
67 Pierre Gourou, "La géographie comme 'divertissement'?" 53; Gourou, *Terres de bonne espérance*, 14.
68 On the former, see Herman Lebovics, *True France: The Wars over Cultural Identity, 1900–1950* (Ithaca, NY: Cornell University Press, 1992). On the latter, see Fuyuki Kurasawa, "Primitiveness and the flight from the modernity: Sociology and the

avant-garde in inter-war France," *Economy and Society* 32, no. 1 (2003): 7–28; Emmanuelle Saada, "Sociability in the imperial republic," in Margrit Pernau and Helge Jordheim eds., *Civilising Emotions: Concepts in Nineteenth-Century Asia and Europe* (Oxford: Oxford University Press, 2015), 63–82.
69 Marc Matera and Susan Kingsley, *The Global 1930s: The International Decade* (Abingdon: Routledge, 2017), 35.
70 Kelly Enright, *A Maximum of Wilderness: The Jungle in the American Imagination* (Charlottesville: University of Virginia Press, 2012), 100.
71 Elizabeth Ezra, *The Colonial Unconscious: Race and Culture in Interwar France* (Ithaca, NY: Cornell University Press, 2000), 5. Also see Philoppe Dewitte, *Les mouvements nègres en France, 1919–1939* (Paris: L'Harmattan, 1985).
72 See Patricia A. Morton, *Hybrid Modernities: Archictecture and Representation at the 1931 Colonial Exposition, Paris* (Cambridge, MA: MIT Press, 2000); Didier Grandsart, *Paris 1931: Revoir l'exposition coloniale* (Paris: Editions FVW, 2010); Jody Blake, "The truth about the colonies, 1931: Art indigène in the service of the revolution," *Oxford Art Journal* 25, no. 1 (2002): 35–58.
73 Pierre Gourou, *Le Tonkin* (Paris: École française d'Extrême-Orient, 1931), 344.
74 Claude Lévi-Strauss, *Tristes tropiques* (Paris: Plon, 1955), 1.
75 On this message see Morton, *Hybrid Modernities*, 1–15; Nicola Cooper, *France in Indochina: Colonial Encounters* (Oxford: Berg, 2001), 74–76.
76 ANOM: FM EE/11/4453/8/GOU Pierre Gourou dossier.
77 Claval, *Géographies et géographes*, 297; Pierre Gourou, "Les divers types de villages du delta tonkinois et leur répartition," and "Présentation d'une carte des densités de la population du Tonkin," in Union géographique internationale ed., *Comptes rendus du Congrès international Paris 1931* vol.3 (Paris: Armand Colin, 1931), 487–490; 580–582.
78 Pierre Gourou and Jean Gottmann, "Albert Demangeon," *Bulletin de la Societé languedocienne de Géographie* 12, no. 1 (1941): 1–16, at 8.
79 Gourou and Gottmann, "Albert Demangeon," 9.
80 Omnia El Shakry, *The Great Social Laboratory: Subjects of Knowledge in Colonial and Postcolonial Egypt* (Stanford, CA: Stanford University Press, 2007), 104.
81 See Paul Claval, "Playing with mirrors: The British empire according to Albert Demangeon," in Anne Godlewska and Neil Smith eds., *Geography and Empire* (Oxford: Blackwell Publishers, 1994), 228–243.
82 See Mark Cleary, "Retour à la terre: Peasantist discourse in rural France c. 1930–1950," in Iain S. Black and Robin A. Butlin eds., *Place, Culture and Identity: Essays in Historical Geography in Honour of Alan R.H. Baker* (Laval: Les presses de l'Université Laval, 2001), 235–254.
83 Jean-Louis Tissier, "Rendez-vous à Uriage (1940–1942). La fonction du terrain au temps de la Révolution nationale," in Guy Baudelle, Marie-Claire Robic and Marie-Vic Ozouf-Marignier eds., *Géographes en pratiques (1870–1945). Le terrain, le livre, la cité* (Rennes: PU Rennes, 2001), 342–351.
84 Pierre Pasquier, 1930, cited in Brocheux and Hémery *Indochina*, 108.
85 Such concerns are captured in Henri Lefebvre, "Problèmes de sociologie rurale: La communauté paysanne et ses problèmes historico-sociologiques," *Cahiers Internationaux de Sociologie* 6, no. 1 (1949): 78–100, and the first volume (1947) of his *Critique of Everyday Life*, to which we shall return.
86 Gourou, *Les paysans*, 388.
87 Gourou, *Les paysans*, 387.
88 Gourou, *Les paysans*, 225.
89 Henri Lefebvre, *Critique of Everyday Life* one-volume edition (London: Verso, 2014), 224.
90 Lefebvre, *Critique of Everyday Life*, 225.
91 Lefebvre, *Critique of Everyday Life*, 229.

92 Lefebvre, *Critique of Everyday Life*, 230.
93 David Biggs, "Arial photography and colonial discourse on the agricultural crisis in late-colonial Indochina, 1930–1945," in Christina Folke Ax, Niels Brimnes, Niklas Thode Jensen and Karen Oslund eds., *Cultivating the Colonies: Colonial States and Their Environmental Legacies* (Athens, OH: Ohio University Press, 2011), 109–132, at 116.
94 David Biggs, *Quagmire: Nation-Building and Nature in the Mekong Delta* (Seattle: University of Washington Press, 2010), 106–107.
95 Gourou, *Les paysans*, 220. On the longer-term impress of Gourou's characterisation on anthropologists and political scientists, see Samuel L. Popkin, *The Rational Peasant: The Political Economy of Rural Society in Vietnam* (Berkeley: University of California Press, 1979), 260–261.
96 Gourou, *Les paysans*, 226.
97 Kleinen, Interview with Pierre Gourou, n.p.
98 Paul Vidal de la Blache, *Principes de géographie humaine* avec Emmanuel de Martonne (Paris: Armand Colin, 1922), 15; *Tableau de la géographie de la France* (Paris: Armand Colin, 1903).
99 Gourou, *Les paysans*, 15.
100 Gourou, *Les paysans*, 111.
101 Gourou, *Les paysans*, 110.
102 Gourou, *Riz et civilisation*, 135.
103 Gourou, *Les paysans*, 11.
104 Jacques Marseille, *Empire colonial et capitalisme français: Histoire d'un divorce* (Paris: Albin Michel, 1984).
105 Vidal cited in Paul Claval, *Histoire de la géographie française de 1870 à nos jours* (Paris: Nathan-Université, 1998), 103. Also see André-Louis Sanguin, *Vidal de la Blache. Un genie de la géographie* (Paris: Belin, 1993); Bréelle, "The regional discourse of French geography," 60–67.
106 See Emmanuelle Sibeud, *Une science impériale pour l'Afrique? La construction des savoirs africanistes en France, 1878–1930* (Paris: l'EHSS, 2002).
107 Gourou, *Les paysans*, 388.
108 Gourou, *Les paysans*, 133.
109 Gourou, *Les paysans*, 575–576.
110 Charles Robequain, "Réception de M. Pierre Gourou à l'Académie des sciences coloniales," *Compte-rendu mensuel des séances de l'Académie des Sciences Coloniales* 15, no. 2 (1955): 63–68, at 67.
111 Gourou, *Les paysans*, 272; 569–576.
112 Gourou, *Les paysans*, 56; Kleinen, Interview with Gourou, n.p.
113 V. Raoul Girardet, *L'idée coloniale en France de 1871 à 1962* (Paris: La Table Ronde, 1972), 15.
114 Gourou, *Les paysans*, 62.
115 Gourou, *Les paysans*, 226.
116 Gourou, *Les paysans*, 575.
117 Gourou, *Les paysans*, 71–72.
118 Gourou, *Les paysans*, 82.
119 Gourou, *Les paysans*, 83.
120 Yi-Fu Tuan, *Romantic Geography: In Search of the Sublime Landscape* (Madison: University of Wisconsin Press, 2014).
121 Henri Copin, *L'Indochine française des années 20 à 1954. Exotisme et altérité* (Paris: L'Harmattan, 1996), 9. Also see Patrick Laude, *Exotisme indochinois et poésie: étude sur l'oeuvre poétique d'Alfred Droin, Jeanne Leuba et Albert de Pouvourville* (Paris: Sudestasie, 1990), who similarly draws a distinction between an exotic literature on Indochina that revelled in the break with a known metropolitan universe and

romanticises distant otherness, and one that is anchored in and tries to face up to colonial realities and pressures.
122 Georges Condominas, Interview with the authors.
123 Michel Bruneau, "Pierre Gourou (1900–1999). Géographie et civilisations," *L'Homme* 153 (2000): 1–25, at 20.
124 Georges Courade, Interview with the authors.
125 Gourou, *Terres de bonne espérance*, 29.
126 Gourou, *Terres de bonne espérance*, 38.
127 Jean-Pierre Raison, Interview with the authors.
128 David N. Livingstone, "The spaces of knowledge: Contributions towards a historical geography of science," *Environment and Planning D: Society and Space* 13, no. 1 (1995): 5–34.
129 Bréelle, Interview with Gourou, 334.
130 The expression is discussed in Nguyen Phuong Ngoc, "Paul Mus et les 'annamitisants' vietnamiens de l'É Française d'Éxtrême-Orient," in David Chandler and Christopher E. Goscha eds., *L'espace d'un regard: Paul Mus et l'Asie (1902–1969)* (Paris: Les Indes Savantes, 2006), 151–171.
131 Pierre Singaravélou, *L'EFEO ou l'institution des marges (1898–1956)* (Paris: L'Harmattan, 1999).
132 John Kleinen, "Nguyen Van Khoan (1890–1975): An odd man out of Vietnamese anthropology?" *Moussons* 24 (2014), online http://moussons.revues.org/3047.
133 Kleinen, "Nguyen Van Khoan."
134 Gourou, *Terres de bonne espérance*, 19.
135 Paul Mus, *L'angle de l'Asie* (Paris: Hermann, 1977), 18.
136 Gourou, "Itinéraire," 8–9.
137 Christophe Cony and Michel Ledet, *L'aviation française en Indochine: Des origines à 1945* (Outreay: Lela Presse, 2012).
138 Cordemoy cited in Gregory Charles Seltzer, "The hopes and the realities of aviation in French Indochina, 1919–1940," 1 Unpublished PhD dissertation, History, University of Kentucky, 2017; and see Seltzer, "The hopes and realities of aviation," 165.
139 Gunn, *Rice Wars*, 27–31.
140 Sophie Quinn-Judge, "Through the glass darkly: Reading the history of the Vietnamese Communist Party, 1945–1975," in Mark Bradley and Marily B. Young eds., *Making Sense of the Vietnam Wars: Local, National, and Transnational Perspectives* (Oxford: Oxford University Press, 2008), 111–134, at 114–116.
141 Gourou, *Les paysans*, 575, 250.
142 John T. McAlister and Paul Mus, *The Vietnamese and Their Revolution* (New York: Harper & Row, 1970), 31. McAlister served as translator.
143 Said, *Orientalism*, 300.
144 IAO: Fonds Paul Mus.
145 Susan Bayly, "Conceptualizing resistance and revolution in Vietnam: Paul Mus' understanding of colonialism in crisis," *Journal of Vietnamese Studies* 4, no. 1 (2009): 192–205, at 200. For a similar argument, see Frederick McHale, *Print and Power: Confucianism, Communism, and Buddhism in the Making of Modern Vietnam* (Honolulu: University of Hawaii Press, 2004), 18–21, 100–102.
146 Bayly, "Conceptualizing resistance and revolution in Vietnam," 199.
147 Frances FitzGerald, *Fire in the Lake: The Vietnamese and the Americans in Vietnam* (Boston: Little, Brown and Co., 1972), 51; also 25–30, 179–180, 444 and 481. FitzGerald studied briefly under Mus, dedicates the book to him, and derives her title from Mus's reflections on Vietnamese history.
148 FitzGerald, *Fire in the Lake*, 179.
149 Tuong Vu, "Vietnamese political studies and debates on Vietnamese nationalism," *Journal of Vietnamese Studies* 2, no. 2 (2007): 175–230, at 221.

134 *Romancing the tropics*

150 Nick Turse, *Kill Anything that Moves: The Real American War in Vietnam* (New York: Henry Holt and Co., 2013).
151 Developing many of Gourou's observations, Mus argued that Vietnamese "nationalism and communism, the programs and roles of the political parties, or similar questions can only be appraised from the standpoint of the villages. Since time immemorial these villages have become the key to the social structure of the country and to its outlook on life." Paul Mus, "The role of the village in Vietnamese politics," *Pacific Affairs* (1949): 265–272, at 265. For an overview of how the field of Vietnamese Studies has recently been re-energised by the question of the 'traditional village,' see Nguyen Tuan Anh and Annuska Derks, "Vietnamese villages in the context of globalization," *Social Sciences Information Review* 7, no. 2 (2013): 20–33; and Hai Hong Nguyen, *Political Dynamics of Grassroots Democracy in Vietnam* (New York: Palgrave Macmillan, 2016), 17–30; Trần Hữu Quang and Nguyễn Nghị, "Reframing the 'Traditional' Vietnamese Village: From peasant to farmer society in the Mekong Delta," *Moussons* 28, no. 1 (2016): 61–88. The wider body of post-war American social science literature with which much of this Vietnamese Studies literature connects stretches from Eric Wolf's 1950s typology of peasant communities as 'closed/corporate' and 'open/interactive,' to the critical frameworks of Samuel Popkin and James Scott. Eric Wolf, "Closed corporate peasant communities in Mesoamerica and Central Java," *Western Journal of Anthropology* 13, no. 1 (1957): 1–18; Popkin, *The Rational Peasant*; and James C. Scott, *Decoding Subaltern Politics: Ideology Disguise, and Resistance in Agrarian Politics* (London and New York: Routledge, 2013), Ch. 3.
152 Phan Huy, Le Tu Chi and Nguyen Duc Nginh, cited in Bréelle, "The regional discourse of French geography," 200–201.
153 Gourou, *Les paysans*, 352.
154 Paul Lévy, "Paul Mus (1902–1969)," *École pratique des hautes études, Section des sciences religieuses* 78 (1969): 82–86.
155 See Gouou, *Les paysans*, 653–658, for this table (assortment) of figures.
156 Gourou, *Les paysans*, 164, 171, 185, 228, 248.
157 Also see Bréelle, "The regional discourse of French geography," 158–163.
158 Bernard-Philippe Grsolier, *Indochine, carrefour des arts* (Paris: Albin Michel, 1961), 6.
159 Jean Bruhnes, *La géographie humaine* (Paris: Librarie Félix Alcan, 1925), 60.
160 Marie-Claire Robic, "From the sky to the ground: The aerial view and the ideal of the vue raisonnée in geography during the 1920s," in Mark Dorrian and Frédéric Pousin eds., *Seeing from Above: The Aerial View in Visual Culture* (London: I. B. Tauris, 2012), 163–187, at 163–164; also see Gourou, *Les paysans*, 164.
161 Robic, "From the sky to the ground," 185 (emphasis in original); and see Gourou, *Les paysans*, 171, 185.
162 Matt Dyce, "Canada and the photograph and the map: Aerial photography, geographical vision and the state," *Journal of Historical Geography* 39, no. 1 (2013): 69–84, at 73.
163 Dyce, "Canada and the photograph and the map," 79. W.T. Lee's 1922 *The Face of the Earth as Seen From the Air* was a key text.
164 Martin Jay and Sumathi Ramaswamy, "Introduction," in Martin Jay and Sumathi Ramaswamy eds., *Empires of Vision: A Reader* (Durham, NC: Duke University Press, 2014), 1–43, at 35.
165 Haffner, *A View From Above*, 20.
166 Gourou, *Les paysans*, 29.
167 Gourou, *Les paysans*, 29–30.
168 An issue taken up in Christopher E. Goscha, "Annam and Vietnam in the new Indochinese space, 1887–1945," in Stein Tonnesson and Hans Antlov eds., *Asian Forms of the Nation* (Copenhagen: NIAS, 1996), 131–150.
169 Gourou, *Les paysans*, 139–142, 402–404, 451, 572. Kleinen, Interview with Gourou, n.p.
170 Gourou, *Les paysans*, 555.

171 Yves Lacoste, *Paysages politiques: Braudel, Gracq, Reclus* (Paris: Librarie Générale Française, 1990), 60–61.
172 Gouvernement Général de l'Indochine, *Service Géographique de l'Indochine* (Hanoi: Imprimerie d'extrême-Orient, 1931); Aéronautique Militaire de l'Indochine, *Historique de l'aéronautique d'Indochine* (Hanoi: Imprimerie d'extrême-Orient, 1930).
173 Louis Finot, "Review of Atlas de l'Indochine," *Bulletin de l'École française d'Extrême-Orient* 20, no. 1 (1920): 69–71 at 69.
174 Kapil Raj, *Relocating Modern Science: Circulation and the Construction of Knowledge in South Asia and Europe, 1650–1900* (London: Palgrave Macmillan, 2007), 82.
175 Biggs, "Aerial photography," 112–114.
176 Biggs, "Aerial photography," 114.
177 Martin Thomas, *Empires of Intelligence: Security Services and Colonial Disorder after 1914* (Berkeley: University of California Press, 2008), 24.
178 Felix Driver and Luciana Martins, "Views and visions of the tropical world," in Felix Driver and Luciana Martins eds., *Tropical Visions in an Age of Empire* (Chicago: University of Chicago Press, 2005), 14.
179 Haffner, *A View from Above*, 20.
180 Gourou, *Les paysans*, 554.
181 Gourou, *Les paysans*, 243.
182 Gourou, *Les paysans*, 237–238.
183 Gourou, *Les paysans*, 225.
184 Gourou, *Les paysans*, 238.
185 Griaulle cited in James Clifford, *The Predicament of Culture: Twentieth Century Ethnography, Literature, and Art* (Cambridge, MA: Harvard University Press, 1988), 68.
186 Clifford, *The Predicament of Culture*, 69.
187 Clifford, *The Predicament of Culture*, 81.
188 IAO: Fonds Paul Mus – Notes (marginalia) on *Les paysans du delta tonkinois*. This quote from, and note in the margins of, *Les paysans* is on page 250. Mus had the Japanese in mind when making this remark about the occupier forcing the villagers to cut their bamboo hedge.
189 Olivier Orain, *De plain – pied dans le monde, Ecriture et réalisme dans la géographie française au XXe siècle* (Paris: L'Harmattan, 2009).
190 Orain, *De plain – pied dans le monde*, 46.
191 Edward W. Said, "Impossible histories," *Harper's Magazine* (July 2002): 18–20, at 18.
192 Cited in Helen Tilley, "Global histories, vernacular science, and African genealogies; or, is the history of science ready for the world?" *Isis* 101, no. 1 (2010): 110–119, at 111.
193 Kleinen, Interview with Gourou, n.p.
194 On which, see John K. Whitmore, "Cartography in Vietnam," in J.B. Harley and David Woodward eds., *The History of Cartography*. Vol. II, Book 2. *Cartography in the Traditional East and Southeast Asian Societies* (Chicago: University of Chicago Press, 1994), 478–511.
195 Gourou, *Les Paysans*, 176, also 256, 271, 271, 277.
196 Roger Lévy, "Review of Pierre Gourou, L'Asie," *Politique étrangère* 19, no. 6 (1953): 525–526, at 526.
197 The quotes and details in this paragraph are drawn from Bréelle, Interview with Gourou, 333–339, and Kleinen, Interview with Gourou, n.p.
198 Especially Albert Demangeon's "Enquêtes regionals: Types de questionnaires," *Annales de Géographie* 97 (1909): 78–81.
199 Kleinen, Interview with Gourou, n.p.
200 Gourou, "La géographie comme 'divertissement'?" 61. He knew the French resident of the province and used local (provincial) French and Vietnamese archives in Ha Dong and Nam Dinh, where he got a sense of how the Vietnamese administration was dependent on the French hierarchy and village Mandarins, particularly when it came to checking census figures.

201 Charles Robequain, *L'Évolution economique de l'Indochine française* (Paris: Institute of Pacific Relations, 1939), 251.
202 Nguyen Van Khoan, "Essai sur le dinh et le culte de génie tutélaire des villages au Tonkin," *Bulletin de l'É Française d'Extrême-Orient* 30 (1930): 107–139, at 117.
203 Gourou, *Les paysans*, 226.
204 Gourou, *Les paysans*, 257–260.
205 Gourou, *Less paysans*, 255, 276.
206 Gourou, *Les paysans*, 263.
207 Michel Foucault, *Power/Knowledge: Selected Interviews and Other Essays, 1972–1977* Colin Gordon ed. (New York: Pantheon, 1980), 90, 99, 114.
208 Joseph Rouse, "Power/knowledge," in Gary Gutting ed., *The Cambridge Companion to Foucault* (Cambridge: Cambridge University Press, 1994), 92–114, at 107–108.
209 Gilberte Bray, Interview with the authors.
210 Gourou, "La géographie comme 'divertissement'?" 72.
211 Philip Taylor, "Introduction," in Philip Taylor ed., *Connected and Disconnected in Vietman: Remaking Social Relations in a Post-Socialist Nation* (Canberra: ANU Press, 2016), 1–41, at 26.
212 Gourou, *Les paysans*, 473, 578.
213 Mumford, *Technics and Civilisation*, 371–372.
214 Mumford, *Technics and Civilisation*, 371–372.
215 Federico Ferretti, "Geographies of peace and the teaching of internationalism: Marie-Thérèse Maurette and Paul Dupuy in the Geneva International School (1924–1948)," *Transactions of the Institute of British Geographers* 41 (2016): 570–584, at 570–574.
216 Arnold, "Europe, technology, and colonialism in the 20th century," 96.
217 David Hunt, "Village culture and the Vietnamese revolution," *Past & Present* 94 (1982): 131–157, at 156.
218 Susan Sontag, *On Photography* (London: Penguin, 1977), 55. On Sontag's critique of aerial and ground photography, and its connections to Foucault, see Elise Morrison, *Discipline and Desire: Surveillance Technologies in Performance* (Ann Arbor: University of Michigan Press, 2016), 230–267.
219 Gourou, *Les paysans*, 569.
220 Edward W. Said, *The World, the Text, and the Critic* (Cambridge, MA: Harvard University Press, 1983), 216.
221 For example, *Direction générale de l'instruction publique (DGIP): Le service de l'instruction publique en Indochine* (Hanoi: Imprimerie d'Extrême-Orient, 1930). This and other public instruction manuals and directives were distributed at the Exposition Coloniale.

4 Networking the tropics

Qu'est-ce que le monde tropical?

In the same month that Pierre Gourou's short article on the Ecole des Beaux-Art de Hanoi appeared in the 1949 *France Illustration* special issue we have studied (see Chapter 2), he published an altogether different – sweeping and ambitious – essay, entitled "Qu'est-ce que le monde tropical?" (What is the tropical world?), in France's leading history journal, *Annales: Economies, sociétés, civilisations*. Gourou had been elected to a Chaire d'Étude du monde tropical at the Collège de France in 1947, and this was the published version of his inaugural lecture. He proceeded in broad strokes and with a different kind of survey than the one that led to his regional monograph on the peasants of the Tonkin Delta. The study of the tropics, he noted, was based on an understanding of both the uniqueness of the tropical world compared to the temperate world, and the many differences between the *pays chauds et pluvieux* (hot and wet lands) that characterised the *zone tropicale*.[1] "Here, in total, is a singularly different nature from our own" he proclaimed.[2] The tropical world "is generally unfavourable" to human existence and "mostly inhabited by peoples with arrested civilisations and sometimes ones in regression," he continued on the first point and in the matter-of-fact way that was his calling card; and if tropical lands were to have a bright future, it would not come from trying to turn them into temperate ones. It was also important to think about "inter-tropical contrast," and particularly about the marked differences in *niveaux de vie* (living standards) between the "highly civilised cultures" and "densely populated lands" of tropical Asia, and the "retarded" (*attardé*) state of other parts of the tropics, which, on the whole, were sparsely populated, and where a combination of an enervating climate, fragile soils, detrimental agricultural practices, and debilitating disease "weighed heavily" on living standards and were not easily solved by the applications of modern science.[3]

Gourou wrote about the tropics in sweeping terms, but also, we sense from this essay, with more directly political overtones than before. A concern with what he liked to call *niveaux de vie* and *amélioration*, and what an Anglophone literature more commonly termed 'development' and 'progress,' had long been embedded in colonial discourse, and by the late nineteenth century with the idea of colonialism as a civilising mission common to both British and French imperial traditions.

Gourou made passing reference to myriad colonial projects of *mise en valeur* over the 30 million square kilometre expanse of the tropics, and to their many shortcomings, and used this French expression as a concept metaphor for colonialism in general. His concern with living standards – albeit chiefly with the food, health, population and poverty, rather than freedom, education and literacy, aspects of this construct – attained heightened importance during the 1930s and 1940s as the world became beset by economic and imperial crisis, and eventually war. During the first half of the twentieth century there was a pronounced increase in the generation and use of standards of living statistics by Western colonial powers and international organisations such as the League of Nations.[4] Before and after World War II Britain and France saw its putative concern with the welfare of its colonial subjects, and thus a scientific and policy interest in food, agriculture, trade, medicine and education, as a means of averting demands for independence and of rebooting empire, whereas the United States saw its professed concern with agricultural modernisation and international trade as both evidence of its professed liberal ideology of universal betterment, and, after the war and particularly in Asia, as a means of containing a looming communist threat.[5]

"Qu'est-ce que le monde tropical?" was published in the year that United States (U.S.). President Harry Truman made his famous Inaugural Speech touting a "bold new program for making the benefits of our scientific advances and industrial progress available for the improvement and growth of underdeveloped nations."[6] The U.S. saw itself as an international leader in this development drive, and saw food supply, and agricultural and population growth, as critical to how revolutionary and Cold War dynamics would play out in Asia.[7] The quest "to promote social progress and better standards of life in larger freedom" was also cornerstone of the 1945 Charter of the United Nations (UN), and in 1946 Gourou was enlisted as a UN consultant regarding living standards in the tropical world.[8] He was charged with designing and implementing the UN's "research and experimental programme for the improvement of agriculture and standards of life in tropical regions," and from 1954 as part of a wider multinational and multidisciplinary panel of "designated specialists" constituting the "humid tropics research programme" established by the United Nations Educational, Scientific and Cultural Organization (UNESCO), culminated in a 1960 report.[9] Between 1945 and 1975 Gourou produced a total of seven reports for UNESCO in this applied vein. But this concern with living standards also infuses his 1949 essay and underscores the idea that by then science and politics were seen as two sides of a tropical coin, each providing a framework for understanding the other.[10] For Gourou, these politics were as much about protecting traditional *encadrements* in many parts of the tropics from Western encroachment as they were about fuelling development, and he was not alone in insisting that the drive to secure adequate levels of subsistence and boost agricultural yields, which was a central plank of UN aspiration, had environmentally specific meanings, moorings and fetters. Matters of food, hunger and poverty could not be tackled without also considering ones of climate, terrain, soil and land use.

Re-scaling tropicality

We shall say more in Chapter 6 about how "Qu'est-ce que le monde tropical?" opened out on to a post-war world of development and decolonisation. Our suggestion in this chapter is that this essay also needs to be read retrospectively, through a set of relations and exigencies that got Gourou from his 1930s fieldwork on Indochina to his more expansive understanding of a *zone tropicale* after World War II.[11] Little attention has been paid to the role that international organisations, and the veneration of expertise, played in the genesis of Gourou's tropical geography, and how this evolution in his work between 1936 and 1949 coincided with *la fin des voyages* (again to borrow Claude Lévi-Strauss's adage for how *Tristes Tropiques*, the ultimate anti-travel travel narrative, came to fruition): a recoil from overseas fieldwork and immersion back in France in the politics of empire and internationalism.

In this chapter we track Gourou's switch in orientation from the local to the zonal (a scaling up of his tropicality), and concomitant (if momentary) shift from a field-based to a desk-bound tropical geography, and suggest that it was shaped by three dynamics: first, his own career path, beginning with his move back to France from Indochina in 1936; second, the creation of new institutional and international networks, and specialist fields, during the interwar years which exalted expertise and comparative analysis, and promoted scholarly and advisory networks and exchanges that crossed disciplines, institutions and borders; and third, the growing significance of bio-politics (a politics of life and death) within a triad of comparison-network-expertise.

The details pertaining to the first of these dynamics – the ballast of standard intellectual biographies – are vital but cannot be fully understood unless they are connected to the other two, and we use the title 'networking the tropics' as a shorthand here for these entanglements. During the 1930s and 1940s the tropics were put in circulation – particular geographies and nations were placed in dynamic debates that both intensified and dispersed understanding of their own specificity, and were pursued in work that was at once more specialist and interdisciplinary, and competitive and collaborative (in national, imperial and disciplinary terms) than before.[12] That circulation stemmed in good measure from new forms of academic mobility and institutional networking, as well as from corporate exploitation and foreign direct investment (and with science as a handmaid of capitalist experimentation and resource extraction), and took a particular twist during the World War II, when, as Robert Fletcher of the U.S. Weather Bureau noted, "a remarkable increase in flying in equatorial regions" allowed "tropical meteorology" to come into much sharper focus and placed the "general circulation model" of the equatorial atmosphere on a stronger empirical footing.[13]

We shall start by developing the point we made in Chapter 1 about the need to release Gourou from the constrictions of disciplinary history and resituate his work and tropicality in a wider web of connections that stretch from the rice paddies of the Tonkin Delta to Parisian hubs of intellectual and political debate and to international gatherings farther afield. We suggest that it was through these networks and the webs of patronage that characterised French institutions (including geography) during this era that his status as an expert was secured. We then examine how it was through these networks that his disciplinary interest in population density became politicised as a matter of 'overpopulation' and how his work was brought to the attention of some prominent international organisations. In the next chapter we explore how Gourou's comparative interests, and ultimately the idea of a *zone tropicale* and the stigma of tropicality associated with it, was fostered not only by these networks but also by the exigencies of World War II.

To put the argument we shall now take up front: Gourou did not tour the Tonkin Delta alone, but with some well-connected French figures (as well as Vietnamese assistants), and it was with their patronage, and especially that of his supervisor, Albert Demangeon, that his expertise was secured and he then became involved in the Guernut Commission of France's Popular Front Government, which endeavoured to adjudicate complaints of colonial abuse and exploitation with specialist advisers and on-the-spot inspection teams. Gourou's concern with land and population also attracted the interest of France's formative foreign affairs think-tank, the Centre d'études de politique étrangère (CEPE, established 1935), the transnational (albeit New York–based) Institute of Pacific Relations (IPR, established 1925), and latterly the UN. During the 1930s the tropics became caught up in a momentous, if vexed and short-lived, growth in different forms of internationalism – scientific collaboration, international political cooperation, imperial and anti-imperial internationalism, and anti-colonial social movements – and Gourou navigated the idealist and realist (utopian and pragmatic, pure and applied) frameworks of international relations that cross-cut the different organisations with which he became involved.

Gourou pursued the idea that since "scientific inquiry is by its very nature a universal enterprise . . . political and national divisions have no meaning," as Daniel Gorman puts it ('ultimate meaning' is perhaps a more precise way of putting it), the tropical world could be viewed as a reality set apart from particular public and private, corporate and colonial, national and international, designs on it.[14] Gorman and a larger literature sees interwar internationalism as an attempt to advance science and secure peace, and as involving a "significant shift from politicians to experts" as key advisers and decision-makers. We pursue the call embedded in this literature, and postcolonial studies, to "explore geographies and nations as they exist in dynamic relation to each other" rather than in fixed hierarchies of metropole and colony, temperate and tropical, modern and traditional, while recognising the drive of centralised and metropolitan institutions to deploy (however incompletely or unsuccessfully) such categories.[15]

The interwar 'tropics' might be deemed an artifice of internationalism (and its snags and snares) and we spend some time in this chapter piecing together how Gourou became connected to the reformist agenda of the French government and an imperial internationalism in order to eschew a hazy association of knowledge with power. It is important to think carefully about precisely how particular scholars and intellectuals became involved in networks of state and imperial power.

How witting or unwitting was their involvement? How did networks of expertise and political influence beckon individual academics, and help to boost (or hamper) their careers and shape (or hinder) their research choices?

Such questions point to the importance of thinking about the lived intellectual and institutional spaces within which knowledge is produced and its meanings are created and conveyed. We use the verb 'networking' to capture the idea that Gourou's tropicality worked in and across a range of international venues (committees, commissions, congresses, organisations, centres, offices) that brought together a plethora of individuals and groups (scholars, politicians, advisers, ministers, pundits, activists), who pursued diverse research, educational and conference programmes and collaborations, and generated a wide array of reports, proposals and publications.[16] Dynamism and eclecticism were the watchwords of this international moment, which spans the middle five decades of the twentieth century, although the 1930s is our focus here, and the 1940s in the next chapter. The networks and spaces of internationalism in which Gourou worked were shaped in decisive ways by what Tamson Pietsch describes (although perhaps stretching the point a little too far) as an "empire of scholars."[17]

Furthermore, given the varieties of life – nationalities, ethnicities, disciplines, political identities and social connections – that convened at such venues, it is not surprising that life itself (bodies, speech, energies, excitements, the nature of human existence) became a matter of central concern. Primitivism and exoticism were part of this internationalism and they arose, in a general sense, from what Sigmund Freud in his famous 1930 essay "Civilization and its Discontents" diagnosed as the struggle between two "mutually opposing" forces: "the phenomenon of life" (how it might be enhanced, preserved, dissected, shared, understood) and "an instinct of death" (enmity, war, barbarity, degeneration and self-destruction).[18]

One does not have to be either a fan or an aficionado of Freud to grasp the gravity of his observation that during the interwar decades 'life' and 'death' attained a new importance. As Foucault later argued, and inverting Freud's argument, the death drive might be deemed the silent agent and arbiter of life, menacing life by promoting it, expanding it, monitoring it, and thus subjecting it to a political rationality – what Foucault termed a 'bio-politics,' which revolved around the administration of life and populations, and the drive "to ensure, sustain, and multiply life, to put this life in order," and with power "situated and exercised at the level of life, the species, the race, and the large-scale phenomena of population."[19] Foucault coined the expression 'biopower' to describe a form of power that does not simply repress or interdict – and in Freud's terms with a degree of happiness traded for a portion of security through the repression and regulation of sexuality – but "exerts a positive influence on life, that endeavours to administer, optimize, and multiply it, subjecting it to precise controls and comprehensive regulations."[20]

Gourou was drawn into, and helped to fashion, a bio-political framing of the tropics in which bodies and populations were rendered "fruits of the exercise of power and control," as Pedro Pereira describes this form of tropicality.[21] Gourou pronounced on how the tropical lives and futures of the rural peasant population of Indochina might be improved, and identified 'overpopulation,' and attempts to mitigate its effects through migration and the colonisation of unused uplands, as a key concern.

House of cards

In "Qu'est-ce que le monde tropical?" Gourou proclaimed "While remaining strictly scientific, if not a uniquely disinterested discipline ... [geography] can put in place a better exploitation of the planet." Latter-day critiques of Gourou that pounce on statements like this as proof of a bogus objectivity have assumed that he regarded the terms "disinterested" and "exploitation" as mutually exclusive, and that his work was flawed because he was ultimately unable to aspire to the former without leaving behind the moral approbation associated with the latter. Yet Gourou did not see these terms as incongruous in 1949, and if we are to gauge how or whether he should be labelled an accomplice to Western power we need to enquire further into the ways and circumstances in which his work became politically meaningful. By 'political' we do not just mean partisan, but also as being enmeshed in spheres of patronage and influence. We see academic identities and knowledges as constituted rather than pre-given entities that are rendered through material and social networks and assemblages, and seek to open up the sometimes decided and *cristallin* but also sometimes fluid and opaque interplay of text and context in Gourou's path from *Les paysans* to *Les pays tropicaux*.

In his *Rule of Experts* Timothy Mitchell examines the important role that practices of modern expertise have played in shaping the power of science and the modern state, and how, from the late nineteenth century, such practices came to infuse diverse vocations and academic disciplines, including geography. He argues that during the interwar years a string of rural and anti-colonial uprisings, from Libya, Egypt and Palestine to India and Indochina, were integral to the constitution of "peasant studies" as a "new field of expertise," and he invokes Gourou's study of the peasants of the Tonkin Delta as exemplary in this regard.[22] Gourou undertook most of his fieldwork in the wake of the 1930 mutiny of the Vietnamese garrison at Yên Bái in northern Vietnam, which was engineered by the non-communist Vietnamese Nationalist Party (Viet Nan Quoc Dan Dang, formed in 1927), the 1931 Xo Viet Nghe-Tinh strikes and 'red soviets' fashioned by the Indochinese Communist Party, and a wider and rising tide of anti-colonial resistance.[23]

Between May 1930 and 1931 there were more than 50 separate demonstrations, marches and attacks in Indochina, chiefly on provincial land and tax offices, and over an extensive and exploitative taxation and *corvée* system. The French imposed monopolies on salt, alcohol (rice wine) and opium, which were mainstays of the peasant subsistence economy, and monetised taxes on land and wages, forcing many poor peasants into new rounds of seasonal labour migration (working on plantations and in mines and factories, and as tenant farmers) to generate the *piastre* they needed to pay taxes and support traditional consumption practices. Rice hoarding, land alienation, eviction by mandarins and colonial authorities, absentee landlordism, and labour exploitation were also major sources of peasant grievance. In their survey *The Peasant Question, 1937–1938* Truong Chinh and Gourou's former pupil and research assistant Võ Nguyên Giáp named all of this the abusive and divisive underside of France's policy of *mise en valeur* – ransacking the peasantry to finance vain French projects of modernisation and line the pockets of French *colons* – and with labourers in both the Tonkin and Mekong Deltas working over 12 hours per day in the stifling humidity of the summer monsoon for a measly three *piastres* per month.[24]

By the end of 1931, Résident supérieur du Tonkin Auguste Tholance was able to tell a Vietnamese provincial council of representatives that "The methodical purge carried out following the odious events in Yen Bey, has been pursued ceaselessly for 18 months. . . . The France of the Marne and Verdun would never let such a minority of revolutionaries and trouble-makers threaten her civilising mission in this country" or allow French power to be undermined.[25] Gourou would not have been blind to what was going on. Pierre Brocheux and Daniel Hémery describe the French terror of early 1930s Vietnam that put down these rebellions as "the years of the first great mass repression of the twentieth century," and Anthony Low adds that during these years there were other "revolutions in monsoon Asia" and brutal colonial state reprisals and repressions that portended more concerted anti-colonial and nationalist movements.[26]

While there were was less peasant rioting and colonial violence far out into the Tonkin countryside, Gourou was aware of what was happening with respect to the land and tax offices, where, at that time, he was busy collecting data, and in *Les paysans* he comments in passing on the deleterious impact of French policies.[27] If one of the only problems he encountered was peasant suspicion that he was a French tax collector in disguise, or an alcohol regulator rooting out illegal distilleries, it was, he noted, because French officials were often the only French people peasants met.[28] Tonkin peasants were diligent tax-payers, he continued, but struggled to repay their debts.[29] Alcohol distilling was forbidden by the French but an activity that many peasants were forced into due to heavy colonial head taxes on rice, and this was "one of the most important areas of dispute between the peasants and the authorities," and a cause of some of the rioting in the Tonkin Delta, Gourou acknowledged.[30] But such remarks are guarded and fleeting. As John Kleinen surmises, French scholars "tacitly agreed to a policy of '*autocensure*' (self-censorship) instead of publishing works . . . which could be seen as criticism of the colonial state."[31]

Yet however much Gourou sought to lose himself in the majestic tropical spectacle he beheld, and did not address colonial violence and resistance directly, he was not blind to the colonial anxiety and disillusionment that marked the years in which he conducted his fieldwork, and this aura of angst formed an important political underside to his text and how it was read. The French colonial regime was in control, but many French (and British) observers, and even the French colonial security and intelligence service, the Sûreté Générale de l'Indochine, pointed to the hollowness of the colonial regime's pronouncements about the return of order. The Yên Bái uprising sparked an important anti-colonial literature, and French commentators and colonial officials latched on to Gourou's detailed examination of poverty and population density as a means of understanding the sources of peasant discontent.[32]

The peasants of the Tonkin Delta were more cushioned from the global economic depression of the 1930s which impacted more severely on their Annam and Cochinchina counterparts, whose lives were more dependent on exports and the price of rice, and other commodities. In Cochinchina, especially, where rice prices fell below the cost of production, landlords started to move to the cities, and left tenant farmers in the problematic position of wanting to reduce production but having still to pay rent.[33] As Irene Nørlund details, it was "middle-sized peasants" on plots of between 1.8 and 18 hectares who suffered most, and the Tonkin Delta was spared from the worst effects of plummeting rice prices and exports because it had a much smaller peasant population in this category (less than 10 per cent, occupying less than 30 per cent of the rice lands, and with sharecropping the most common form of renting), and a peasant population that, as Gourou explained, led a more self-sufficient, if "miserable," existence.[34] Indeed, "for the poorest section of the peasantry, who did not produce enough for their own consumption, falling rice prices had less of an impact and were even an advantage."[35]

From 1931 onwards, numerous commentators and texts addressed the rice problem. Notable among these was Paul Bernard's *Nouveaux aspects du problème économique indochinois* where this powerful French banker attacks the anti-developmental character of French colonial policies and proposes to improve the situation by sorting out rural debts, promoting industrialisation, and attacking what he described as the "cancer of over-population" through mass emigration from the Tonkin.[36] Here and in much else written around this time, Gourou's contribution to knowledge and debate is amply registered, although more in relation to the political implications of poverty and colonial abuse than in terms of the economics of rice production. Bernard surmised that *Les paysans* was "an overwhelming witness to the misery which is rampant throughout Tonkin," and worried about how this misery might foster rebellion.[37] There is a potently pessimistic underside to the last – "beauties of the delta" – section of *Les paysans*, and Andrew Hardy suggests that it stemmed from "too much data": "the work's very scale constituted both its strength and its weakness. It is as if, by writing down this complex society of villages, Gourou created a towering house of cards, fragile and immobile."[38] As Mitchell notes, however, given Indochina's great remove from Parisian centres of decision-making, data mattered, and Gourou's sharply worded "diagnoses of the peasant condition" and advice to the colonial authorities "of the 'delicate' task they faced in preserving the existing Vietnamese 'moral and social' system that is so meticulously described, along with the peasants' 'strikingly wretched material conditions'" had considerable influence.[39]

Ecological vulnerability, and prospects of famine and peasant rebellion, were not far from Gourou's mind when he observed: "Only this traditional civilisation can give the people the happiness they deserve; outside that there is only disorder and despair."[40] In any event, historians have deemed these observations prophetic and acutely political.[41] Gourou did not simply produce the Tonkin Delta as an exotic object of study. *Les paysans* also pointed to new of subjects of power and resistance – of poverty, gloom, exploitation and peasant agitation; possibly a house of cards.

Patronage and expertise

But how did Gourou become an expert? The question is complex, in good measure since 'expertise' has different meanings and trajectories in different settings, but also because the answer does not just revolve around his data. The word 'expert' comes from the Latin *experiri*, which translates as 'experienced in,' and for Isaac Newton and Albert Einstein (who both professed on the subject), an expert, in this root sense, is someone who is not simply skilled in a particular area and can demonstrate that they have accrued experience in that area and can use that accumulated experience to deepen or revolutionise a field or discipline, and but can also make their contribution to the stock of knowledge look effortless, as if it appeared to "flow spontaneously from experience itself," as Einstein put it.[42] While areas of expertise and their relationship with particular types of institutions varied, the process of becoming an expert has some core characteristics: experts are deemed to excel in their area by seeing more meaningful patterns in it than amateurs, immersing themselves in the problems associated with it for longer, deploying the skills that characterise it more incisively than others, communicating the meanings and findings that come from it more precisely and persuasively than others, and, as a result, to have the support and validation of a wider community.[43]

Brian Balogh notes that in 1930s and 1940s America expertise connoted bureaucratisation and efficiency, and organisations were conceived (broadly following the sociology of Max Weber and later Talcott Parsons) as "rational and efficient actors," and professionals were treated as "independent and objective experts who brought unique skills to the problems they faced."[44] *Les paysans* conforms to the latter of these two images – the once vaunted modern scientific canon that true knowledge is created by a sovereign (independent and skilled) observer who speaks for the other and enjoins that other to bow before the observer's superior gaze – and was hailed in these terms by those who reviewed his books. One deemed it the "masterwork" of an "exceptional scholar"; another as a "very rare work" displaying the author's "warm human interest"; and yet another as "one of the best theses ever passed at the Sorbonne" and "rightly deserving" of the medal of the Société de géographie commerciale et d'études coloniales for that year, and a *mention très honorable* at the Sorbonne.[45] Gourou kept clippings of these and other flattering reviews.

As this suggests, it would be a mistake to think either that original and specialist works like Gourou's were produced and read as the detached works of lone scholars, or that the organisations that fêted authors as experts were always or completely rational and efficient. *Les paysans* was lauded within a system of intellectual production and circulation, and a specific political climate. It was published in Paris by the École française d'Extrême-Orient (EFEO), and while Gourou had good reason to claim that his work was the original product of his own imagination and labours, his contention that he did not speak to other scholars about what he was doing is contradicted by ample testimony.[46] In 1955, for instance, Charles Robqequain recalled his frequent conversations with Gourou "under the banyan tree at the EFEO in Hanoi," and lamented: "We could [then] move from one end of the country to the other, without weapons, without fear. . . . The Indochinese Union was a reality."[47] In fact, Gourou had toured the Tonkin Delta with a number of influential French scholars, including Jacques Weulersse, Paul Mus and Paul Rivet, and such connections were crucial to how he later became incorporated into metropolitan and international academic and political networks. In 1947 a further vital link in these networks, Lucien Febvre, noted that Rivet was especially important. Rivet provided an ebullient Preface to *Les pays tropicaux*, and not only valued his time with Gourou in Indochina but also confided how the geographer appeared downhearted and weary and "in need of encouragement to complete his valuable work."[48]

Hugh Clout also underscores the significance of patronage – relationships of guidance and allegiance between professor and student – in the professionalisation of French geography.[49] Gourou's supervisor was Albert Demangeon, one of the figureheads of French geography. He agreed to be Gourou's patron when they met in Paris in 1931. Gourou had taken "the fast route" back to Paris: a six-week journey from Haiphong, via Hong Kong, to Vancouver on the Canadian Pacific Steamship Company's Empress of Asia, and then by train to Montreal and New York, and on to Le Havre by steamer.[50] Gourou regarded his supervisor as "very kind," but Demangeon's guidance was limited to a convivial annual letter to his student in Indochina, and the completed thesis presented to him early in 1936 was the first piece of Gourou's field research that he had seen.[51] But the very well-connected and prolific Demangeon was supportive from the outset and instrumental in launching his student's academic career, first of all as *chargé de cours* at the Université libre de Bruxelles in 1936. He also wrote a glowing review of *Les paysans* in the prestigious *Annales de Géographie*. "The book that Mr Gourou has just dedicated to peasants of the Tonkin Delta," Demangeon remarked, "is remarkable in both method and spirit, and represents a highly original understanding and explanation of the human community that peoples the area. It took a huge amount of work on his part to collect documentation of rare value and in a country whose climate is not always favorable to the European."[52]

In France and elsewhere, book reviews, and the facility to oversee or control the review apparatus of journals, played an important role in moulding academic authority and interpretative communities. Leading French geographers and historians such as Demangeon, Gourou and Febvre were copious reviewers of books and conference proceedings; and in France *savoir-faire* (expertise) revolved around these forms of patronage and leverage, and the intellectual tastes they cultivated or left out to dry, and did not simply rest on the nature and extent of one's fieldwork or the acuity of one's observations and arguments. Moreover, while French geography at this time was in some respects set in its ways, and the expertise that came from it rested on its distinct identity as a subject that dealt with the relations between people and environment, geographers had a strong sense of public service and were keen to bring their disciplinary wares to political arenas of debate and policymaking. Demangeon had been a member of the committee that helped to define France's approach to the Versailles Peace Conference in 1919 and during the interwar years was heavily involved in the development of the secondary school geography curriculum.

Colonial reform and international networks

These 'networks' helped to boost Gourou's reputation as a scholar and specialist. But the political climate in which *Les paysans* appeared was a deciding factor too. While it is unclear whether Gourou wrote the conclusion to *Les paysans* – with its observations about peasant misery – before or after he had left Indochina, publication of the book coincided with the election of the left-leaning Popular Front coalition government under Léon Blum (of the socialist Section Française de l'Internationale Ouvrière, SFIO) in May 1936. Gourou's musings about how the peasant situation might be handled need to be read in the context of the pledges that the Popular Front made to undertake a programme of colonial reform, and not least to alleviate rural poverty. Spurred on by a general strike in France during the summer of 1936 marking workers' high hopes for radical change, Blum's government immediately set about drafting a comprehensive package of domestic social legislation.[53] But also concerned by economic exploitation in France's overseas possessions, the Popular Front promised a parliamentary commission to investigate colonial abuses, instilling hope among North Vietnamese nationalists and communists, and the various political movements in Cochinchina that formed an Indochinese Congress (people's assembly) to negotiate colonial reforms with the French, that positive changes in the region were at hand.[54]

Gourou became part of what Gary Wilder terms the "reformist network" and "administrative-scientific complex" of French colonial humanism, and what David Fisher calls an "interclass alliance" of artists, writers, intellectuals, celebrities, politicians and business leaders that grappled with domestic and colonial pressures.[55] Gourou became immersed in a number of specific spaces of knowledge production comprising the network Wilder describes, and they shaped the direction of his work in some unanticipated ways.

Blum appointed his SFIO colleague and lawyer Marius Moutet Minister of Colonies, and Moutet sought from Blum three other ministers who supported his vision of 'democratic colonisation' and 'human development' – Maurice Viollette (the former Governor-General of Algeria), Pierre Viénot (a civil servant on the Middle Eastern Desk at the Quai d'Orsay), and Charles-André Julien (a colonial historian of the Arab world and anti-colonial intellectual). Having joined an array of prominent writers and intellectuals (such as Andrée Viollis, André Malraux and Paul Rivet) who criticised France's brutal reprisal of the Yên Bái uprising, and helped to form the Committee of Amnesty and Defence of the Indochinese People, Moutet was viewed as a friend within Vietnamese political circles, and within a few months of taking office the new Popular Front government had eased restrictions on the Vietnamese press, granted amnesty to hundreds of political prisoners, and was drafting a labour code restricting child labour and banishing forced labour.[56] Moutet also used the juxtaposition of the poverty-stricken Vietnamese peasant and the opulent lives led by French rubber planters and colonists (constituting just 0.2 per cent of Indochina's population during the 1930s), along with the shortcomings of French famine prevention commissions and relief programmes in Indochina, to underscore the need for what he termed a *colonisation altruiste*. France had a moral duty and paternalistic responsibility to improve the lot of its indigenous colonial subjects, he declared in a June 24 circular to colonial Governor-Generals calling them to a conference on the subject of colonial reform; and increasing the peasant's standard of living was the primary means by which France's colonies would form a "complex whole" with the mother country.[57]

Early in 1937 Moutet established the Office de l'Alimentation Indigène in French Indochina and began to extend to the colony many of the labour and welfare reforms being implemented at home. Press freedom and literacy (in both *quoc ngu* and French) in Indochina also flourished under the Popular Front, especially in Saigon and among the urban classes.[58] However, French politicians soon started to argue over the political meaning and economic burden of colonial obligation, and part of what lay behind Moutet's "complex whole" was a concern shared by all political parties that an abrupt overhaul, if not dismantling, of the French empire would cause more harm than good. They also still feared the oral transmission of revolutionary ideas. Pierre Brocheux and Daniel Hémery ascertain (from secret government documents) that what (ironically) Moutet and the Popular Front cabinet feared most was that famine and poverty would spur communism in the region, and that Vietnamese perception of France's inadequate colonial response to cycles of flood and drought would be key.[59]

While the SFIO and the French Communist Party (PCF, which was not formally part of the Popular Front coalition but supported the government) were openly critical of many facets of France's colonial system when in opposition, they had shifted their position by the summer of 1936, seeing France's overseas empire as a continuing source of national strength, and arguing that it constituted a key buffer against economic insecurity and social unrest at home (fuelled by the increasingly deleterious international impact of the Great Depression), and the rise of fascism on its borders. Geoffrey Gunn notes that the French Popular Front government "facilitated the legalization of the various communist parties for the first and only time in the colonies . . . [which] led to a unique cohabitation between adherents of the Third and Fourth Internationals in standing up to French colonialism."[60]

Yet colonial reform rather than revolution was the political watchword of the day (and in Britain as well as France), and the PCF supported this approach, and not least because it was in line with the Seventh Congress of the Communist International convened in Moscow in July 1935, which resolved that a new alliance of democratic and nationalist forces – a united anti-fascist front – was needed to combat Nazi Germany and Japan, and that the goal of fomenting proletarian revolution and constituting soviets in the colonies should play second fiddle to this front.[61] Between July 1936 and March 1937, various leftist groups in Indochina also strove to create a people's assembly in Saigon to negotiate colonial reforms with the French, but the colonial authorities held out for the arrival of a committee of inquiry (colonial commission) from France.[62]

Colonial reform proved difficult to implement, however, and that commission did not materialise. The Minister of Labour, Justin Godard, was sent instead, in January 1937, and his trip was overshadowed by the fallout from a ruinous cycle of drought and food shortage affecting southern Tonkin, northern Annan and much of Laos in the winter of 1936–1937, and then floods in the north and south the following summer, along with strikes in the Tonkin coal mines and by tradespeople in the cities in Hanoi, Haiphong and Nam Dinh, and attacks on private property.[63] The colonial authorities scrambled to respond to the agrarian crisis with a set of emergency measures, and urban unrest diverted attention from reform. All of this, along with criticism of Moutet's plans, contributed to the political downfall of Blum in June 1937 (although the Popular Front lingered until the autumn of 1938). Moutet's reforms were resisted by the French Senate (partly because it was fiscally hamstrung), and by colonial Governor-Generals and French *colons* who begrudged his dictatorial tone. Tony Chafer and Amanda Sackur surmise that if the Popular Front was a defining moment in the history of French colonialism, it was in the way it made "the contradictions inherent in the [French] colonial project . . . clearly visible" for the first time.[64] In fact, the young geographer Jacques Weulersse ascertained these contradictions before that in a notebook he wrote during his trip around the world in 1931 comparing the French and British colonial systems, although it was little known at the time and only published, as *Noirs et blancs*, decades later.[65] Gourou hosted him during his sojourn in Indochina, and it is interesting to compare the vigour with which Weulersse questioned French policy compared with Gourou's muted remarks in *Les paysans*.

As we saw in Chapter 2, the idea of colonial altruism was profoundly contradictory. Debates raged, for example, about whether the granting of workers' rights was compatible with colonial domination. Attempts at reform were also hampered by the geography of empire – by marked differences in colonial conditions (resources and land use, styles of colonial rule, and labour practices) – and as Vietnamese nationalists quickly grasped there was a large chasm between the rhetoric and reality of colonial reform. By the end of 1937, Gunn observes, the French Popular Front government had become acutely aware of "the fragility of the agrarian-hydraulic pact into which they had entered" with both the Vietnamese elite and French *colons*, as a means of expanding agriculture, boosting exports, and seeing irrigation projects as a way of pacifying the masses by promising a basic level of subsistence and ameliorating the effects of flood, drought and famine.[66]

A larger body of postcolonial theory now reads such instances of reformist impasse as indicative of imperial Europe's persistent failure to live up to its universalist (liberal, humanist, emancipatory) pretensions when the crunch came.[67] Reform was overdetermined by binaries of civilisation and savagery, and modernity and tradition, which installed Europe as the fount and pinnacle of development and arbiter of what counted as right, normal and true (and what did not), and configured colonial subjects as 'not yet' ready for independence and self-rule. It was in keeping with this epistemological and political malfunction that interwar colonial reformers like Moutet could express shock at colonial oppression and injustice yet continue to defend empire as a bastion of order and beacon of improvement, thus impeding the very possibility of progress and liberation that his colonial humanism pronounced. Moutet drew sharp distinctions between 'civilised' French workers and 'backward' Tonkin peasants, and Martin Thomas suggests that colonial reform was more "a clutching at obvious straws of manpower and materials" in the face domestic uncertainty than "a reasoned assessment of the colonies' importance to French economic, military, and international power."[68]

Intellectuals played an intriguing part in this story and dynamic. Fisher notes that they provided the French Popular Front coalition "with moral authority, prestige, ideological legitimacy, a rhetoric of hope," and writes of Parisian and provincial intellectuals being called upon by government to offer their expertise regarding colonial questions.[69] Moutet sought academic counsel on a routine basis, and we might see his establishment of a commission of inquiry into the conditions and aspirations of France's colonial populations, chaired by Henri-Alfred Guernut (the radical Deputy for Aisne and former Minister of Education), as further evidence of his embrace of this alliance.[70]

The Guernut Commission was approved by the French Senate on 30 January 1937, started its work in July 1937, and called forth 37 experts from a broad cross-section of French and colonial society, including intellectuals such as Rivet (a socialist and the founder of the Comité de vigilance des intellectuels antifascistes) and critics of French colonialism such as the novelist André Gide, whose "1926–1927 Congo journey began as a pleasure jaunt into the exotic and the erotic," as Walter Putnam notes, but ended up one bewitched by his witnessing of violence and atrocity, and spawning a determination to "write the wrongs" of empire.[71] The Commission's remit was to focus firstly on social and economic conditions, and only latterly on political problems, and to offer practical solutions. Academics played a central part as both specialists and pragmatists, and the plan was that they would solicit information using questionnaires, and offer advice and solutions in the form of monographs on particular colonial regions and questions. The commission was divided into three working groups, dealing with: (1) Tunisia and Morocco; (2) America, Central Africa, Madagascar and La Réunion; and (3) Indochina, French India and Oceania. Gourou served the third group, chaired by Victor Basch (a Sorbonne Professor of Philosophy and founding member of the League Against Imperialism) during the spring of 1937.

Moutet regarded famine, poverty, hunger and population as colonial problems that pervaded France's colonial empire, and argued that if famine relief was to work, indigenous elites as well as academic experts needed to be more closely involved in the fashioning of context-specific measures.[72] Gourou's concluding remarks in *Les paysans* provided an obvious touchstone in this regard, and Robequain was one of the first to say so. In his lengthy review of the book in the *Bulletin* of the EFEO he explained that it might be regarded as a consummately detailed response to the "pressing need" for detailed studies of colonial regions and problems, and that politicians should heed Gourou's observation that the French faced as a very "distressing prospect," and his advice that colonial "remedies" to peasant poverty that sought "the overthrow of the values of the traditional world of the peasant would herald a misery worse than all others."[73] "The moral and social stability of the region, the complex of customs and traditions which enables the peasant to endure his awfully wretched material existence, must in no way be threatened," Gourou had written; and he asked: "what would become of a people who reflected on their own infernal wretchedness?"[74] The implication was clear: peasants might train their thoughts on the roots of their misery and identify France's colonial presence as a primary cause.

Gourou accepted the arguments made by French *colons* and colonial administrators that French rule in Indochina had yielded some improvements in rural infrastructure (dikes, roads, irrigation systems), and in December 1936 the interim Governor-General for Indochina, Auguste Silvestre, trumpeted these improvements in a letter to Moutet defending his colonial government's record on poverty and famine relief, and claiming that rural standards of living were gradually rising.[75] However, as many other observers insisted, the vast majority of Indochina's population was desperately poor and it had proved tremendously difficult to improve their lot. As Robequain viewed matters, the Vietnamese peasant "just manages to live from day to day, and at most he has extremely limited reserves."[76]

Moutet sought to fathom the causes of what he regarded as this 'backwardness.' His initial plan was to send a team to Indochina to liaise with colonial officials and listen to Vietnamese concerns. But this failed to materialise, fuelling the ire of the Indochinese Congress, which had compiled a list of demands to present to commissioners. The Labour Minister, Justin Godart, was dispatched instead to undertake a more limited (three-month) appraisal, but he was greeted in Saigon in January 1938 with strikes and protests over Moutet's failure to implement his promised labour code, prompting the Minister of Colonies to telegraph colonial officials insisting that they maintain public order at all costs.[77] It was partly for this reason that Moutet resolved that the commission should do its work from Paris and look there for expertise.

In the records pertaining to the Commission there is a letter to Moutet dated 17 October 1936, postmarked Hanoi, and thus presumably from Silvestre, providing "official [government] testimony of satisfaction" regarding Gourou and emphasising his "devotion" to his teaching and research.[78] However, Gourou was already acquainted with Moutet; he had written to him twice in July 1936, possibly in response to being asked about his ability to assist the government, outlining his research on the Tonkin Delta, noting his intention to take up a teaching post in Brussels, and explaining that while he had not resigned himself to the thought of never returning to Indochina, he might at that moment usefully place his "competence" at the disposal of the Ministry.[79]

It is also likely that Gourou was recommended to Moutet by Rivet, who had close links with the Popular Front. Rivet had toured Indochina from December 1931 to April 1932, collecting objects for the Musée d'Ethnographie du Trocadéro (MET) in Paris and a planned ethnological museum at Dalat, and had toured the Tonkin Delta with Gourou towards the end of his trip. As Alice Conklin shows, he had also used his trip to enlist the support of a range of – younger and older, amateur and professional, and metropolitan and colonial – "experts" to expedite his vision of a global network of ethnological research, which centred on the MET and Institut d'Ethnologie in Paris, and was discussed in detail in 1937 at a Congress organised by the Association Colonies-Sciences in Paris.[80] Rivet had sought the assistance of the EFEO in Hanoi as part of his preparations for the Congress, which he hoped would facilitate the professionalisation of ethnology as a field-based research discipline with metropolitan-trained scholars. However, he realised that the EFEO's mission was primarily archaeological and philological rather than ethnological, and knew from one of its young talents, Paul Mus, that attempts to expand or reorient its remit would be viewed with diffidence. Rivet held Gourou's work in high regard because of its palpably ethnological dimension. He had studied Gourou's 1931 *Le Tonkin* and was fascinated by the geographer's discussion of village life and traditional handicrafts. Rivet also knew that Gourou had also been reviewing books on French colonial policy in the *Bulletin de l'Ecole française d'Extrême-Orient* since the late 1920s, and had gained a reputation for fair-mindedness. In short, Rivet deemed the geographer a key Indochina expert.[81]

As Demangeon intimated, the CEPE (Centre for the study of foreign relations) – the French equivalent of the U.S. Council on Foreign Relations and the Britain's Royal Institute of International Affairs – which was created in 1935 with Roger Lévy as director, was also significant in Gourou's inculcation into this French network.[82] Lévy was an international affairs specialist who had been the Far Eastern editor of the weekly magazine *L'Europe Nouvelle* and secretary general of the Comité d'études des problèmes du Pacifique (CEPP).[83] The CEPE's remit was to commission and publish detailed and impartial studies of international affairs (including – centrally – colonial policy) in articles, pamphlets and scholarly monographs, and to provide a bridge between universities, specialist research institutes and government. It extended French efforts – forged, for instance, by the École Coloniale (created in 1887) and Institut d'Ethnologie (created in 1925) – to articulate academic, colonial and governmental concerns, and rationalise the colonial service.

Such initiatives were supported by government (especially the Ministry of Colonies and Ministry of Public Education), and, as Wilder shows, were about more than the ideological justification of French rule.[84] Academics and administrators such as Gourou and Rivet, and ethnologists such as Marcel Mauss, Michel Leiris and Marcel Griaulle, insisted that foreign inquiry was not made and measured simply for the state and government policy. At the same time, they grasped that expert academic knowledge played an important strategic function in policy-making. Research institutes like Lévy's CEPE were inventive and supple knowledge networks that were at once tied to and free from government and disciplinary oversight, and experts like Gourou and Rivet both followed disciplinary protocols (their reputations rested on them) yet deemed themselves free to take inquiry in new directions. For instance, Rivet – who in 1928 (and along with Georges-Henri

Rivière) was put in charge of reorganising the Musée d'Ethnographie (located in the Trocadéro, and where the Cubists discovered Negro art) into the more scientifically and systematically organised Musée de l'Homme (inaugurated in 1938 in the Palais de Chaillot) – helped to bring anti-colonial thought and activism, and the racism inherent in German fascism, to the heart of governmental debate.[85]

Gourou had reviewed Lévy's 1935 survey *Extrême-Orient et Pacifique* in the *Bulletin* of the EFEO and the two of them struck up a friendship in 1937. One of the prime initiatives that Lévy had instituted at the CEPE was to bring scholars, writers, politicians and diplomats together to identify and debate international issues, and the centre invested Gourou's work with new governmental and international overtones. Demangeon, along with André Siegfried (who held a chair of economic and political geography at the Collège de France), sat on the CEPE's administrative council, attesting to the high regard in which geographical knowledge was held, and of the seven research groups covering different parts of the world moulded by Lévy the one devoted to the Far East and Pacific was the first named, attesting to the centrality of Indochina in French political debate.

Lévy also looked to forge connections that would give France a stake in the internationalism of the day, and as means of better understanding, and hopefully countering, the rise of fascism in Germany and Japan. He took the first French delegation to attend an IPR conference, in Banff, Canada, in 1933, and in 1935 appealed to the IPR's International Secretariat in New York to make his *Comité* its French National Council, which it duly did. By 1937 the work of the CEPE had become closely intertwined with that of the IPR, although impending war in Europe and German occupation prevented French delegations from making it to the IPR conferences of 1939 and 1942.[86] The IPR was the first large non-governmental international organisation concerned with Pacific issues. It was founded in 1925, based first in Honolulu and from 1934 in New York, funded by powerful American business and philanthropic interests such as the Rockefeller Foundation and Carnegie Endowment for Peace, and was comprised of autonomous national councils that fed into a central secretariat, and representatives from which met every two-to-three-years at 10- to 14-day international conferences held in luxurious hotels, with different national councils taking it in turns to host the event. Priscilla Roberts has described the IPR as "perhaps the most visible and impressive international foreign policy think tank [of the interwar era]. . . representing the most ambitious and probably most idealistic contemporary effort to establish a transnational knowledge network."[87] Governmental officials prevailed in some national councils, but academics served in many of too, as foreign policy experts.

The CEPE, and its periodical *Politique Étrangère*, which was allied with the IPR's journal *Pacific Affairs*, tackled a wide range of international affairs questions,[88] and it was through Lévy that Gourou and Robequian became valued members of the IPR's Pacific research group and French national council. Gourou's 1936 monograph was quickly recognised as the most important work published on Indochina during this period, and it soon became linked to what Tomoko Akami describes the as "internationalization" of the Pacific as an area of study and foreign policy concern.[89] The IPR published monographs (through its Inquiry Series)

154 *Networking the tropics*

and pamphlets, and ran area and country training programmes for the military. Gourou's next book, *L'utilisation du sol en Indochine française* (1940), along with Robequain's *L'évolution économique de l'Indochine française* (1939) were both commissioned by the CEPE and published by the IPR.

The Pacific was not just being internationalised by such initiatives, however. In Gourou's hands it was also being tropicalised: identified once more as 'monsoon Asia.' From 1937 through to the 1950s Gourou's research was moulded by the remit of the CEPE and Lévy asked his trusted colleague to join the French delegations that attended important IPR conferences in 1945 (at Hot Springs, Virginia), 1947 (at Stratford-on-Avon, England, which Gourou chaired), and 1950 (at Lucknow, India).[90] The CEPE prompted Gourou to develop a broader interest in questions of population and land use in the Far East, and he became more and more interested in how 'monsoon Asia' was shaped by extraneous (by which he often meant sub-tropical) Chinese and Indian influences. André Burguière surmises that while French academics had been ambivalent about serving governments and ideological causes during the 1920s, by the mid-1930s and with the onset of global economic depression, growing colonial unrest, and a looming fascist threat that stretched to Japan, they had become much keener to offer their services.[91]

Both before and after World War II, the IPR and Lévy's CEPE articulated a liberal internationalism which had not abandoned faith in empire as a progressive and altruistic instrument of economic and social change, although it was an internationalism guided more by the realisation that France was passing through a difficult period than by the idealism that had characterised the 1920s.[92] Geographers' growing awareness of the interdependence of different parts of the world, and the problems this raised about the occupation and organisation of space, constituted part of this realisation.

An increasingly international and networked world was at hand, and geography responded to it by flagging the importance of comparison and synthesis.[93] Geography insisted on its autonomy, even purity, but also tried to express its legitimacy by demonstrating its political utility.[94]

Bio-tropicality: the Guernut Commission

Gourou produced a 32-page report for the Guernut Commission, in which his academic interest in population density morphs into a concern with *surpopulation* (overpopulation; overcrowding), and where he underscores the fragility of France's agrarian and hydraulic projects in Indochina. As Alison Bashford notes, overpopulation (and population decline) formed one of several "spaces" (along with demography, eugenics, and bodily and reproductive rights) through which twentieth-century "population politics and expertise" have been understood.[95] It was a 'space' that haunted imperial powers, and, for reasons later theorised by Foucault, it was at once a code word for empire's death drive – its self-defeating attempt to control and rule enormous and distant peoples and lands with limited resources and rods of violence and division – and a cipher for empire's attempt to avert this drive by imagining and managing remote and ill-understood environments and occupants as territories and populations to be enumerated, classified, regulated, rearranged, medicalised and governmentalised.[96] Colonial life was constituted by the omnipotence of death (disease, destruction, fatality and ruin), and exposed both colonial racism and the fragility of colonial control.

"[I]f much scholarship posits that a mid-century world 'overpopulation' discourse displaced an interwar national 'depopulation' discourse," Bashford continues, "we find book after book on overpopulation written in the earlier period sitting alongside the 'degeneration' and 'depopulation' tomes that historians typically focus on."[97] Colonial explanations of the relationship between poverty and overpopulation often sought to absolve colonial powers of blame, and Bashford shows that doctrines about "human-climate interaction" – a bio-tropicality, if you will – were part and parcel of that absolution. The physical and psychological health and well-being of tropical bodies and populations was examined "along axes of environmental difference, human difference, and comparative adaptability," Bashford observes, and in regions such as Indochina with problems of white acclimatisation and the colonisation of 'unused' land compounded by the spectre of large, impoverished and potentially seditious peasant populations.[98] "In the context of an accelerating population growth, and the pressing planetary limit that so many acknowledged, the tropical question was at once physiological, geographical, and deeply political. It was no longer just an imperial issue, but an international one."[99]

Gourou occupied an important, if complex, position in these debates and concerns. On the one hand, and as Van Nguyen-Marshall relates, he "was one of the first researchers to conduct surveys of rural and urban living conditions in Indochina, and provided valuable ammunition for critics of French colonialism."[100] He began his report to the Guernut Commission by reiterating one of his concluding remarks in *Les paysans* that while the social evolution wrought by French colonialism should not be opposed *per se*, it should be pursued cautiously in the peasant countryside; that "often local improvements . . . would reap a better return than vast enterprises."[101] He went on to identify "overpopulation" as a key facet of "the undernourishment of the inhabitants" of the Tonkin Delta, and both in this report and his *L'utilisation du sol*, he used bio-medical language to address the issue at hand: "It is certain that the Tonkin peasants are miserable and that their misery must be alleviated. But the task is a delicate one, because the Tonkin Delta is an ancient country with limited, if complex, resources, where many centuries of human activity have forged a subtle and multi-layered organism. We cannot just cut into the flesh there, and any attack on one of the layers could have the most grievous consequences for the whole."[102] His invocation, both here in his report and in *L'utilisation du sol*, of a fleshy and "pathological" demographic situation is intriguing.[103] For as Sokhieng Au shows, during the interwar years surgical practices and metaphors of "cutting the flesh," and recourse to a medical language of 'pathologies' and 'cancers' to talk about populations, attained a broader significance in European colonial empires as means of thinking about colonial populations as bodies with shapes (geo-bodies), anatomies and ailments, as well as "fantastic consumption stories" (cannibalism) specifically associated with the tropics, which scientists and experts could diagnose in order to heal, improve and perfect.[104] Tropical medicine was the most important site in the development of bodily practices and representations of incision, but by no means the only one.

On the other hand, and as the Australian political scientist MacMahon Ball reflected in 1952 on this bio-political scene:

> A good deal has been written about the poverty and misery of the Indochinese people, but M. Pierre Gourou, a leading authority on the economy of the region, suggests caution in this regard. He points out that processions of starving people and hordes of beggars are not seen in Indochina as in some other parts of eastern Asia, that the people are not brutalized serfs but free men eager to improve their lot. The deep misery that produces inertia and dumb fatalism is not evident in Indochina.[105]

Ball went on to underscore the significance of Gourou's geographical approach to the relations between poverty, colonial control and peasant unrest in the Far East. As Gourou had explained, "Poverty in Indochina is associated with different circumstances in different areas. In Tonkin in the north it is linked with an uneven distribution of population in Indochina as a whole and the overcrowding of the Red River Delta. In Cochinchina in the south it is linked with the unequal distribution of the cultivated land."[106]

During the course of the 1930s the issue of overpopulation began to permeate the *Annales de Géographie* (and American and British geography too), and in 1938 Demangeon wrote a global overview of the subject.[107] The cut that Gourou made into the question was different from that of either Demangeon, or Robequain, who was a supporter of the colonial regime.[108] Robequain declared that the question of overpopulation, so carefully unpacked by Gourou, was the greatest problem facing the French, and "the one on whose solution depend all others."[109] He took the Malthusian line that there were natural limits to population growth in Indochina (limited resources and a lack of space), and also put overcrowding in the north and the demographic "disequilibrium" between the north and south down to France's stabilising, modernising and medical influence. Paradoxically, he argued, the French had created poverty by trying to boost peasant living standards, and sat on the horns of a dilemma, particularly in the north, where most of the rebellions had occurred. "This is a difficult problem to solve and one of the white man's greatest burdens. Will he not be worn out in his double attempt to increase the native's life span and feed him better?"[110] All this had done was to swell the "so-called 'proletarian' class in Indochina;" his proposed solution was birth control.[111]

Gourou saw the question of overpopulation differently: as an inevitable and interminable upshot of the peasant agricultural civilisation he cherished. From the mid-1930s through to the 1950 IPR paper on "The economic problems of monsoon Asia" he presented at Lucknow, he took a consistent line:

> Dense rural populations are not caused by lack of space.... The fundamental causes of the increase in the rural population are to be found in the nature of their agricultural techniques and in their dietary regime.... There is a

natural tie, and one difficult to break, between the living standard and the agricultural techniques employed. It is the latter that are responsible for the low level of consumption, not poverty that has forced the peasant to adopt these techniques.[112]

Nearly one third of Indochina's 22 million people lived in the northern delta region on 2 per cent of the colony's land surface, he reported to the Guernut Commission, and with average population densities of 430 km^2 (and in some parts of the delta over 1000 km^2) – "a swarm of humanity," as he described it, surpassing densities in the most compactly inhabited industrial regions of Europe.[113] Over 60 per cent of the Tonkin peasant families possessed or worked less than 0.36 hectares (one *mau*) of land, and their average income was one sixth of that of their French counterparts. The vast majority of Indochina's cultivated land was devoted to rice production, he continued, and the tiny plots of land and village-based labour made for a precarious existence, driving many off the land and in search of waged employment.

Gourou outlined a raft of measures that might be adopted to alleviate hunger and raise living standards in Tonkin: improvements in drainage and irrigation; the use of plant and mineral fertilisers; the introduction of new seeds and plant stock; the development of agricultural credit agencies and rural cooperatives; the preservation of community land and property rights and traditions; the development of handicraft industry; and "the possible development of a new peasant mentality." A long list, and one based, he suggested, on his detailed knowledge of the region – to which, he confided in his letter to Moutet, he hoped to return. In conclusion, he argued that "all of these measures, if they are applied in a judicious and unhurried fashion, will produce the desired results: the development of production and improvement in the standard of living of the peasant."

However, he ended with the following message:

> We should not have any illusions on this subject. We can hope to soften considerably, by the means laid out above, the condition of the Tonkin population, but it would be dangerous to believe that we could transform their lifestyle. The Tonkin peasants taken as a whole are poor and will remain so; but we will have achieved a magnificent result if we have in a modest way reduced the instability of their conditions of existence and warded off the danger of famine. We should not hide the fact that the extreme density the Tonkin population has reached is a disease without radical cure, and that this population grows annually at 10 to 15 per 1000. In order to confront this problem, we must therefore use all means available; that is why we have outlined here a category of these means of action, of limited but certain effectiveness. Finally, I cannot end this report without emphasising once more the need for careful statistical surveys.[114]

The above list of recommendations is very similar to the one arrived at by Godart for Indochina as a whole.[115] In fact, the section of Godart's report that deals with the "peasant question" relies heavily on Gourou's "magisterial" study.[116] Robequain similarly argued that agricultural credit and mutual associations were needed to combat the "usurious interest rates" on agricultural and industrial initiatives imposed by Vietnamese creditors, and that any overhaul of "inferior methods of [peasant] cultivation" that had arisen from the peasant's "chronic state of want and his need for immediate returns on his labour" needed to be undertaken cautiously.[117]

Gourou's contribution needs to be placed in this encompassing, politically driven debate about the relations between population, land and colonial power. He did not discover the problem of 'overpopulation,' and others saw it in different ways. Around this time Bernard and another prominent colonial banker, René Bouvier, declared overpopulation a social and political time bomb and discussed the merits of outmigration to the highlands north and west of the Tonkin Delta, and other parts of Indochina.[118] Like Gourou, the agronomist and future Third World ecological campaigner, René Dumont, addressed the socio-economic problems facing the Tonkinese peasant, without denouncing France's colonial presence *per se* and arguing that "development should not be hurried."[119] Dumont singled out taxation and usury for special criticism. Money was scarce in the delta, and with the drop in rice prices taxation had become too heavy, he thought. Villages were taxed by number of inhabitants and plots of land, but this often did not correspond to their material conditions, and as a result, the villages often did not co-operate with the colonial administration. In addition to criticising the taxation system, Dumont appealed to banks and usurers to understand that it was not in their interest to ruin the Tonkinese peasant.

And like Gourou, the demographer Grégoire Khérian stressed in the first of two articles on the population "disequilibrium" between north and south, the need for careful statistical analysis, and argued that colonial policies which undermined the region's complex and fragile rural stability would be catastrophic.[120] Demangeon underlined the broad interest in the problem in his review of *Les paysans*, and Andrew Hardy has since remarked that economic depression and political agitation in Indochina "concentrated French minds like never before" and "overpopulation received almost obsessive attention."[121] In 1936–1937 alone, the issue was debated in five books and eight articles, and with a great deal of bombast and rhetoric about the strengths and weaknesses of French colonial policy, and debate rumbled on into the 1940s, with further studies by French and Vietnamese scholars.[122] Silvestre (and others) saw out-migration from the delta as a remedy, but questioned its efficacy – as did Godart and Gourou – on the grounds that it would be difficult to achieve because peasants were strongly attached to their ancestral villages. Hardy adds that how one defined terms like 'overpopulation' and 'disequilibrium' was part and parcel of the problem and debate.[123]

However, Gourou's tropical take on the problem – the primacy he accorded to monsoon relations between society and environment, and the crippling judgement that they locked the peasant into a particular existence – left a decisive mark on the debate. As Martin Thomas explains:

> landholding statistics indicated that in several northern provinces peasant agriculture was unviable. Farmers could neither feed their families from

privately owned paddy nor from communally farmed village plots. Hence, the rapidity of Vietnamese proletarianization that so terrified the Popular Front's commissioners. Where French reformers dwelt on agricultural modernization, irrigation schemes and improved rural credit facilities, leading ICP members Truong Chinh and Võ Nguyên Giáp were the first to ascribe the plight of Vietnam's peasantry to colonial land seizures, punitively high taxes and the oppressive demands of plantation owners.[124]

Land use, population density and poverty were keys to understanding and debating colonial violence and unrest.

Aware that Moutet saw overpopulation as a cause of poverty, and poverty as the lever the communists were using to win peasant hearts and minds, Gourou knew that he needed to make some concrete recommendations, and the tardy ones he did make were more in keeping with how the reform process played out than with Moutet's utopian rhetoric. In a February 1937 circular to Governor-Generals imploring France to revamp its 'civilising mission,' for instance, Moutet noted: "Enslaved for millennia by the caprice of the forces that are beyond them . . . the native has necessarily taken the habit of submitting to them without even dreaming of defeating them, and living day-by-day carefree of a future to which he has resigned himself in advance," and that the "true grandeur of our civilising role" lay in "teaching" the native "how to provide for himself."[125] This image of millennial enslavement was echoed by Gourou, and like Godart he thought that impoverished peasants striving to improve their lot were being hampered by counter-productive colonial policies. Gourou and Godart found themselves caught between what Hardy sees as the "urgency (of discourse) and conservatism (of action)."[126]

The Guernut Commission came too late to be of any direct use to the Popular Front (the Commission folded in July 1938), and as Nguyen-Marshall concludes, the process of colonial reform in Indochina was soon "supplanted by the bureaucratic imperative to follow directives and to produce results on paper."[127] Young communists like Giáp – who undertook research for Gourou's minor thesis on Annamite houses, and who had presented Godart with a report on "the peasant question" which drew on this research – felt betrayed and began to argue that the question of 'population' was less significant than that of the radically unequal distribution of land, labour and capital.[128] However, Gourou's configuration of population as a political problem shaped how his work was read, how his status as an expert was secured, and how his thinking and work developed over the next 10 years. In both his Guernut report and subsequent writing his pessimism regarding France's ability to alleviate the peasants' "misery" is the watchword and echoes the political uncertainties of the 1930s.[129]

While Gourou's work is commonly viewed as a corrective to racism (and other determinisms), the debate about poverty in which he became embroiled cannot be detached from the wider promulgation of 'population' (birth and death rates, life expectancy, illness, diet, habitation) as a colonial object of analysis, target of intervention, and matter of security (for the French, concerned about the spread of peasant unrest) during this period – and with the squalor caused by overpopulation deemed a 'disease' for which the French imperial state sought a 'cure.' Gourou deployed such bio-medical metaphors in his Guernut report and *L'utilisation du sol*, and as Foucault famously argued, such language was germane to the advance of the modern state and disciplines like geography into matters of life and death, fostering a regime of "biopower" that revolved, by the twentieth century, around the idea of race.[130] The Guernut Commission undertook detailed surveys of the 'métis problem' in the French empire, and growing French administrative interest during the 1930s in migrants, children and marginalised populations made racial categorisation central to the functioning of colonial power.[131]

Comparison-network-expertise

Let us pause for a moment to reflect on the direction in which these entanglements had started to take Gourou and the discourse of tropicality. In the last chapter, we dwelt, with the caption 'the data delta,' on how Western knowledge about foreign lands was translated into colonial power through what Bernard Cohn termed "investigative modalities"[132] – historiographic, observational/travel, survey, enumerative, museological and surveillance modalities that were central, he argued, to the exercise and legitimation of British power in India. Benedict Anderson added that in Southeast Asia the "nexus" of the census, the map and the museum "profoundly shaped the way in which the colonial state imagined its dominion – the nature of the human beings it ruled, the geography of its domain, and the legitimacy of its ancestry."[133]

France had a long tradition of scientific research on its colonies, and by the nineteenth century much of it was organised by centralised metropolitan research institutions.[134] The drive to annex and exploit territory, people and resources yielded a range of amateur and professional knowledge which was produced by a diverse cast of explorers, soldiers, merchants, missionaries, surveyors, administrators, travellers and scholars who either had very prescribed aims or for whom the dissemination of knowledge was a by-product of their main employment. By the late nineteenth century, however, there was a growing concern within elite circles with the need for knowledge and projects of investigation that were more consistent with the standards of scholarly production within emerging disciplines and institutions, either within or parallel to the university.

These "colonial sciences," as Pierre Singaravélou terms them, have been the subject to intense critical scrutiny. They were handmaids of imperial expansion and colonial dominance, and with what counted as science and progress in knowledge set up in Eurocentric terms, through what Johannes Fabian described as a "denial of coevalness." Western knowledge was extolled, the role of native informants and indigenous knowledge was downgraded (as mythic and superstitious), and non-Western lands were deemed suppliers of 'data' (as well as of resources), which the Westerner took home and made meaningful.[135]

Yet as Fabian and now a large critical literature shows, Westerners had to work hard to keep up the pretence that their observations were the sole founts or sure-footed providers of knowledge and truth. Western and colonial knowledges were produced and worked through complex, and sometimes conflicting, networks that brought scholars into diverse political, academic and institutional circles, and scholars such as Helen Tilley have suggested with reference to early twentieth-century Africa that the term 'colonial science' should be approached with caution. As Alice Conklin reflects on Tilley's scenario:

> all scientific research circulates both locally and globally in ways that its producers cannot control – even when this research is sponsored by imperial governments seeking solutions to problems of colonial governance. From this perspective, defining any "science" as specifically "colonial" obscures more than it illuminates. Her [Tilley's] point is not that "good" science triumphed over "bad" science in Britain's African colonies, but that the outcome of the appeal to science was never absolutely predetermined by the fact of empire. Professionalizing scientists in the field could and often did maintain their distance from policy-making: their training encouraged them to look for the very kind of complexity in human societies that overburdened administrators or their superiors did not have time to consider.[136]

In the U.S., for example, 'the jungle' was as much a product of popular culture (travel writing, film, environmental thought, and tourism) as it was of corporate capitalism and American empire – "a tropical extension of the American wilderness ideal," writes Paul Sutter, "a place of great natural diversity as well as a testing ground for scientific explorers," albeit still as a realm designed to be consumed, both directly and at 'home' (in zoos, museums, exhibitions, food halls and restaurants).[137]

While Gourou did not have any direct connection with the colonial state, his work and that of the EFEO was implicated in France's colonial project by virtue of the fact that such research was made possible by, and gained meaning and authority within, the colonial system initiated by Governor-General Paul Doumer (1897–1904), who created a centralised administration and primary export sector, awarded land grants to French settlers, and created an extensive network of roads, dikes, schools, prisons and research institutes.[138] Indigenous societies and environments were of particular concern to colonial governments, and a new generation of engineers and scientists (including geographers) were either tasked or expected to bring precision and expertise to French colonial understanding. This arm of *mis en valeur* in Indochina was meant to assuage the bad reputation that Tonkin had attained in France as a costly possession in both money and blood; and what Julia Waters suggests with respect to Marguerite Duras's commiseration with the misery of the peasants of the Mekong Delta also pertains to Gourou's Tonkin peasants and their misery: "the link established between native peasants, their environment and universal, natural cycles not only dehumanises and de-individualises them, but also, as in imperialist ideology, serves to justify the coloniser's – and the reader's – complacent, passive, spectatorial role in the face of the colonised subjects' suffering."[139]

However, one of the criticisms levelled at this scientific outreach was that it could just as easily backfire – reveal some uncomfortable truths and stoke anti-colonial sentiment.

Les paysans became caught up in this problem, and as Singaravélou signifies, if the "colonial sciences" were simply instruments of othering and control that produced stereotyped visions of colonised territories and peoples, they would not have lived for long. In spite of the problematic imperial context in which they arose, they performed, by the virtue of the novelty of the objects of knowledge they constituted and fluidity of the imperial discourses and colonial settings in which they operated, an epistemological decentring and made it possible to invent new fields, methods and orientations. Singaravélou sees oral history, tropical geography and legal anthropology as three noteworthy, if inadvertent, inventions and suggests that they had polyvalent meanings that straddled science and sentiment, and involved both acquiescence to and criticism of empire.

Gourou was not blind to the opportunities opened up by Doumer's network of écoles and instituts, and nor to the political meaning that his interests had in mid-1930s France. Connecting these two sites and sides of his 'colonial science,' and the wider eventualities of 'epistemological decentring' and 'unintended consequences' pursued by Singaravélou and Tilley, was a modality (a triad) of comparison-network-expertise. With increasing vim during the interwar years, values and practices of comparison and expertise hinged on the idea of the modern self as coming into being through movement and circulation across different lands, peoples and knowledge domains. In the late nineteenth and early twentieth centuries, what Deborah Neill describes as "networks in tropical medicine" – collaborations among doctors and scientists that strove to get past national and imperial rivalries and tackle major tropical diseases such as sleeping sickness in sub-Saharan Africa – were key to the development of this medical speciality and fostered a shared 'Europeanness,' albeit one, she points out, that was still underpinned by a common imperial belief in the scientific and racial superiority of the European.[140] Hannah le Roux writes in similar terms about post–World War II "networks of tropical architecture."[141]

These two modalities of 'census-survey-map-museum' and 'comparison-network-expertise' – the one immersive, the other expansive; one shaping nineteenth-century, and the other twentieth-century, empire – shared what Emily Rosenberg describes as "a scientific, positivistic faith":

> if sufficient data on a particular question could be collected, the information could then be ordered, analysed, and used to solve problems in the natural and social worlds. In this great task, divisions of religion, ideology, and even national loyalties might be put aside as specialists constructed common understandings of statistics and society. . . . Those who constructed circuits of expert knowledge claimed they could assemble observations from around the globe and extend their expertise by using far-flung networks in which localized 'facts' could be compared, tested, confirmed, and connected together.[142]

By the early twentieth century there were clear affinities between these far-flung imperial networks and the discourses of Orientalism and tropicality. However, the complicity of science in colonialism should not be treated as a "foregone conclusion," Conklin remarks; and it is also possible to write about "dead white scientists attached to empire in the interwar era without in any way eulogizing or apologizing for them."[143]

The drive to create what Rosenberg describes as "transnational epistemic communities" was deemed pivotal to the survival of empire, played a key role in the rise of multinational corporations (such as Unilever and The United Fruit Company), venerated comparative and systematic modes of inquiry and inspection, and was imbued with rhetorics of improvement and rationality. It was thought that "Taking diverse environments into account in a systematic fashion might hold the key to the "future welfare and progress of mankind," as Tilley writes of the treatment of Africa as a cohesive (physical, biological and human) entity during this period, and with different branches of science (ecology, ethnology, geography, medicine, meteorology, sociology, zoology) working in tandem to overcome the pervasive criticism that business suffered from a lack of scientific understanding.[144]

From the late nineteenth century, imperial powers and the League of Nations "increasingly supported scientific coordinating conferences in specific territories and metropolitan centers," and at which specialists from diverse backgrounds "were able to pool information, consider common problems, and share ideas and strategies. Inadvertently, these conferences [also] created a forum for individuals to criticise the status quo."[145] Tilley lists 19 interwar British imperial coordinating conferences between 1920 and 1940, and before the Guernut Commission the French held a large Imperial Conference between December 1934 and April 1935, which was the apogee of a campaign led by the business community and France's Minister of Colonies, Pierre Laval, to place the organisation of the imperial economy on a rational footing, and involved an array of scholars, government ministries, technical advisers, bankers and pundits. Lauren Janes noted that the Great Depression "spurred a new emphasis on coordinating and planning the imperial economy so that Greater France could be self-sufficient in the protectionist global economy of the 1930s."[146] This networking of welfare and progress came with its own – sometimes old, sometimes new – conceits, limits and delusions.

The interwar era was a boom time in the articulation of scientific theory and practice in France – a boom fuelled less by the dream of peace and progress than by the need to explain, predict and offset calamity and colonial unrest through the creation of technological and scientific links between France and its colonies. An ebullient *Journal des économistes* reviewer noted that *Les paysans* was "an excellent working tool" for the economist on account of its abundant tables, illustrations, coloured cards, as well as detailed descriptions, and would be welcomed by readers far and wide.[147] Hugh Clout adds that the field excursions accompanying international geography conferences "provide[d] important opportunities to transmit geographical knowledge" and bring doctoral students into the discipline's patronage networks.[148] And from 1936, Gourou fostered cooperation between

French, Belgian and Portuguese colonial authorities in their efforts to gauge tropical conditions in their respective African possessions.

However, these networks and communities were often not as supple, cohesive or international as is sometimes supposed. Western experts and metropolitan centres "dominate the circuitry," Rosenberg stresses, and scholars clung tightly to their right to be believed and to the remit of their own disciplines.[149] Tilley notes that many scientists "doubted that a 'centralized' approach would yield the sort of knowledge they needed to do their jobs well. The drive to localize knowledge was just as strong an imperial imperative as the desire to standardize and systematize it."[150] Nor was communication between disciplines seamless. The head of the archaeology department at the EFEO, Henri Parmentier, wrote to Gourou congratulating him on "the ingenious idea of photographing and redrawing the maps of the colonial state to give the impression of real village densities," but continued: "I lament to see that you use none of my lectures when you deal with the particularly exciting architectural issues covered in your work . . . although they are treated with remarkable accuracy and no less clarity by someone who is not guided directly by his normal preparation to study such subjects. . . . Give my compliments to Inguimberty for the assistance he gave you. Yet it is frustrating for an architect to see an academic and a painter do the work that architects neglect."[151]

Reflecting on how tropical geography evolved in France between the mid-1930s and 1945, Paul Claval notes that "Because there was no central organisation for overseas scientific research prior to the Second World War," French knowledge of tropical problems developed independently in Indochina (with Gourou and Robequain), the Americas (and with the French geographer Pierre Monbeig helping to establish the first departments of geography in Brazil), and Africa (with Jacques Richard-Molard, who worked in the highlands of Guineas in 1941–1942 with bursaries provided to young geographers by the Vichy regime).[152] At the same time, one of Demangeon's students, Monbeig, at the University of São Paulo (and where Braudel and Lévi-Strauss also taught) between 1935 and 1946, became President of the Association of Brazilian Geographers, played a significant role in the Brazilian *Conselho Nacional de Geografia* (national survey), which was a multidisciplinary venture, and brought the work of American geographer Isaiah Bowman to the attention of geographers there. And from 1945, when he headed the geography section at the Institut Français d'Afrique Noire (IFAN) in Dakar (established 1940), "Richard-Molard gathered together a dynamic group of geographers, financed their research and organised their fieldwork [across tropical French Africa], generally in cooperation with geologists, pedologists, botanists and ethnographers."[153] Monbeig and Richard-Molard created their own unique networks, the former on the back of the Brazilian state's desire to internationalise its educational system, particularly by attracting French scholars, and the latter in connection with France's attempt to reboot its overseas research.

The positivistic faith of scholars such as Gourou, Monbeig and Richard-Molard did not banish exoticism and primitivism: in some respects quite the reverse. They were all highly literate scholars, and as Sam Rohdie notes, like the artists and writers to whom they frequently turned for inspiration, they found positivism a means of dealing with "the inadequacy of words when faced with the immensity of what could be seen. "The Romantic challenge to Neo-Classicism and its idealisations ... was in part posed by a new realism and attention to new social, ethnic and geographical realities for which travel and Orienalism [and tropicality] were important stimulations. This was not without scientific echoes in a new materialist and Positivist science ... with its stress on direct observation and visualisation and within which the subject of geography assumed pride of place."[154] In other words, it was meaningful for there to be both a poetics and a politics to Gourou's writing about the Tonkin peasants, and for him to invoke the two on the same page; or to Monbeig's poetic evocation of the Amazon rainforest and political broadsides against the sin of its deforestation. Richard-Molard imbibed the work of French ethnographers and anti-colonial thinkers and went on to develop a new conception of black African culture – one with a distinct geographical identity.[155]

Gourou's *Les paysans* was lauded in this vein, as a book with a politics and poetics that "rightly deprecates any sort of 'progress' which would give the Tongkinese peasant the 'blessings' of modern machinery to ameliorate their present laborious lives," as the reviewer for *The Geographical Journal* put it; and "robbing them of their work, it would leave them (Gourou says) 'nothing to do but die discretely.'"[156] Another reviewer reflected that through his "keen feeling for the beauty of the delta," Gourou had concomitantly and rightly asserted that "the ultimate caution should be used in introducing elements of change lest the delicate balance of their ancient economy be upset."[157]

However, Gourou's work became increasingly cast in a dead-pan die, and not without a wider significance on other counts. For during the interwar years geography's factual orientation became central to the nascent academic and policy field of international relations because the discipline's concern with questions of climate, terrain, land use, population and borders was deemed pragmatic.[158] Gourou described his inaugural lecture as the Collège de France as an "enumerative summary" of the key differences between "tropical experience" and "European experience," and one finds in *Les pays tropicaux* few of the aesthetic flourishes one encounters in *Les paysans*.[159]

Our argument here has been that metropolitan-colonial networks and systems of expertise were pivotal to the creation of this outlook and Gourou's concern with questions of population. We shall see in the next chapter that wartime experience played its part too. But we shall finish this chapter by saying a little more about how this interwar 'networking' of the tropics presaged another a key element of Gourou's tropicality: comparative analysis.

The imperial power of comparison

By the end of 1937 Gourou had positioned himself as a voice of academic authority in what he saw as the "stormy debates taking place around him," and he found in the lives and work of his friends Lévy and Rivet humanist values to which he clung tightly too (and others have since associated with him): an innate curiosity about the world; clear judgement and single-mindedness; a combination of scientific rigour and aesthetic finesse; warmth and generosity; and, overall the aspiration to be "not an Olympian scholar, but a man of research, of action and of heart."[160] These were not just personal traits. They were also social traits emanating from a particular era and political conjuncture, and were integral to the make-up and standing of the network of which Gourou was part. Lévy, we might recall, described Gourou as a "man of vast culture," and for whom "the landscape is in charge."[161] During the late 1930s and into World War II the nestling of his tropicality within the agendas and conceits of colonial reform developed in two directions. First, Gourou developed new comparative lines of inquiry, and second, he thought more deeply about the relationship between scholarship and politics.

His growing comparative interest in the Far East culminated in two general geographical accounts *L'utilisation du sol*, which we have already encountered, and *La terre et l'homme en Extrême-Orient* (1940). Both books might be seen as a natural progression from his regional monograph on the Tonkin Delta, and as ventures expedited by the recognition that it was difficult for him to return to Indochina to undertake further fieldwork. French geography, of course, had its own strong traditions of global as well as regional study, but as the above discussion also suggests, comparative inquiry was embedded in the research orientations of the IPR, CEPE and Guernut Commission, and in the intellectual outlooks of Lévy, Rivet and the humanist-reformist assemblage they fostered.[162] Gourou's lengthy (466-page) *L'utilisation du sol* is a detailed, technical study based on a synthesis of land, population and labour statistics pertaining to Indochina's three territories and Vietnam's three regions.[163] Gourou was asked by the CEPE and IPR to focus on indigenous Vietnamese land use and traditional economic activity, while Robequain was asked to produce a companion volume on the effects of French colonisation and questions of economic development.

L'utilisation du sol extends and reinforces the points Gourou made in his Guernut report. A 1:2,500,000 scale map of villages the Tonkin Delta (a level of detail unsurpassed in overseas fieldwork at this time), along with four 1:500,000 maps for the other Vietnamese provinces, were photographic reductions of maps produced by the Service Géographique de l'Indochine and extended the 'birds-eye view' that had allowed Gourou to envision the Tonkin Delta as an intricate and beautiful patchwork of villages, fields and dikes (see Chapter 3). But the task he had been set also encouraged what became a more enduring intellectual concern: how the problems facing Indochina (and other tropical regions) had neither colonial nor physical origins, but stemmed from the uneven geographical distribution of population and land use, and the millennial struggle of peasants seeking to shape 'vegetal civilisations' in begrudging lands. This is what reviewers saw in Gourou's *La terre et l'homme*, and the most astute of them knew that his geography could be used in an apologist fashion to defuse a charged political situation by revealing the age-old roots and texture of many Far East problems.[164]

La terre et l'homme paints a vast canvas and corrects what the British geographer Dudley Stamp saw as a deficiency in *L'utilisation du sol*: a reluctance to step beyond "linguistic and political boundaries" and take advantage of "excellent scientific work" done on other parts of Asia.[165] Gourou draws Japan, Korea, China and Indochina into a unitary analytical frame and seeks to envision a peasant world devoid of political and imperial turbulence. Concerned with "the peasant framework" of this vast region, he writes:

> The traditional peasantries of the Far East existed in a solid framework comprising the family, the clan, the village and the state.... The dominant characteristic of the human geography of the Far East up to 1940 was the existence of densely packed peasantries strewn over large spaces. This goes back to the dawn of written history, and it cannot be explained without referring to the efficiency of the civilisation and its 'peopling' quality.

Gourou underscores the importance of a "hierarchy of political institutions" (from village level upwards) in making this framework durable, and argues that the Far East is a climatic and cultural unit, and that its landscape is "like a parchment that still bears the scratches of its ancient text."[166]

We can find in this comparative vision the inklings of Gourou's zonal tropicality. Gourou sought order and harmony in the peasant landscapes of the Far East, and as Lévy implied, by locating this region in 'monsoon Asia' rather than in 'the Orient' he implicitly challenged Orientalist constructions, which had placed China in the classical Orient and Indochina in 'Southeast Asia' (the *Extrême-Orient* in French).[167] Gourou started to argue that the tropical world had its own 'peculiar' and 'surprising' geography. These adjectives are made to do a good deal of work in *Les pays tropicaux*, but the problematic of difference they conjure with also runs through *La terre et l'homme:*

> The fragmentation, the minute landholdings, the absence of pastoralism, and the dominance of wet rice cultivation all result in a rural landscape very different from those of Western Europe. The latter are also peasant landscapes, but of a quite different nature. In the Far East one looks in vain for the Normandy grasslands, the hedges and trees of the bocage, the vineyards, and the open fields with the strip of cultivation that recalls the medieval three-field system. The European, surprised by a new world, will experience many disappointments if he expects to find the landscapes of his childhood. . . . To appreciate the beauty of the Far Eastern rural landscape, one must forget the West completely and accept the different standards of this corner of Asia.[168]

Forget the West, but with the West as the yardstick of comparison: this was the double-standard of Gourou's tropicality as it emanated from this book. Imperialism, industrialisation, urbanisation and revolution are effaced from the Far Eastern scene Gourou paints, and its 'different standards' are described through direct comparison with idealised, and by implication 'normal,' European landscapes.[169] It was no longer the differences between the French and Vietnamese that were significant; increasingly, it was the differences between a temperate world and a tropical one. And in these passages Gourou's temperate references points – the bocage, the vineyards, a medieval world – are not represented as dreary and

168 *Networking the tropics*

degraded, as the French peasants and *pays* evoked in *Les paysans* are made to appear, but as ancient and picturesque (see Chapter 3).

But let us think some more about Gourou's comparative orientation. Raewyn Connell has argued that the zest for comparative research in the social sciences has strong colonial roots. The Western research institutions and networks we have considered, she would say, "rested on a one-way flow of information" and "displaced imperial power over the colonized into... [an] abstract space of difference" that naturalised and sometimes concealed colonialism and Western influence.[170] While even some of Gourou's staunchest critics see his use of comparison as one of the strengths of his oeuvre, his work made the sovereign gaze of the Western academic upon far-off lands paramount and side-lined the role that inter-Asian networks and actors played in the production of knowledge. The metropolitan networks through which Gourou, Godart and their expert advice travelled eschewed Vietnamese networks, such as the Indochinese Congress and the EFEO, and enabled colonial specifics – the problem of overpopulation in Tonkin – to be placed in a wider, comparative space of French and Western governmental reason.

While the Guernut Commission did not ultimately deliver the fully fledged picture of the needs and aspirations of France's diverse and diffuse indigenous colonial population that it promised, it was modelled on a style of inquiry that, as scholars such as Bashford, Rosenberg and Tilley (already cited) and Jennifer Robinson argue, facilitated the transfer of ideas, lessons and experiments between different colonial regions.[171] It is partly on account of the comparative association of social decay with population decline that Richard Overy has dubbed the interwar years Britain's "morbid years."[172] Robinson observes that it was in the 1930s that we find the first links in "a chain of expert advisory committees" connecting Britain, France, Germany, Holland and the United States that expedited a "combination of pragmatic technical and historically specific colonial dimensions to comparative research projects" aimed at supporting and reforming colonial empires.[173] She argues that an underlying assumption in this comparative research vision, which turned Africa, particularly, into an object of expert knowledge and site of state intervention, was the teleology of development, with the West positioned as the standard-bearer of modernity and progress.

Already in *La terre et l'homme*, and more strongly after World War II, Gourou's work brushed against the grain of this teleology because he was sceptical about the role that Western science could play in transforming and managing Far Eastern landscapes. Demangeon was sceptical too. In a lengthy review of Australian geographer Grenfell Price's 1939 *White Settlers in the Tropics*, which was one of the most geographically comprehensive surveys of this "problem," as he saw it (albeit a study that ignored Indochina), Demangeon suggested that it was not easy to find progress in the tropics because of a host of problems – climate, soil, disease, subsistence, and "racial mixing and policy" – facing white settlers and Western corporations.[174] The year before, in a short article in the French mass circulation magazine *Le Monde Colonial Illustré*, where (to our knowledge) he used the expression "les pays tropicaux," Gourou questioned two sentiments that he thought applied not just to Germany's distorted view of the tropics but also to a growing comparative literature (and that were given a full airing by Price): that environment had a determining influence on race; and that many of the problems of white settlement in the tropics would be abated, if not eradicated, by science and technology.[175]

However, Césaire was one of the first to realise that Gourou's comparisons belonged to another dimension of Western dominance: the manipulation of meaning, an intellectual imperialism with Western experts ascribing traits of difference to specific peoples, environments and natures. Ironically, it was not until Clifford Geertz's famous pair of studies of Indonesia, *Agricultural Involution* and *Peddlers and Princes*, were published in 1963, and with scholarly scrutiny of the Western biases in development thinking then in the air, that Gourou's *La terre et l'homme* started to be mined for its comparative insights into what Geertz termed "involution" – the possibility of agrarian change and growth through an intensification of traditional practices.[176]

Gourou's response to what Demangeon had dubbed "the decline of Europe" after World War I was different to that of other interwar European and American geographers.[177] While Demangeon and others fuelled debate within and beyond French geography about the dangers of nationalism and empire, the need for an economically integrated federal Europe, and about how Europe should be conceived as a space of interaction and connection rather than division and isolation between peoples and regions, Gourou found in the Far East autonomous "plant-based civilisations" unlike any in Europe, and argued that to fathom them one needed an "an intimate knowledge of general problems."[178] For him this meant eschewing the European (post-Enlightenment) idea of civilisation as progress and adopting a comparative perspective on human-environment interactions that rejected physical determinism and was concerned with the *longue durée*. Elements of this perspective were Eurocentric: it was Gourou doing the representing. But it might also be suggested, as we have in part here, that an appeal to the 'different standards' of the Far East and tropical world at least promised a less insidious geography of difference – even if, ultimately, it did not deliver it.

In this chapter we have also been concerned with how to relate Gourou to both the history of geography and the internationalism of the interwar years. While it would be misleading to suggest that his work has to this point been treated largely as if it can be separated from the historical and political circumstances in which he worked, it is fair to say that many commentaries have a disciplinary hue. It is Gourou's contribution to a field of study, and the development of his own body of ideas, that has generally been of concern. While context is not expunged from such accounts, it is often reduced to background noise. We have sought to show that more might be done with the *interplay* of text and context. Gourou belonged to an interdisciplinary and international intellectual culture, and many of the questions he took up, and to which he gave a distinctive inflection, had a broad political resonance in interwar France and Europe.

At the end of her study of interwar American geography, Susan Schulten concludes that "geographical knowledge – in its resistance to change and its effort to make sense of the world – has operated conservatively . . . and conservatively

170 *Networking the tropics*

in its service to and defense of national power."[179] While we can see elements of this judgement at work in interwar French geography (particularly through its adherence to Vidalian ways and the patronage networks that shored up this tradition), there was also much experimentation. We see the increasingly comparative bent in Gourou's work as significant in this respect. While French geographers operated at a global as well as regional scale, few worked in the explicitly zonal manner that Gourou came to popularise. As we now seek to show, the rise of Nazi Germany, the occupation of France, and Gourou's experience of war are vital (if to date missing) links in understanding of the evolution of his tropical geography and post-war tropicality.

Notes

1 Pierre Gourou, "Qu'est-ce que le monde tropical?" *Annales: Économies, Sociétés, Civilisations* 4, no. 2 (1949): 140–148, at 140.
2 Gourou, "Qu'est-ce que le monde tropical?" 141.
3 Gourou, "Qu'est-ce que le monde tropical?" 145.
4 See, for example, Anne Booth, "Measuring living standards in different colonial systems: Some evidence from South East Asia, 1900–1942," *Modern Asian Studies* 46, no. 5 (2012): 1145–1181; Wendy Way, *A New Idea Each Day: How Food and Agriculture Came Together in One International Organisation* (Canberra: ANU Press, 2013); Priscilla Roberts, "The Institute of Pacific Relations: Pan-Pacific and pan-Asian visions of international order," *International Politics* (2017), https://doi.org/10.1057/s41311-017-0108-y.
5 Corey Ross, *Ecology and Power in the Age of Empire: Europe and the Transformation of the Tropical World* (Oxford: Oxford University Press, 2017), 380–390. Rodolphe de Koninck, "Southeast Asian agriculture post-1960," in Chia Lin Sien ed., *Southeast Asia Transformed: A Geography of Change* (Singapore: Institute of Southeast Asian Studies, 2003), 191–230.
6 Cited in John A. Simon and Michael W. Miller, "Development assistance: Rationale and applications," in Derek S. Reveron, Nikolas K. Gvosdev and John A. Cloud eds., *The Oxford Handbook of U.S. National Security* (Oxford: Oxford University Press, 2018), Ch. 11. For the post-war era see, for example, Benn Steil, *The Battle of Bretton Woods: John Maynard Keynes, Harry Dexter White, and the Making of a New World Order* (Princeton: Princeton University Press, 2013).
7 See Nick Cullather, *The Hungry World: America's Cold War Battle Against Poverty in Asia* (Cambridge, MA: Harvard University Press, 2010).
8 United Nations, *Charter of the United Nations and Statute of the International Court of Justice* (San Francisco: United Nations, 1945), 2.
9 General Conference, UNESCO House, Paris, 20 November–10 December 1946 (Paris: UNESCO, 1947), 130.
10 Pierre Gourou, "Un programme de recherches et d'expériences en vue du relèvement de la production agricole et des niveaux de vie dans les régions tropicales," *Tiers-Monde* 1, no. 3 (1960): 373–385.
11 See Clark A. Miller, "'An effective instrument of peace': Scientific cooperation as an instrument of U.S. Foreign policy, 1938–1950," *Osiris* 21, no. 2 (2006): 133–160.
12 See A. D. Roberts, "The British empire in tropical Africa: A review of the literature to the 1960s," in Robin Winks ed., *The Oxford History of the British Empire* volume V: *Historiography* (Oxford: Oxford University Press, 1999), 463–485.
13 Robert D. Fletcher, "The general circulation of the tropical and equatorial atmosphere," *Journal of Meteorology* 2, no. 3 (1945): 167–174, at 167.

14 Daniel Gorman, *International Cooperation in the Early Twentieth Century* (London: Bloomsbury, 2017), 194.
15 Roberta Bivins, "Coming 'home' to (post)colonial medicine: Treating tropical bodies in post-war Britain," *Social History of Medicine* 26, no. 1 (2013): 1–20, at 2.
16 Michael Riemens, "International academic cooperation on international relations in the interwar period: The International Studies Conference," *Review of International Studies* 37, no. 2 (2011): 911–928.
17 Tamson Pietsch, *Empire of Scholars: Universities, Networks and the British Academic World, 1850–1939* (Manchester: Manchester University Press, 2013).
18 Sigmund Freud, *The Standard Edition of the Complete Psychological Works of Sigmund Freud*, Volume 21 *(1927–1931)* James E. Strachey ed. and trans. (London: Hogarth Press, 1981), 118–119, 208.
19 Michel Foucault, *The History of Sexuality: An Introduction,* trans. James Hurley (London: Allen Lane, 1979), 137–138.
20 Freud, *Standard Edition*, 21, 115; Foucault, *History of Sexuality*, 137–138.
21 Pedro Paulo Gomes Pereira, "In and around life: Biopolitics in the tropics," *Vibrant: Virtual Brazilian Anthropology* 10, no. 2 (2013): 13–37, at 18.
22 Timothy Mitchell, *Rule of Experts: Egypt, Techno-politics, Modernity* (Berkeley: University of California Press, 2002), 124.
23 See William Duiker, "The red soviets of Nghe-Tinh: An early communist rebellion in Vietnam," *Journal of Southeast Asian Studies* 4, no. 2 (1973): 186–198. These were the first serious rebellions against French rule since 1917. The French used 'Nghe-Tinh' as a shorthand for the provinces of Nghe An and Ha Tinh. In 1930, peasant villages in the Tonkin Delta were bombed by French aeroplanes and thousands of nationalists were imprisoned or executed, crippling the Nationalist Party and allowing the ICP to gain political ground in the north. See Ngo Vin Long, *Before the Revolution: The Vietnamese Peasants Under the French* (Cambridge, MA: MIT Press, 1973), 4–41.
24 Martin J. Murray, "The development of capitalism and the making of the working class in colonial Indochina, 1870–1940," in B. Munslow and H. Finch eds., *Proletarianisation in the Third World* (London: Croom Helm, 1985), 216–233, at 227; Truong Chinh and Vo Nguyen Giáp, *The Peasant Question, 1937–1938* (Ithaca, NY: Cornell University Press, 1974), 29.
25 Auguste Tholance, "La situation politique et économique du Tonkin," *Revue du Pacifique* 2, no. 11 (1931): 650–660, at 653–654. His successor reiterated roughly the same message: Pierre-André Pagès, "La situation politique et financière du Tonkin," *La Revue du Pacifique* 1, no. 1 (1933): 27–40.
26 Martin Thomas, "French empire elites and the politics of economic obligation in the interwar years," *The Historical Journal* 52, no. 4 (2009): 989–1016. Pierre Brocheux and Daniel Hémery, *Indochina: An Ambiguous Colonization, 1858–1954* trans. Ly Lan Dill-Klein (Berkeley: University of California Press, 2009), 319. "When indeed one looks about one in the mid-1930s for the major figures of a decade or so later, not only does one see Mao cooped up in Yennan and Ho in exile in Moscow, but Nehru in Derha Dun prison and Sukarno banished to flores." D.A. Low, *Eclipse of Empire* (Cambridge: Cambridge University Press, 1991), 29.
27 Pierre Gourou, *Les paysans du delta tonkinois. Etude de géographie humaine* (Paris: l'École française d'Extrême-Orient, Les Editions d'Art et d'Histoire, 1936), 378, 478–479.
28 Pierre Gourou, *Terres de bonne espérance, le monde tropical* (Paris: Plon, 1982), 12.
29 Gourou, *Les paysans*, 378.
30 Gourou, *Les paysans*, 479.
31 John Kleinen, "The village as pretext: Ethnographic praxis and the colonial state in Vietnam," in Jan Breman, A. Saith and P. Kloos eds., *The Village in Asia Revisited* (New Delhi and Oxford: Oxford University Press, 1997), 353–395, at 358.
32 For example, Louis Roubaud, *Viet-Nam: La tragédie indochinoise* (Paris: Valois, 1931).

33 Irene Nørlund, "Rice and the colonial lobby: The economic crisis in French Indochina in the 1920s and 1930s," in Peter Boomgaard and Ian Brown eds., *Weathering the Storm: The Economies of Southeast Asia in the 1930s Depression* (Singapore: Institute of Southeast Asian Studies, 2000), 198–228.
34 Nørlund, "Rice and the colonial lobby," 202.
35 Nørlund, "Rice and the colonial lobby," 203.
36 Paul Bernard, *Nouveaux aspects du problème économique indochinois* (Paris: Fernand Sorlot, 1937), 113–120.
37 Bernard, cited in Andrew Hardy, *Red Hills: Migrants and the State in the Highlands of Vietnam* (Copenhagen: NIAS Press, 2003), 81.
38 Hardy, *Red Hills*, 81.
39 Mitchell, *Rule of Experts*, 124.
40 Gourou, *Les paysans*, 576. Mitchell's study ranges across accountancy, anthropology, economics, engineering, politics and surveying as well as geography. He shows how Gourou's 1936 study influenced the pioneer of Egyptian peasant studies, Henry Ayrout, and underpinned French and Vietnamese scholarship on peasant society. The latter point was reaffirmed to us in interviews with French and Vietnamese scholars. Sylvia Fanchette, Interview with the authors; Olivier Tessier, Interview with the authors; Dào Thê Tuân, Interview with the authors. As we discuss in the last chapter, a Vietnamese translation of Gourou's work appeared in 2003.
41 Joseph Buttinger, *Vietnam: A Dragon Embattled*, 2 vols. (London: Pall Mall Press, 1967), I, 172–222; Paul Isoart, *Le phénomène national vietnamien: De l'indépendance unitaire à l'indépendance fractionée* (Paris: Librairie Général de Droit et de Jurisprudence, 1961), 178–184; 261–272.
42 Einstein, cited in Thomas M. Skovholt, Matthew Hanson, Len Jennings and Tabitha Grier, "A brief history of expertise," in Thomas M. Skovholt and Len Jennings eds., *Master Therapists: Exploring Expertise in Therapy and Counseling*, 10th Anniversary Edition (Oxford: Oxford University Press, 2016), 1–15, at 5.
43 M.T.H. Chi, R. Glaser and M.J. Farr eds., *The Nature of Expertise* (Hillsdale, NJ: Lawrence Erlbaum Associates, 1988).
44 Brian Balogh, *Chain Reaction: Expert Debate and Public Participation in American Commercial Nuclear Power, 1945–1975* (Cambridge: Cambridge University Press, 1991), 2
45 Anon., "Review of *Les paysans du delta tonkinois*," *Les Nouvelles littéraires* (1937): 2; H.W.P., "Review – *Les paysans du delta tonkinois*," *Geographical Journal* 89, no. 5 (1937): 665–667; L. Ravéreau, "Review, *Les paysans du delta tonkinois*," *Review économique française* 24 November in 1937, clippings in AULB: Fonds Pierre Gourou PP153.
46 A minor thesis was also required. Gourou wrote his on Central Vietnamese houses, and the illustrations for both works (including many house plans) were produced by the artist Joseph Inguimberty, who moved to Hanoi in 1925, having been hired by Victor Tardieu to head the painting department at the newly established École des Beaux-Arts de l'Indochine (see Chapter 2). Attainment of the state doctorate did not automatically qualify one to be elected to a chair in a French state-run university.
47 Charles Robequain, "Réception de M. Pierre Gourou à l'Académie des Sciences Coloniales," *Compte-rendu mensuel des séances de l'Académie des Sciences Coloniales* XV (1955): 66.
48 Paul Rivet, Preface to *Les pays tropicaux*, cited in ACDF: Lucien Febvre, Assemblée des professeurs du 16 février 1947.
49 Hugh Clout, "Professorial patronage and the formation of French geographical knowledge: A bio-bibliographical exploration of one hundred non-metropolitan regional monographs, 1893–1969," *Cybergeo: European Journal of Geography* 549 (2011), http://cybergeo.revues.org/24203?lang=en.

50 ANOM: FM EE/11/4453/8/GOU – 25 July 1930.
51 Dany Bréelle, Interview with Pierre Gourou, Bruxelles 29 August 1995, Appendix H in Dany Bréelle, "The regional discourse of French geography: The theses of Charles Robequain and Pierre Gourou in the context of Indochina," Unpublished PhD dissertation, Geography, Flinders University, 2003. Clout adds that in the interwar years the professor-student patronage relationship was often loose, and especially for students undertaking overseas fieldwork. Gourou was in Paris in 1931 to attend the Paris Exposition in the Bois de Vincennes, for which he had written a work of popular geography, *Le Tonkin*, to accompany the Indochina exhibit. Gourou's patron, Demangeon, was heavily involved in organising the human geography activities at the International Geographical Congress meeting held at the Sorbonne that year. See Hugh Clout, "Albert Demangeon, 1872–1940: Pioneer of La Géographie Humaine," *Scottish Geographical Journal* 119, no. 1 (2003): 1–24.
52 Albert Demangeon, "Les paysans du delta tonkinois," *Annales de Géographie* 46, no. 262 (1937): 404–407, at 404.
53 See Martin Thomas, *The French Empire Between the Wars: Imperialism, Politics and Society* (Manchester: Manchester University Press, 2005), 277–288. The package included a 40-hour working week, minimum wage legislation, paid holidays, improved sickness and unemployment benefits, the legal right to strike, and new collective bargaining rights. Many of these measures and concessions were articulated in the Matignon Accords signed on 7 June 1936. While some of these accords were subsequently diluted or rescinded, they are still seen as a pillar of social rights legislation in France.
54 Sud Chonchirdsin, "The Indochinese Congress (May 1936–March 1937): False hope of Vietnamese nationalists," *Journal of Southeast Asian Studies* 30, no. 2 (1999): 338–346. By the 1930s, French Indochina was comprised of Cambodia, Laos and the Vietnamese territories of Tonkin (North), Annam (Central) and Cochinchina (South).
55 Gary Wilder, *The French Imperial Nation-State: Negritude and Colonial Humanism Between the Two World Wars* (Chicago: University of Chicago Press, 2005), 61–65; David Fisher, *Romain Rolland and the Politics of Intellectual Engagement* (Berkeley: University of California Press, 1988), 236.
56 H. K. Khánh, *Vietnamese Communism 1925–1945* (Ithaca, NY: Cornell University Press, 1982), 209. Moutet had formerly been a radical in the League of the Rights of Man, and the legal sponsor of the Vietnamese anti-colonial politician Phan Chu Trinh. Another Vietnamese nationalist, Phan Van Truong, dedicated his 1922 *Essai sur le code Gia Long* (on French colonial penal codes) to Moutet, calling him an "*indigénophilie*." Truong Buu Lam, *Colonialism Experienced: Vietnamese Writings on Colonialism, 1900–1931* (Michigan: University of Michigan Press, 2000), 37.
57 Moutet cited in Panivong Norindr, "The Popular Front's colonial policies in Indochina: Reassessing the Popular Front's '*colonisation altruiste*'," in Tony Chafer and Amanda Sackur eds., *French Colonial Empire and the Popular Front: Hope and Disillusion* (London: St Martin's Press, 1999), 230–248, at 232.
58 See Philippe M.F. Peycam, *The Birth of Vietnamese Political Journalism* (New York: Columbia University Press, 2012).
59 Brocheux and Hémery, *Indochina*, 235, 328–335.
60 Geoffrey C. Gunn, *Rice Wars in Colonial Vietnam: The Great Famine and the Viet Minh Road to Power* (New York: Rowan & Littlefield, 2014), 133.
61 See William Duiker, *Ho Chi Minh: A Life* (New York: Theia Books, 2000), 223–229.
62 Gunn, *Rice Wars*, 115.
63 Gunn, *Rice Wars*, 129.
64 Tony Chafer and Amanda Sackur, "Introduction," in Tony Chafer and Amanda Sackur eds., *French Colonial Empire and the Popular Front: Hope and Disillusion* (London: St Martin's Press, 1999), 1–32, at 27.

65 Jacques Weulerrse, *Noirs et blancs: A travers l'Afrique nouvelle: de Dakar au Cap* (Paris: CTHS, 1995), viii.
66 Gunn, *Rice Wars*, 134.
67 See, especially, Dipesh Chakarabarty, *Provincialising Europe: Postcolonial Thought and Historical Difference* (Princeton, NJ: Princeton University Press, 2000).
68 Thomas, *The French Empire Between the Wars*, 101–105. Norindr also alights on the paradox that in Indochina the French revenue raised to fund colonial reforms was to be based on a spurious French levy on opium and gambling.
69 Fisher, *Romain Rolland and the Politics of Intellectual Engagement*, 236.
70 Cited in Thomas, *The French Empire Between the Wars*, 287.
71 Walter Putnam, "Writing the wrongs of French Colonial Africa: *Voyage au Congo* and *Le Retour du Tchad*," in Tom Conner ed., *André Gide's Politics: Rebellion and Ambivalence* (London: Palgrave Macmillan, 2000), 89–110, at 89.
72 It was not just the French, of course, who grappled with the colonial question of famine, either out of a genuine humanitarian concern or with a more cynical interest in how this 'problem' tarnished the imperial reputation of the mother country.
73 Charles Robequain, "Les Paysans du Delta tonkinois," *Bulletin de l'Ecole française d'Extrême-Orient* 36 (1936): 491–497, at 493–494.
74 Gourou, *Les paysans*, 578.
75 See Van Nguyen-Marshall, *In Search of Moral Authority: The Discourse on Poverty, Poor Relief and Charity in French Colonial Vietnam* (Vancouver: UBC Press, 2008), 122–124.
76 Charles Robequain, *The Economic Development of French Indo-China* trans. I. Ward (New York: Oxford University Press, 1944), 240.
77 Although the account of his journey published in 1994 has photographs of him being greeted with due colonial ceremony wherever he went.
78 ANOM: FM/GUERNUT Carton 2 (Reports and correspondence); FM EE/11/3881/11/GOU (Gourou's personal dossier – 1); ANOM: FM EE/4453/8/GOU (Gourou personal dossier – 2)
79 Pierre Gourou to Marius Moutet 6 juillet 1936, cited in Trinh Van Thao, *L'école française en Indochine* (Paris: Karthala, 1995), 253.
80 Alice Conklin, *In the Museum of Man: Race, Anthropology, and Empire in France, 1859–1950* (Ithaca, NY and London: Cornell University Press, 2013), 222. Also see Nélia Dias, "Rivet's mission in colonial Indochina (1931–1932) or the failure to create an ethnographic museum," *History and Anthropology* 25, no. 2 (2014): 189–207; and Nélia Dias, "From French Indochina to Paris and back again: The circulation of objects, people, and information, 1900–1932," *Museum & Society* 13, no. 1 (2015): 7–21.
81 Pierre Gourou, "In memoriam – Paul Rivet," *Annales: Économies, Sociétés, Civilisations* 14, no. 1 (1959): 200–204, at 200.
82 Albert Demangeon, "Le centre d'études de politique étrangère," *Annales de Géographie* 45, no. 258 (1936): 646–648.
83 Pierre Gourou, "Le souvenir de Roger Lévy," *Politique étrangère* 43, no. 3 (1978): 347–349.
84 Wilder, *The French Imperial Nation-State*, 59–68.
85 Christine Laurière, "Paul Rivet (1876–1958), Le savant et le politique," *Nuevo Mundo* (2007), http://nuevomundo.revues.org/3365.
86 Not least through Lévy's own work. He collaborated with French and North American historians Guy Lacam and Andrew Roth to produce *French Interests and Policies in the Far East* for the IPR in 1941. Much of the volume was about Japan, and it was published a month before the Japanese attack on the American military base at Pearl Harbour.

87 Roberts, "The Institute of Pacific Relations," n.p.; J. B. Condcliffe ed., *Problems of the Pacific, 1929: Proceedings of the Third Conference of the Institute of Pacific Relations, Nara and Kyoto, Japan, October 23 to November 9, 1929*, vol. 3. (Chicago: University of Chicago Press, 1930); Jacques Vernant, "Roger Lévy," *Politique étrangère* 43, no. (1978): 351–353.

88 The Centre (which changed its name to the *Institut français de relations internationales* in 1979) was similar to the Royal Institute of International Affairs based at Chatham House, London (1919–present), and the New York-based Council on Foreign Relations (1921–present). All of them were non-profit non-governmental organisations dealing with questions of national and international significance. *Politique Étrangère* (issued every two months) was similar to the Royal Institute's periodical *International Affairs* and the Council's *Foreign Affairs*. All three publications remain leading international relations periodicals to this day. The following institutions were involved in the creation and functioning of the Centre: l'Université de Paris; l'École libre des Sciences politiques; la Commission française de coordination des Hautes Études internationales; la Bibliothèque de Documentation internationale contemporaine; le Centre de Documentation sociale de l'École Normale Supérieure le Groupe d'études diplomatiques; Comité d'Études des problèmes du Pacifique; and the Comité d'Études de l'Europe centrale.

89 Tomoko Akami, *Institutionalising the Pacific: The United States, Japan, and the Institute of Pacific Relations in War and Peace, 1919–1945* (London and New York: Routledge, 2002).

90 The Hot Springs and Lucknow meetings were also attended by Charles Robequain.

91 André Burguière, *The Annales School: An Intellectual History* trans. J. Todd (Ithaca, NY: Cornell University Press, 2009), 29–35.

92 Richard Little, "Historiography and international relations," *Review of International Studies* 25, no. 2 (1999): 291–299.

93 See Jean-Baptiste Arrault, "Penser à l'échelle du Monde: Histoire conceptuelle de la mondialisation en géographie (fin du XIXe siècle/entre-deux-guerres)," Géographie. Université Panthéon-Sorbonne – Paris I, 2007, https://tel.archives-ouvertes.fr/tel-00261467/document.

94 See Ronald Hubscher, "Historiens, géographes et paysans," *Ruralia* (1999), online http://ruralia.revues.org/87.

95 Alison Bashford, "Nation, empire, globe: The spaces of population debate in the interwar years," *Comparative Studies in Society and History* 49, no. 1 (2007): 170–201, at 170.

96 See David Arnold, *Colonizing the Body: State Medicine and Epidemic Disease in Nineteenth-century India* (Berkeley: University of California Press, 1993); Ann Laura Stoler, *Race and the Education of Desire: Foucault's History of Sexuality and the Colonial Order of Things* (Durham, NC: Duke University Press, 1995); and Stephen Legg, "Foucault's population geographies: Classifications, biopolitics and governmental spaces," *Population, Space and Place* 11, no. 3 (2005): 137–156.

97 Bashford, "Nation, empire, globe," 172.

98 Alison Bashford, *Global Population: History, Geopolitics, and Life on Earth* (New York: Columbia University Press, 2014), 147.

99 Bashford, *Global Population*, 147–148.

100 Nguyen-Marshall, *In Search of Moral Authority*, 55.

101 ANOM: FM/GUERNUT carton 2; ANOM: FM EE/11/3881/11/GOU; Gourou, *Les paysans*, 107, 577.

102 ANOM: FM EE/11/3881/11/GOU; Pierre Gourou, *L'utilisation du sol en Indochine française* (New York and Paris: Institute of Pacific Relations; Centre d'études de politique étrangère, 1940), 431.

176 *Networking the tropics*

103 ANOM: FM EE/11/3881/11/GOU; ANOM: FM EE/11/4453/8/GOU.
104 Sokhieng Au, "Cutting the flesh: Surgery, autopsy and cannibalism in the Belgian Congo," *Medical History* 61, no. 2 (2017): 295–313, at 296.
105 MacMahon Ball, "Nationalism and communism in Vietnam," *Far Eastern Survey* 21, no. 3 (1952): 21–27, at 22.
106 Ball, "Nationalism and communism in Vietnam," 22.
107 See Arrault, "Penser à l'échelle du Monde," 393 and *passim*.
108 See Paul Marres, "Le problème du surpeuplement dans l'Indochine française et en Extrême-Orient, d'après Pierre Gourou et Charles Robequain," *Annales de Géographie* 51, no. 285 (1942): 52–57.
109 Robequain, *Economic Development of Indochina*, 344.
110 Robequain, *Economic Development of Indochina*, 344–345.
111 Robequain, *Economic Development of Indochina*, 85; also see also see Nguyen-Marshall, *In Search of Moral Authority*, 52–58.
112 UBCA: Institute of Pacific Relations Fonds, 86–11: Pierre Gourou, "The economic problems of monsoon Asia and the example of Japan," 1950, 2–3.
113 ANOM: FM EE/11/3881/11/GOU; ANOM: FM EE/11/4453/8/GOU.
114 ANOM: FM EE/11/3881/11/GOU.
115 Justin Godart, *Rapport de mission en Indochine, 1er janvier – 14 mars 1937* (Paris: L'Harmattan, 1994), 172–179.
116 Godart, *Rapport de mission en Indochine*, 107.
117 Robequain, *Economic Development of Indochina*, 168, 240.
118 Paul Bernard, *Nouveaux aspects du problème économique indochinois* (Paris: Fernand Sorlot, 1937), 113–120, 174–176. On Bouvier, see Andrew Hardy, *Red Hills: Migrants and the State in the Highlands of Vietnam* (Singapore: Institute of Southeast Asian Studies, 2003), 76–82.
119 René Dumont, *La Culture du riz dans le delta tonkinois* (Hanoi: Société d'éditions géographiques, maritimes et coloniales, 1935), 394.
120 Grégoire Khérian, "Le Probleme démographique en Indochine," *Revue Indochinoise Juridique et Economique* 1–2 (1937): 7–28. And see Hardy, *Red Hills*, 76–82.
121 Albert Demangeon, "Les paysans du delta tonkinois d'après P. Gourou," *Annales de Géographie* 46, no. 262 (1937): 404–407; Hardy, *Red Hills*, 76.
122 Charles Robequain, *L'évolution économique de l'Indochine française* (Paris: Hartmann/IPR, 1939), 82–86.
123 Hardy, *Red Hills*, 78–84.
124 Martin Thomas, *Violence and Colonial Order: Police, Workers and Protest in the European Colonial Empires, 1918–1940* (Cambridge: Cambridge University Press, 2012), 144.
125 Moutet cited in Nguyen-Marshall, *In Search of Moral Authority*, 123.
126 Hardy, *Red Hills*, 82.
127 Nguyen-Marshall, *In Search of Moral Authority*, 126.
128 Pierre Gourou, *Esquisse d'une étude de l'habitation annamite dans l'Annam septentrional et central du Thanh Hoá au Binh Dinh* (Paris: École française d'Extrême-Orient. Les Editions d'Art et d'Histoire, 1936).
129 Gourou read a lot of philosophy, and while we do not know whether he read anything of the Frankfurt School (and possibly not, as he generally bucked against German writing) there are parallels between his assessment of poverty and the anti-developmental inclinations of the Tonkin peasantry and Theodor Adorno and Max Horkheimer's refutation of the idea that the humanisation of nature over time through labour carries within it an emancipatory logic (a way out of scarcity) – a view they developed during the 1930s and 1940s. See Seyla Benhabib, *Critique, Norm and Utopia: A Study of the Foundations of Critical Theory* (New York: Columbia University Press, 1986), 11.

130 Gourou, *L'utilisation*, 421–433. On Foucault, biopower and population, see Stephen Legg, "Foucault's population geographies: Classifications, biopolitics and governmental spaces," *Population, Space and Place* 11, no. 3 (2005): 137–156; Michel Foucault, *Security, Territory, Population: Lectures at the Collège de France, 1977–1978* (Basingstoke: Palgrave Macmillan, 2007); Nikolas Rose, "The politics of life itself," *Theory, Culture and Society* 28, no. 6 (2001): 1–30.
131 On this language, see Emmanuel Saada, "Race and sociological reason in the Republic: Inquiries on the Métis in the French Empire (1908–1937)," *International Sociology* 17, no. 2 (2002): 361–391; and Michael Vann, "Caricaturing 'the colonial good life' in French Indochina," *European Comic Art* 2 no. 1 (2009): 83–108.
132 Bernard S. Cohn, *Colonialism and its Forms of Knowledge: The British in India* (Cambridge: Cambridge University Press, 1995), 11.
133 Benedict Anderson, *Imagined Communities: Reflections on the Origins and Spread of Nationalism* Revised edition (London: Verso, 1991), 243.
134 See, for example, Numa Broc, "Les grandes missions scientifiques françaises au XIXe siècle (Morée, Algérie, Mexique) et leaurs travaux géographiques," *Revue d'histoire des ceicnes* 23 (1981): 319–350; James E. McClellan III and François Regourd, "The colonial machine: French science and colonization in the Ancien Régime," *Osiris* 15, no. 1 (2001): 31–50; Lewis Pyenson, *Civilizing Mission: Exact Sciences and French Overseas Expansion, 1830–1940* (Baltimore and London: The Johns Hopkins University Press, 1993); Paul Claval, "Colonial experience and the development of tropical geography in France," *Singapore Journal of Tropical Geography* 26, no. 3 (2005): 289–303.
135 Pierre Singaravélou, *Professer l'empire: Les "sciences coloniales" en France sous la IIIe République* (Paris: Sorbonne, 2011). Johannes Fabian, *Time and the Other: How Anthropology Makes It Object* (New York: Columbia University Press, 1983), 31, 148.
136 Alice Conklin, "What is colonial science?" *Books&Ideas.net*, 31 January 2013, 3–4. Also see Kapil Raj, "La Construction de l'empire de la géographie: L'Odyssée des arpenteurs de Sa Très Gracieuse Majesté la reine Victoria, en Asie centrale," *Annales. Histoire, Sciences Sociales* 52, no. 5 (1997): 1153–1180; Simon Schaffer, Lissa Roberts, Kapil Raj and James Delbourgo eds., *The Brokered World: Go-Betweens and Global Intelligence, 1770–1820* (Sagamore Beach, MA: Washington Publishing International, 2009); and Patrick Petitjean, "Science and the 'civilizing mission': France and the colonial enterprise," in Benediky Stutchey ed., *Science Across the European Empires – 1800–1950* (Oxford: Oxford University Press, 2005), 107–128.
137 Paul S. Sutter, "The tropics: A brief history of an environmental imaginary," in Andrew C. Isenberg ed., *The Oxford Handbook of Environmental History* (Oxford: Oxford University Press, 2014), 178–198.
138 See, for example, Christophe Bonneuil, *Des savants pour l'empire: La structuration des recherches scientifiques coloniales au temps de 'la mise en valeur des colonies françaises' 1917–1945* (Paris: ORSTOM, 1991).
139 Julia Waters, "Marguerite Duras and colonialist discourse: An intertextual reading of *L'Empire français* and *Un barrage contre le Pacifique*," *Forum for Modern Language Studies* 39, no. 3 (2003): 254–266, at 263.
140 Deborah J. Neill, *Networks in Tropical Architecture: Internationalism, Colonialism, and the Rise of a Medical Speciality, 1890–1930* (Stanford, CA: Stanford University Press, 2012).
141 Hannah le Roux, "The networks of tropical architecture," *Journal of Architecture* 8, no. 3 (2003): 337–354.
142 Emily S. Rosenberg, "Transnational currents in a shrinking world," in Emily S. Rosenberg ed., *A World Connecting 1870–1945* (Cambridge, MA: Harvard University Press), 919.

143 Conklin, "What is colonial science?" 7.
144 Tilley, *Africa as a Living Laboratory*, 116.
145 Tilley, *Africa as a Living Laboratory*, 131.
146 Lauren Janes, *Colonial Food in Interwar Paris: The Taste of Empire* (London: Bloomsbury, 2016), 99. Also see M. Aso, "How nature works: Business, ecology and rubber plantations in colonial Southeast Asia, 1919–1939," in Michitake Aso in Frank Uekötter eds., *Comparing Apples, Oranges and Cotton: Environmental Histories of the Global Plantation* (Frankfurt am Main and New York: Campus Verlag, 2014), 195–220.
147 M.C. Notices bibliographiques, *Journal des économistes*, janvier-février no. 1 (1937): 108–109.
148 Hugh Clout, "French Geographers under international gaze: Regional excursions for the XIIIth International Geographical Congress, 1931," *Belgeo* Online 1–2 (2012), http://journals.openedition.org/belgeo/6248, doi:10.4000/belgeo.6248.
149 Rosenberg, "Transnational currents in a shrinking world," 920.
150 Tilley, *Africa as a Living Laboratory*, 130.
151 AULB: Fonds Pierre Gourou PP153 H. Parmentier, c. hon du Service archéologique de l'EFEO to Pierre Gourou, Phnom Penh, 8 July 1937.
152 Claval, "Colonial experience," 297.
153 Claval, "Colonial experience," 297; also Pierre Monbeig, *Ensaios de geografia humana brasileira* (São Paulo: Livraria Martins, 1940), and Jacques Richard-Molard, *Afrique Occidentale Française* (Paris: Berger-Levrault, 1949).
154 Sam Rohdie, "Geography, photography, the cinema," *Screening the Past* 12, no. 1 (2014): 1–28, at 4.
155 Jacques Richard-Molard, *Problèmes humains en Afrique Occidentale* (Paris: Présence Africaine, 1958), 116.
156 H.P.W., Review *Les paysans du delta tonkinois*, 467. Cuttings in AULB: Fonds Pierre Gourou PP153.
157 Anon, "The Tongking Delta and the Annamite House," *The Geographical Review* 27, no. 3 (1937): 519–520.
158 Lucian Ashworth, "Mapping a new world: Geography and the interwar study of international relations," *International Studies Quarterly* 57, no. 1 (2013): 138–149.
159 ACDF: Pierre Gourou "Leçon inaugurale du cours de Pierre Gourou, Collège de France, 4 décembre 1947."
160 Gourou, "Le souvenir de Roger Lévy," 353 Gourou reflected: "I was seduced by the extent of his knowledge of the East, by his vast culture, immense reading, his sense of the beautiful, a remarkable aesthetic erudition, his courtesy, a great benevolence; the latter led him to ask me to contribute to the Centre my *l'utilisation du sol*," Gourou, "Le souvenir de Roger Lévy," 347.
161 Roger Lévy, "Pierre Gourou: Pour une géographie humaine," *Politique étrangère* 38, no. 5 (1973): 647–648.
162 In France there were various Géographie Universelle (global geography) projects. The most significant of them were spearheaded by Élisée Reclus (1876–1894) and by Vidal de la Blache's students (1927–1948). Gourou's land use survey was not unique – although his extensive use of aerial photography was. Between 1932 and 1938 the British geographer Dudley Stamp oversaw the production of a land utilisation map of Great Britain.
163 As was Gourou's 1947 *L'avenir de l'Indochine*. An English translation of this essay was published in the IPR's periodical *Pacific Affairs* in 1947 and Gourou presented his argument at the IPR's meeting in Stratford-on-Avon in that year.
164 Shannon McCune, "The diversity of Indochina's physical geography," *Far Eastern Quarterly* 6, no. 4 (1947): 334–344.
165 Dudley Stamp, "Review *L'utilisation du sol en Indochine Française*," *The Geographical Journal* 96, no. 2 (1940): 131–132.

166 Pierre Gourou, *Man and Land in the Far East*, trans. S. H. Beaver (London: Longman, 1975), 145, 1. We quote from the English edition because the translation cannot be bettered.
167 Roger Lévy, "Pierre Gourou. L'Asie," *Politique étrangère* 19, no. 6 (1953): 525–526.
168 Pierre Gourou, *Les pays tropicaux: Principes d'une geographie humaine et economique* (Paris: Presses Universitaires de France, 1947), 2, 6, 145, 161; Gourou, *Man and Land in the Far East*, 88–89.
169 The subject of 'disappointment' later became a focal point of Claude Lévi-Strauss's *Tristes Tropiques*.
170 Raewyn Connell, *Southern Theory: The Global Dynamics of Knowledge in Social Science* (London: Allen and Unwin, 2007), 12–16.
171 Jennifer Robinson, "Comparisons: Colonial or cosmopolitan?" *Singapore Journal of Tropical Geography* 32, no. 1 (2011): 125–140.
172 Richard Overy, *The Morbid Years: Britain and the Crisis of Civilisation 1919–1939* (London: Allen Lane, 2009).
173 Robinson, "Comparisons," 132.
174 Albert Demangeon, "La colonisation blanche sous les tropiques," *Annales de Géographie* 49, no. 278–279 (1940): 98–105, at 101–102. Grenfell Price, *White Settlers in the Tropics* (New York: American Geographical Society, 1939).
175 Pierre Gourou, "La colonisation blanche dans les pays tropicaux au regard de la science," *Le Monde Colonial Illustré* 16, no. 1 (1938): 3–4.
176 See, for example, W.F. Wertheim, "Peasants, peddlers and princes in Indonesia: A review article," *Pacific Affairs* 37, no. 3 (1964): 307–311. This rekindling of interest in Gourou's book was instrumental in its re-issue in France by Flammarion in 1972, with a new preface by Gourou, and the publication of an English translation of this second edition in 1975.
177 See Michael Heffernan, *The European Geographical Imagination* Hettner-Lecture 2006 (Stuttgart: Franz Steiner Verlag, 2007), 41–60.
178 Pierre Gourou, "Un livre sur l'Amérique," *Annales de Géographie* 60, no. 319 (1951): 128–129, at 128.
179 Susan Schulten, *The Geographical Imagination in America, 1880–1950* (Chicago: University of Chicago Press, 2002), 240.

5 *Gourou en guerre*

From empire to war

If Gourou networked his way to the project of tropical geography for which he became best known, it was not with the careerist connotations that this expression carries today. But it was not unwitting either. His immersion in the tensions of empire corresponds to what Daniel Gorman and others see as "an important shift from politicians to experts" as key advisors and decision-makers between 1900 and 1945.[1] However, war itself often gets lost in accounts that accentuate the networked and comparative bents of international cooperation. As we seek to show in this and parts of the next chapter, the exigencies and fallout of World War II were important (and barely remarked upon) influences on how Gourou's project of tropical geography came into focus and attracted attention. There are some overlaps with the last chapter – chiefly in terms of how a set of disciplinary dynamics within French geography were entangled with wider networks of power and spheres of influence. On the whole, however, we plot a different course from 1936 to 1950 than the one that has just preoccupied us by focusing on the coming of war and experience of occupation for Gourou.

Spheres of influence

On 19 May 1936 Demangeon wrote to the Geography Commission of the Université Libre de Bruxelles (ULB) proposing three candidates from the Sorbonne to fill its vacant full-time professorship of human and regional geography: Léon Aufrère (aged 47), who was preparing his doctoral thesis (and never completed it); Pierre George (aged 27), an overt communist, who had completed his two years before; and Gourou, who was to about to defend his. Gourou was Demangeon's leading candidate and the committee's decision to appoint him was unanimous, although he was initially appointed *chargé de cours* rather than professor since he had not yet finished his doctorate.[2] Six weeks later Gourou wrote to Marius Moutet (France's Minister of the Colonies, who had been hailing him to advise the Popular Front Government over population issues in Indochina) explaining why he had applied and accepted the post "with much hesitation" (for he had really wished to return to Indochina). Gourou noted that Demangeon had deemed him "the most worthy, both for general reasons and since I specialise in the question of tropical and colonial geography and would be the best prepared for fruitful studies in the Belgian Congo," and added

> It is a good idea for a Frenchman with solid qualifications, and inspiring complete confidence in his masters, to be proposed to ULB . . . [because] it is known that, without a Frenchman with these qualities Hegenscheidt [the incumbent professor and founder of the Institute of Geography at ULB] would have looked for a successor in Germany, a country where he has some sympathy and where you find some good specialists of the tropical world. It would have been most unfortunate to see escape from French influence a post with a certain importance, and a post where the study of colonial issues will be paramount, and which a Frenchman will carry out in a spirit of Franco-Belgian collaboration – which would obviously not be shared by a German geographer.[3]

As Gourou intimated, Belgium was a liminal zone in the competition between French and German geographers – a tussle which had become increasingly acrimonious and nationalist in tone since the end of War World I, and had been heightened by the fact that the International Geographical Union (IGU, formed in 1922 and replacing the older International Geographical Congresses dating back to 1871) excluded Germany, Austria and Hungary. Gourou acknowledged that the colonial and tropical geography he would teach at ULB was by no means detached from these ideological vistas. Indeed, during the 1930s his master, Demangeon, had denounced German *Geopolitik* (and particularly the work of Karl Haushofer) as a betrayal of Friedrich Ratzel and the German school of geography, and as destabilising the border arrangements instituted by the Versailles Treaty (which, crucially for Demangeon, had returned Alsace to France).[4] Haushofer and others were criticised for construing the state as a 'person' and 'organism' rather than as the dynamic product of interaction between civilisations and their geographical milieu, and for seeing French power and identity as becoming diluted by its contact with its colonial races.

At ULB Gourou began to put together courses on regional geography and the principles of human geography, with a thematic focus on tropical and colonial problems, and focused his reading on equatorial Africa and South America, which he knew least about.[5] A typed copy of his Curriculum Vitae in the Collège de France Archives also lists his associations from 1936 with the École nationale de la France d'Outre-Mer and Institut indochinois pour l'étude de l'homme, initiating a pattern of transit between Belgium and France that lasted until he retired.[6] The former, created in 1934, was a re-named version of the École coloniale (founded in 1889), France's prestigious grande école for training colonial administrators, and between 1936 and 1945 Gourou travelled from Brussels, and during the war years from Montpellier and then Bordeaux, to Paris to deliver classes on colonial Indochina.[7] He was also connected with the Institut indochinois pour l'étude de l'homme, co-founded by Paul Lévy (an ethnologist) and Pierre Huard (an anatomist), as a "corresponding fellow."[8] The Institut sought to bring together French and Vietnamese researchers and had its main offices in the building of the École française d'Extrême-Orient (EFEO) in Hanoi. Its 30 or so full members included leading EFEO scholars Louis Malleret, Paul Mus (who succeeded Delavignette as Director of the École nationale de la France d'Outre-Mer), Nguyễn Văn Khoan, Nguyễn Văn Huyên and Trần Văn Giáp. Gourou knew many of these scholars from his time in Indochina, and that they held his work in high regard.[9]

While these two teaching and research facilities were concerned with the dynamics of empire, growing Franco-German tensions were also much in the intellectual atmosphere of 1930s Paris, and French geography.[10] German geographers responded to Demangeon's attack on their preoccupations (at the ideological core of which was the idea of *Lebensraum*, living space for a 'greater Germany,' which the Nazi Party soon solicited to its expansionist designs) by attacking Gourou's other 'master,' Emmanuel de Martonne, who became the secretary general of the IGU in 1931 and president in 1938.[11] De Martonne initially worked to overcome intellectual grudges between American and German geographers that were rooted in German disgruntlement at the challenge William Morris Davis's "cycle of erosion" theory posed to German intellectual supremacy in physical geography, and readmit German geographers to the IGU.[12] But tensions deepened as Nazi power grew. Nazi ideologues succeeded in getting German geographers to shun the 1931 IGU meeting in Paris, where the American geographer Isaiah Bowman assumed the Union's Presidency, and when the Nazis came to power in 1933 German geography was reorganised around the *Führerprinzip* ('leader principle') under Siegfried Passarge, a geomorphologist. Passarge wrote to de Martonne that "Germany has awakened and will never fall asleep again."[13] De Martonne replied that his letter "confirms the impression that Germany is a gravely ill nation"; and added, "Powers and political ideas pass, but science is eternal."[14] Large and disciplined delegations of German geographers (who had been vetted by the Gestapo) under Ludwig Meeking and Wolfgang Panzer were sent to IGU conferences in Warsaw (1934) and Amsterdam (1938), and when de Martonne took over the presidency, the conflict became increasingly personal and acrimonious.

Gourou attended the Amsterdam meeting with two of his acquaintances from his Indochina years, Charles Robequain and Jacques Weulerrse, and later confided that the occasion was turned into a "very unpleasant congress" by the "unbearable representatives from fascist Germany and Italy" who had come to discuss their "small ideas" concerning *Lebensraum*.[15] "Intensely shocked by the impertinence of the Nazi geographers," he later reflected, "we were immensely concerned without guessing that twenty-four months later, German tanks would reach the banks of the Loire."[16] Indeed, "From 1933 onwards it was difficult to avoid the fear that a megalomaniac drunk on racist myths, and supported by accomplished technicians and the virtuousness of a blinded people, was leading Europe to disaster."[17] Hitler could have been stopped in 1936 (at Munich), he continued, but in France "part of the right admired Germany and Italy, while part of the left ignored the external threat and placed its hopes in Communism. And the League of Nations offered no comfort."[18]

On the eve of Germany's invasion of Poland in September 1939 Demangeon observed that of all the states that had grown on the soil of Europe, none was "more unique, nor more disquieting, than Germany."[19] Gourou was mobilised and joined an army regiment in the Rhône Valley. But just three months later, and with a German Blitzkrieg having failed to materialise in Belgium and France, ULB sought to get Gourou back, and did so with his approval. He wrote to the University's head administrator, Paul Hyman, on 7 December 1939, saying that while he needed to serve his country, "my conscience would be less uneasy if I carried out my university duties than if I continued to lead a monotonous and useless life in a sleepy barracks."[20] He started teaching again in Brussels in February 1940, but then on 13 May 1940 informed ULB administrators that "In accordance with the order given by the French Embassy, I will be obliged to leave . . . and I do not know if I will be able to return before the end of the war."[21] But there was also a strong element of concern, if not desperation, in his decision to act on this order. He sought to flee Brussels with his family before the Nazis arrived at his office or apartment door, and in lieu of being returned to his army post and in need of work he placed himself at the disposal of the French Embassy.[22] Accordingly, in late July he was appointed *chargé de cours de géographie* in the Faculté des lettres de Montpellier, in the *zone libre*, nominally to replace Professor Jules Sion, who had just died.[23] He moved to Bordeaux in January 1942, becoming *maître de conférences (géographie)* in the Faculté des lettres, and was made Professor of Geography there in January 1943.[24]

Upon the fall of Paris to the Germans in May 1940 the French government relocated to Bordeaux, and Marshal Philippe Pétain, the hero of Verdun, opposed the proposal made by Paul Reynaud that the French government quit the mainland and fight the war from the Empire. Pétain's request for an armistice was greeted with approval, and General Charles de Gaulle's departure for London and radio appealing for continued resistance went unnoticed. France was carved up, with the most densely populated, productive and strategically important northern region and western seaboard (including Bordeaux) under direct German occupation (with an enormous tax imposed on the French to fund the occupation). The south and east – the so-called *zone libre* – as well as the Empire, were placed under the control of Pétain's Vichy government. Pétain took revenge not only on the Popular Front but also on the Revolution of 1789. As we saw in Chapter 3, his 'National Revolution' centred on a return to the earth and the virtues of the pious peasant – a rural, quasi-feudal and racial idyll unsullied by communism and cosmopolitanism – and geography was targeted as a subject of national importance. The Vichy regime strengthened the teaching of geography at school level, with emphasis placed at primary level on the study of local communities and their ancestral roots, and at secondary level on France's 'natural regions' and its overseas empire. Pétain was also instrumental in creating a discrete *agrégation* in geography, in 1943, and with de Martonne (who later had to deal with allegations of collaboration) playing a leading hand in its design and implementation.[25]

Collaborationists did not wait for German orders to begin to persecute Jews, and Marc Bloch and Gourou's friend Jean Gottmann were among them. It was in

Montpellier that Gourou met Bloch, and momentarily greeted Rivet (who in 1940 had organised an anti-fascist resistance network at the Musée de l'Homme) as he fled France for Spain and then Colombia to continue his research. And Gottmann, whom Demangeon had initially advised to leave Paris and continue with work on farm structures in western France with Sion before the situation for Jews worsened and he was forced to flee, to New York.[26] New bonds were made, and others renewed, in fraught times as older ones and the international networks through which Gourou's work was circulated quickly fractured and faded from view.[27] Indeed, one of the first things that German troops did upon entering Paris was to launch a campaign against "international associations." On 30 July 1940 a special commando unit of the German SS stormed the IGU, based in the Institut de Géographie in Paris, seized its archive, and after a short investigation concluded (among other things) that "the president E de Martonne has anti-German feelings" and that "the society's documents give interesting indications of the work done behind the scenes in an international scientific society against the German representatives and on the manner in which scholars were sought from other countries to oppose the Germans."[28] The German campaign against the IGU was subsequently waged on two fronts: first, as a personal campaign against de Martonne and French influence at the IGU; and second, as an attempt to expose connections between university geography and French resistance to the German Occupation. When the head of the German Geographical Society, Professor Schneider, came to Paris in May 1942 to convince French geographers to collaborate with the Germans, he was met with suspicion and prevarication.

"We therefore ran head on into catastrophe," Gourou reflected: "It was an abyss in May 1940. For four years we were plunged into shame, humiliation, racist persecutions, fratricidal hatreds, and anxiety in the middle of the night"; and yet "it seemed certain to me that Germany would lose. I was also supported by the loyalty and hard work of students who confronted adversity and whom I could not let down. So I did not lose sight of my research."[29] There are parallels between Gourou's personal response to occupation and Braudel's articulation of how the trauma of war and for him imprisonment created a "wound" that threw historians back on to their "deepest selves" and prompted them to keep the "semistillness" of history – the *longue durée* – in view as a refuge from the storms of the present.[30] For Braudel and Gourou alike, keeping their research in sight was a way of coping with the trauma and privations of war, and no doubt contributed to the strong bond they developed.

Tropiques

Given its focus on rural lifestyles and tradition, geography was ideally placed to support Vichy's peasant fascism. But while de Martonne's involvement in the creation of the geography *agrégation* might be read in this light, Febvre's and Gourou's veneration of peasant land and life should not. As we will show in more detail in Chapter 6, they were concerned with how peasants might evade the clutches of fascism, and how the study of peasant life might be used to contest the politically motivated ends to which the Vidalian 'land and life' tradition was being put.

Interestingly, between 1941 and 1945 Aimé Césaire, his wife Suzanne Roussi, and René Menil co-edited a subversive literary magazine, *Tropiques*, in Martinique under the prying eyes of French censors. The magazine brought surrealism and *négritude* to the French Antilles, and had a complex relationship with metropolitan primitivism, which, as we have shown, envisioned the tropics as a scene and source of redemption for the pathologies of modernity (see Chapter 3). One of the overarching aims of the essays, poems and polemics in this journal was to question the way Vichy France had woven images of tropical grandeur and excess into its patriarchal imagining of a 'greater France' and fascist *Volk*.[31] Tropical flora and fauna appear in many of Césaire's contributions, as both emblems of an imprisoning Western gaze and pointers to an alternative Caribbean identity. Pierre Mabille (friend of the surrealist André Breton) extended this line of enquiry in an essay entitled "La Jungle," observing that "tropical paradises suppose the existence of prisons" and seeing an absolute opposition between "the jungle where life explodes everywhere, free and dangerous, the most luxuriant vegetation being ready for all kinds of mixtures, transmutations and trances, and that other sinister jungle where a Führer, perched on a pedestal, watches, along the neo-Greek colonnades of Berlin, the departure of mechanised cohorts that are ready, after having destroyed all other living things, to annihilate themselves in the rigorous parallel lines of endless cemeteries."[32] Similarly, Roussi deployed the expression "*le grand camouflage*" to evoke how, on the one hand, Vichy representations of an idyllic and obedient tropics disguised "the flowers of human debasement" (realities of colonial racism and violence), but how, on the other hand, "Poets who saw the tropical flames [of oppression] fanned by hunger, fear, hate and ferocity" might use this enmity to recover a "beautiful Antilles."[33]

Tropiques thus gave a wartime inflection to the association of tropicality with personal and cultural enrichment through artistic expression, and toyed with exoticism. But Césaire insisted that the journal was insurrectional, and we see it as part of the militant tropicality we looked at in Chapter 2 with reference to the Viet Minh and Indochina War. The journal challenged Vichy's alienating value system, which had turned the Caribbean into a "mute and sterile Earth" where "the tam-tam in the bush" could no longer be heard, Césaire wrote, through the embrace of wonder and symbolic language.[34] However, as Michael Dash suggests, we can also locate in Césaire's (and later Derek Walcott's) re-imagining of a colourful, lush, sonorous Caribbean tropics a major problem for anti-colonial thought: the prospect of only being able to counter colonial discourses like tropicality by using and potentially reinforcing their imagery and binaries.[35] As Césaire himself intimated in *Tropiques*, a radical poetics of tropical knowledge was not easily translated into a concrete project of resistance or social transformation. In other words, tropicality did not just emanate from the West; and while we do not want to underestimate the huge differences in tone and perspective between Gourou and Césaire, they found common ground in their resistance to fascism and Vichy France. However,

Gourou did not fight Vichy power with poetics. His concern was with the practicalities of survival.

The wartime origins of French tropical geography

German occupation placed some basic practical and physical restraints on research, and it is important to note that the ideological relationship between Vichy and disciplines such as geography and history was by no means clear cut.[36] Immediately after the war, Aimé Perpillou observed that "spiritually" as well as "physically" French geographers "found it impossible on account of the war to work effectively even in their own country."[37] He relates that travel was difficult; that it was difficult to access contemporary (cartographic, economic and statistical) data; that censorship prevented the publication of some work (on the physical geography of the French coastline, and French railway system, for example); and that many were restricted to working on projects began in the pre-war years. "It would be too long, indeed impossible," he added, "to list all the geographers, teachers in various Lycées in France, who, while still too young to have made reputations, were arrested, interned, deported, or executed." Many geographers left their posts "to devote themselves to the clandestine activities of the Resistance," providing maps from their own private collections and concealing them from the Germans.[38] Perpillou gives Gourou and Jules Blache special praise for their high-profile contributions to the Resistance.

Some avoided censorship by taking refuge in the putatively innocuous study of physical geography. Others continued to work on human regional geographies and landscapes, and herein resistance to Vichy preoccupations could be expressed in more or less veiled ways, and the traces of defiance and opposition that censors found in what they read were often tenuous. Nor, in every area, did Vichy thought and practice constitute a marked break with what had gone before. A concern with land use (and particularly the exploitation of forests and waste land), and wartime worries about population decline, can be traced back through the interwar years into the nineteenth century. And Vichy ideologues and censors did not monopolise understanding of the deeply political figure of the peasant.[39]

Gourou's story from 1940 to 1945 is certainly one of stealth and survival: of refusing to become what Braudel had called (with reference to himself) "the lost France," and thinking carefully about how best to resist.[40] As one of his young students in Bordeaux, Paul Pélissier remembered his teacher's advice: "we were twenty-year-old kids. He told us not to leave for the Maquis. We had no arms, papers, munitions, nothing. Come when we call you, he said. There was a certain connivance, a good connivance, between Gourou and his students."[41] The Vichy regime considered universities to be generally in opposition to the cult of the leader due to their spirit of independence, and at this juncture Gourou clung tightly to Descartes' aphorism "Once in your life you need to doubt everything."[42] The universities employed only a tiny proportion of French people, however, and were difficult to control.

Gourou fled Brussels on 14 May 1940 (three days after his family) with just a suitcase. With no savings to fall back on, he needed to secure work quickly, and with all of his books and papers still in Brussels he knew it would be hard to resume his research swiftly. His remarkable wartime correspondence with Jean Gottmann underlines both his quotidian struggle to keep his research going, and finance it by teaching, amidst anguish and uncertainly, and the sense of release that this experience afforded him.

In 1941 Febvre confided to a friend that trying to maintain one's regular routines of academic research and writing was "a way to victory against the powers of death" – the death of freedom and the critical spirit resulting from occupation.[43] And as Philippe Burrin writes of the wartime output of the *Annales*, "absorption in the study" was deemed a comfort and "scientific language" a shell that stopped the occupier from peering too closely into one's affairs.[44] Until early 1942, Gourou had a burning desire to flee France and expand his research horizons by going to Rio de Janeiro, where, he had heard, there was an opening.[45] But this and other plans during these dark years were stifled by "the uncertainties and possible dangers for my family," and Gourou settled instead for a "cordial welcome" in the "very stuffy Faculty" at Bordeaux, where he worked on "black Africa," furnishing an "academic interest that everyday life scarcely offers" and a dream of one day "visiting tropical Africa to verify my intellectual constructions."[46] However, food in Bordeaux was in short supply and Gourou left his wife and family in Montpellier and sought to commute whenever he could get a pass. He wrote to Gottmann, who by this time was ensconced in the safe and privileged surroundings of Princeton University: "I envy you because you will be able to work, whilst I just build courses, without any chance for personal fieldwork"; indeed, under the duress of hunger and occupation Gourou started to wonder whether the idea of a "general human geography is not an elusive myth . . . if human geography can be anything other than particular and regional," as it very much was in wartime France.[47]

In a long letter dated 30 August 1942 Gourou writes to Gottmann: "Everything I see confirms in me the idea that you were dead right to leave France; conditions of existence would not have improved for you in the *zone libre*; of course, I'm not talking about the occupied zone where life is clearly unbearable for your Jewish friends." At the same time, Gourou reports that he has been able to "polish" his course on Africa, and fills his friend in on recent developments in their fragmented geographical world: "Blanchard has brought out a brilliant book about the western Alps, full of terrible swipes at Cholley." Gourou found time to read Jean Dresch's thesis on the High Atlas mountain region in Morocco, which he describes as "almost completely morphological" and "terribly boring on the whole." He also muses over de Martonne's retirement and how, with Demangeon now dead too, "Our masters will not have to complain of being eclipsed by their successors."[48] But he ends with a lament: "I'm doing nothing of interest, just accumulating notes and information for future works. And anyway, the shortage of paper makes any publication problematic. The *Annales de géographie* has only brought out one issue in 1942."[49]

188 Gourou en guerre

Frustration at a research life not being led, and anxiety about the dangers and privations that his family were enduring, were Gourou's overriding emotions, and they spilled into his intellectual life. How could they not? It seemed likely that if it had not been for the constraints of war, Gourou would have undertaken research overseas. Yet his wartime experience, particularly at Bordeaux, which had many research connections with tropical Africa, fostered a more general interest in tropical geography. The exigencies of a world at war and a country occupied had crept into his geography. While the services of American, British and German geographers were sought and valued by their country's war machines, both at home and overseas, the implication of Gourou's correspondence with Gottmann is that the war stirred within him a more chaste and nominally detached interest in 'the tropics.'

As early as 1938, when he encountered the 14 German geographers who were dispatched to the IGU meeting in Amsterdam, Gourou was arguing that colonisation was not directly responsible for the condition of the tropical world, and he left that 1938 meeting aghast at the anti-semitic tone of the papers presented by the German geographers about the possibilities of colonisation in the tropical zone by the white race.[50] Gourou's subsequent removal of racial questions of 'whiteness' and 'blackness' from the picture of tropical development had an important personal and political dimension. There was an anti-German element to his stance, and in the longer term his pessimism about the prospects of development in the tropical world can be read as a rebuttal of the utopian hue of much German colonial writing about tropics.

The Bordeaux Liberation Committee

But Gourou's resistance does not end here. The push of the Soviet Red Army into Eastern Europe in June 1941 finally propelled the Communists in France (and elsewhere) into armed resistance. The tide started to turn against Hitler, and Germany's exploitation of France escalated, pushing more and more French people into the resistance camp led by De Gaulle. German occupation of the *zone libre* in November 1942, after the Anglo-American landings in North Africa, dispelled the myth of Vichy as 'shield' protecting the French people from Nazi excesses, and within a year had prompted Gourou to become more directly involved in the French Resistance. Recent scholarship explores how German occupation of the *zone libre* expanded the meaning and repertoire of resistance, and emphasises its specificity, diversity and even disjointed nature, and how it fractured Vichy's essentialist discourse of peasant and land.[51]

On 2 May 1944 Gourou wrote to his cousin Marinette in coldly ironic and acerbic terms about the worsening situation in Bordeaux.

> How you must suffer from being deprived of your radio. You who delighted in listening to the voice of Philippe Henriot [the prominent collaborator, who was assassinated by the Resistance in June 1944], there you are reduced to reading only the rare articles he publishes in the newspapers. We are

threatened with suffering the same fate as you; it is whispered that the radio sets will be seized in the near future. That will unfortunately deprive us of the possibility of receiving at this critical moment the instructions of our Marshall that are the envy of the world.[52]

We have had some "noisy nights," he continued; "the judeo-capitalists came to bomb the aerodrome at Mérignac . . . [and it was] quite impressive; for once the masonico-atlantic airmen's aim was true; and the atlantico-democratic bombs make very big craters." Talk of evacuation from Bordeaux was in the air, but Gourou thought it unlikely because the railway and bridges outside the city were the real targets. "We have had no news from Nelly and Jean [other relatives] since January. They had to leave their home. . . . And we do not know what has become of them since. I am very worried about them."

This was the build-up to the Allied landings in Normandy (the British were seeking bridgeheads further south too) and Gourou was being brought closer to his own form of engagement. In August 1944 he joined the Comité de libération de la Gironde, which was "comprised exclusively of men who, for four years, have fought secretly against German oppression" and who were united by their "patriotric enthusiasm" and the motto "Justice will be done for everyone."[53] The Committee, with René Caillier from Parti républicain as President, was a truly rainbow group with representatives from the Parti socialiste, Résistance rurale, Franc-tireurs et partisan, Libération, Mouvements unis de résistance, Front National, France-Liberté, Presse résistante, Détenus politiques, Parti radical-socialiste and Parti communiste. Gourou was listed a hailing from France-liberté.[54] The Committee had five "commissions" – *municipalités; épuration; ravitaillement; travail, chômage, salaires et prix;* and *presse* – and Gourou was put in charge of food and fuel supply, and of liaising with trade unions wishing to contribute to the commissions.[55]

Although Bordeaux was not as ravaged by war as the French Atlantic ports of La Rochelle, Saint-Nazaire and Lorient further north, it shared many of the difficulties facing France as a whole in the wake of occupation, and then in the aftermath the Liberation. When rationing was introduced in 1940, as what Pétain termed a "painful necessity," the daily bread quota was 350 grams per person, and by 1944 it had slumped to just 250 grams.[56] The League of Nations calculated the typical calorific value of adult rations in France to around 1,200 calories per day over the period 1940–1944, which was significantly lower than in other western European countries, and with bread rationing becoming a focal point of consternation and conflict. The severely cold winter of 1944–1945 put extra strain on the very limited supply of coal, gas and electricity, and convoys of food that managed to overcome this deficiency in fuel, and find ways past destroyed bridges and roads, and through harbours in the Gironde blocked by sunken vessels, were further hampered by labour shortages and German interference. Bakeries were attacked by German garrisons, and the war years were years of galloping inflation, with a rampant black market and rising social conflict over food and resources. Shortages were felt most acutely in rural areas. The freezing winter of 1944–1945 also precipitated a fall in milk production (and, some argue, a rise in infant mortality), and the following year wheat and potato yields fell by 30 per cent across France.[57]

The *comités de libération* that sprang up in French departments and villages towards the end of the war provided the French Resistance with a civil structure that would coordinate with its military arm, the Forces Françaises de l'Intérieur. The *comités* were concerned primarily with the rationing and provisioning of basic commodities, and looked to university academics to help them devise workable planning and distribution systems. Food and fuel were vital to the way urban and rural areas were connected, and to the scope and limits of German and Vichy control over the affairs of town and country. Chris Pearson shows how the natural environment – the use and transformation of forest, farms, coastlines and waste land – was pivotal to how both Vichy envisioned German occupation, to how Germany sought to exploit France for its own ends, and to how resistance was mounted.[58] Vichy saw the French peasantry and the cultivation of the countryside and exploitation of the forests as integral to French national renewal. The German military and Vichy administrative apparatus eyed town and country in more utilitarian ways, requisitioning food and seizing petrol, fixing prices for, and levying duties on, basic commodities (coal, oil, wood, livestock, grain, dairy products and vegetables), fragmenting transportation systems, and thus, in places (the Gironde being one of them), destabilising the rural economy and fuelling the black market. The port of Bordeaux lay at the centre of German attention, and by 1941 had become a key port for trade with Spain and Japan, and was heavily defended.

The Germans took particular interest in the mixed rural economies of northern and western France, and found it more difficult to exploit agricultural districts like the Gironde, where monoculture (viticulture) prevailed.[59] However, they prized French wine and within two months of their arrival in Bordeaux, in June 1940, château and merchant wine cellars had been ransacked, and the looting only stopped with the arrival of a *Weinführer* (wine controller, as the French nicknamed the figure), appointed by Hitler to regulate the sale of wine in each of France's wine-producing areas.[60] In Bordeaux, French opposition to the Germans came, as it generally did elsewhere in France up until the final months of the occupation, not in the form of direct military confrontation (German troops in just the coastal zone numbering 1.4 million by 1944) but in clandestine activity and everyday geographies of resistance – through sabotage, concealment, information-gathering and dissemination, the organisation of supply and escape routes, and black market activity – that aimed to alleviate shortages and ameliorate webs of rumour and deceit spun by the Vichy regime. Over 70 per cent of resisters were under 40 and 85 per cent were male, but women carried much of the burden of everyday resistance.[61]

Gourou's involvement came at the tail end of the occupation. In July 1944, a month after the Allied landings in Normandy, the German *Kriegsmarine* in southwestern France was ordered to quit Bordeaux. The Vichy apparatus of rationing, requisitioning and surveillance was still in place, however, and resistance to it by regional liberation committees still had a big part to play in securing victory. The port was left fairly unscathed (Nazi orders to destroy it, to foil a further allied landing, were foiled) but the harbour and the Gironde was strewn with scuttled and sunken vessels, making the river almost unnavigable, and there were fissures in the transport network on land. Gourou started work for the Bordeaux liberation committee in mid-August and reported to it on the matter of food supply on 12 September 1944, basically arguing that order and planning needed to be restored to the rural economy and its connections with the city. "The situation remains serious because of the shortage of transport and administrative order," he surmised, and an "increase in bread rations would depend on improved relations between the northern departments and the west."[62] He recommended that soldiers confined to barracks should have their civilian ration cards withdrawn, and that Bordeaux would be supplied with adequate quantities of butter and milk if 60 vehicles could be made available. Then, on 21 September, he provided a detailed report on the port and requested that British authorities supply the necessary equipment to clear it of debris.[63] It is not clear how Gourou gathered his information, and whether he toured the Medoc. But it is likely that he was advised on rural conditions by another Bordeaux geographer, Louis Papy, an expert on France's Atlantic coast, and by Alphonse Grange, a Free French special agent to the region from 1942.[64]

From *engagement* to *épuration*

Gourou's remit quickly expands from here. On 26 September 1944, a history and geography lycée teacher, Gabriel Delaunay, of the Mouvements unis de résistance, was elected President of the Bordeaux liberation committee and insisted on Gourou's appointment as Vice-president because of his practical geographical skills.[65] Delaunay asked the Comité National de la Resistance to extend the Bordeaux region to include the departments of Lot-et-Garonne and the two Charentes, and asked Gourou to justify this extension on geographical grounds, which he dutifully did.[66] Then, in October 1944 Gourou arbitrated a dispute over who from the committee would have dinner with De Gaulle when he visited: Gourou asked for a list of those whom the Republic's *commissaire* Maurice Papon had introduced to the General.[67]

In November 1944 Gourou returned to the problem of food supply, providing a detailed report on the worsening meat supply situation, asking again for the port to be unblocked, and suggesting that basic rations in the city would need to be reduced.[68] He also envisioned the development of a barter economy not dissimilar to the one he had seen in operation in large parts of the Tonkin Delta, with meat from within and outside the Medoc region possibly bartered for wine, and charcoal for butter.

In December 1944 he also sought to hasten administrative decentralisation by trying to ascertain how many civil servants were involved in épuration (purging/ cleansing) inquiries, and who to trust to oversee rationing, local markets and the wine production system.[69] However, it was in December too that the committee president announced Gourou's resignation, to attend to "international matters"; he had been chosen to represent France at the "Pacific conference" (organised by the Institute of Pacific Relations – IPR) in the U.S.[70] Gourou was congratulated on this "promotion" and thanked for the "active and constructive role" he had played on the committee.[71]

This was geography in the service of the French Resistance and a way, perhaps, for Gourou to act on his worries for his family. Pélissier recalled that his mentor never spoke or wrote about his involvement in the French Resistance, or of how it related to his emerging tropical geography. However, Gourou's daughter told us that he remembered the U.S. flying fortresses that flew above the city, and was drawn to the acid wit of the writer Marcel Aymé, who satirised the occupation years and the pretentions of the 'resistants.'[72] In other words, Gourou was self-effacing about the whole thing.

The archival evidence we have reviewed is tantalising with regard to the broader question of Gourou's tropical geography. It seems that he was asked to serve the Bordeaux liberation committee because of his professional skills, and he exercised them in connection with rationing, food supply and the organisation of rural economy. While he left no diaries or reflections on his wartime experience, it seems to us that it confirmed his view that organisation and administration were pivotal to all landscapes, economies and *encadrements*. The idea of resistance at work in his work for Bordeaux liberation committee centred on organisation as a practical activity – a kind of 'action within limits,' to use Gramsci's expression, that sprang from a spirit of resistance and an institutional form manufactured to meet a particular need.[73]

It would of course be disingenuous to suggest that it was in Bordeaux in 1944, rather than the Tonkin Delta before or tropical Africa after, that Gourou arrived at the notion of *techniques d'encadrement* (the production and landscape moulding techniques of particular groups and civilisations). Yet his reports on food supply in the Gironde dwell on what he regarded as these basic imperatives of economy and society, and we might find in Gourou's activities the seeds of his grounded and material conception of 'the tropics' as a domain of human-environment interaction, and his concern with the human organisation of space.

Gourou greatly admired Roger Dion's 1930s work on the wine-growing areas of the Loire valley and viewed viticulture as a "progressive and sociable" form of land use that "promoted the liberation of the peasant."[74] He admired the way Dion had challenged the assumption that agricultural success was determined by climatic factors and soil quality, and had probed the significance of historical processes (trade, markets, church and state) and traditions (centuries spent working the soil). While Gourou did not address these ideas directly in his reports on food and fuel supply problems in wartime Bordeaux, we can glean from his remarks about Dion that his tropicality was not simply about the projection of European values and geographical categories. It also involved a concern with practical ensembles and organisational techniques – a geography of "care" and "discerning selection" in the relations between people and land, to use his terms.[75]

On 5 May 1945, and with the liberation of France mostly accomplished, Gourou wrote to the Rector of ULB, ostensibly to say that he would not apply for a vacant teaching post at the Sorbonne and wished to return to Brussels, and made the following observation:

> We await the bells and sirens announcing the end of the war in Europe. But I can't manage to rejoice at this event. You cannot tear yourself away from the horror of the war, this horror that sticks to your skin like a viscous peace [peace of the dead]. You think of friends and colleagues that the Gestapo took away and whom you will never see again. I don't know why, but I am still haunted by the image of the truck taking a load of prisoners under German guard from the sinister Fort du Hâ [Bordeaux] last August. The incredible details we receive from Buchenwald [the largest Nazi concentration camp on German soil] and other places add to the horror of our meditations. Can we draw vengeance from that, and can we rid mankind of its terrible habits? I fear that German bestiality could well be contagious.[76]

He ended by noting that he had been extremely busy since returning from the U.S. presiding over the Conseil académique d'enquête de l'Académie de Bordeaux (CAE, Academic board of enquiry for the Academy of Bordeaux) which was tasked with 'purging' (*épuration*) the education system (from primary to university level) of collaborators.

In January 1945 the Rector of the Académie de Bordeaux was informed that around 250 dossiers would be examined, and that Professor Gourou would lead the enquiry. He did so, intermittently, from November 1944 to September 1945, and the Rector had been warned that progress might be slow because the academics on the board had limited time, some of the dossiers were incomplete, and the council had but a tiny office. Even so, on 24 March 1945 Gourou reported to the Minister of Education in Paris that he had dealt with 238 cases and sent 75 dossiers to Paris to be dealt with by the courts.[77] Indeed, in October 1944 Gourou signed a letter "to the Commissar of the Republic and the Prefect" asking whether "a purge of the supplies service in Bordeaux" was needed, and whether it should include people who had been directly under his supervision.[78] He obviously felt nervous about some of the work given to him. Nevertheless, examples of collaboration that were deemed punishable by Gourou's council included: flirting and drinking with Germans; sleeping with the enemy (termed *collaboration horizontale*); saying that De Gaulle was ridiculous; wearing the Pétainist françisque badge; denouncing the resistance; and asking pupils too much about their parents' radio sets.[79]

Gourou's reputation thus grew in the world of the French Resistance and Liberation, and he was made Officier de la Légion d'Honneur in 1948 for his patriotic endeavour.[80] While "all branches of the Resistance supported the return of the Republic, which implied elected rather than self-proclaimed representatives," Megan Koreman notes, "few resistants had a public reputation on which to build." Indeed, they "had to prove their resistance all over again after the liberation," and they found the local purge of Vichy collaborators an important means of so doing.[81] Many, like Gourou, were not looking to gain office, but were responding to local demands for justice and national renovation. Violent retribution was the exception rather and the rule at national and local level, Koreman continues, and local purge councils like Gourou's sought to gather evidence about collaborators that would then be passed on to the central authorities, where justice would be meted out by the courts. By mid-1945, however, the purge was bifurcating along national and provincial lines, with central government working through legal channels, and local purge councils responding to public demands for revenge as well as for justice and closure on the past. While local councils sought to use legal mechanisms, many of them felt frustrated at the time it took the judicial system (which itself was being purged) to hear cases, and hence resorted to extra-legal methods.

Denunciation (of the Resistance) and treason (against the Republic) were complex crimes that locked informants and collaborators into complex webs of political and personal accusation, revelation and betrayal.[82] Many of the cases reviewed by Gourou's council fell in this convoluted category, and the penalties that local purge councils – including his – used took diverse forms. Penalties such as execution and imprisonment were reserved for the central courts, but local councils could levy fines and sack workers, and these measures were coupled with extra-legal penalties such as job suspension and demotion, and ostracisation by the public.[83] Of actual cases dealt with by Gourou's council: Mademoiselle Roturier, "a primary school teacher and collaborator from 1942," was suspended from teaching for two years for wearing the françisque; and another female primary school teacher was demoted because of her "dubious morality" – for her *collaboration horizontale* – a charge she denied, arguing that she was giving "just simple hospitality to an old officer."[84] Tony Judt points to the frequency with which women were accused of cavorting with Germans and Vichy officials, and adds that while the exchange of sexual services for goods, clothing and protection undoubtedly occurred, "the popularity of the charge and the vindictive pleasure taken in the punishment is a reminder that for men and women alike the occupation was experienced above all as a *humiliation*. Wreaking revenge on fallen women was one way to overcome the discomforting memory of personal and collective powerlessness."[85]

Hot Springs, Dakar and Dalat

We can sense both Gourou's anguish about this process, yet acceptance that justice needed to be done, in his comment to the Rector of ULB about the "viscous peace." But he was also keen to look to the future – to work his way back from death to life – and grasped that the Resistance saw empire as a compensation for the humiliation of 1940.[86] Yet other outlooks were stirring. Gourou later observed that the IPR Conference at Hot Springs Conference jolted him into think about the aftermath of war, and about problems of restoration as well as retribution. Over 150 representatives from 12 countries and 9 national IPR councils attended this meeting, between 6 January and 17 January 1945.[87] That they were willing to travel by air and boat in hazardous wartime conditions attests to the significance attached to the meeting, which was attended by academics and administrators with lengthy first-hand experience of international, and particularly East Asian and Pacific, affairs. And they were rewarded by being hosted by the IPR's American national council at the swanky Homestead Hotel at Hot Springs.[88]

The French delegation was headed by Paul Emile Naggiar, who was a former ambassador to China and Moscow, and Gourou was taken as an expert on Far Eastern land and society. He was involved in two of the four round table discussions instigated at the conference, on "Economic recovery and progress in Pacific countries," and "the future of dependent areas," and although, as Yutaka Sasaki attests, "no verbatim records" of these discussions were kept, a number of key individuals have provided eyewitness accounts. Gourou was one of them; William Holland (the IPR Sectretary) another.[89] "It was by a fortunate twist of fate that I was in New York at the end of 1944," Gourou recalled. "With victory over Nazism assured, it was now a case of applying a post-war political programme that would reconcile the happiness of peoples to the political and moral leadership of the United States."[90] Yet while accepting of the American Council's declaration that it "was not interested in the long-term maintenance of the colonial system and would indeed be disillusioned by any suggestion that the main function of an international system after the wat would be to conserve empires," Gourou was less sure about American talk of "the idea of the 'development' of the liberated colonies."[91] And the newly invited Indian delegation, many of whom were centrally involved in agitating for independence, was unsure too – and also about the large number of officials from Chatham House (which had a staunchly imperial heritage) in the British National Council and conference delegation.

In the paper Gourou presented at the conference, on "The standard of living in the Tonkin Delta and the Far East," he pointed out that in the Far East overpopulation would fetter development, and insisted that "appropriate *techniques d'encadrement*" rather than the imposition of American models "are the key to 'development.'"[92] Indeed, there were "sharp exchanges," William Holland recalled, over the future of Britain's and France's colonial empires, to the extent that many British, French and Dutch delegates left the meeting "quite upset" over American criticism of their colonial records in India and Southeast Asia.[93] Gourou also spurned the American idea that colonialism was chiefly "responsible for the backwardness of the dominated territories."[94] The conference occasioned him to press the broader idea that the problems of the tropical world pre-dated and surpassed European colonisation, and in his inaugural lecture at the Collège de France he noted that the oscillation of scholarship, politics and emotion in his life and work over the previous five years had prompted him to take up "the most urgent issues of our time."[95]

However, it was a further unplanned occurrence that sealed the direction in which his work would go. With the war nearing its end, "a mission to Senegal" to preside over that year's jury for the *baccalauréat* examination at the Institut français d'Afrique noire (IFAN) in Dakar allowed Gourou "to fully regain contact with the tropical world and feel excited again by my geography which had helped to distract and entertain me" during the war years. "Luckily," he continued, a bad reaction to a yellow fever vaccination forced him to convalesce in Dakar for more than three months, interrupting his *épuration* work in Bordeaux and giving him "time to explore the library of IFAN, which was full of British, African and American publications that were unavailable in France. This ample harvest of new information rejuvenated and filled out the documentation which formed the basis of *Les pays tropicaux*."[96] The British literature was particularly strong on sub-Saharan Africa and India, the American literature stronger on the Amazon and American tropics.[97] He also found Dutch and German literature that was mostly new to him.[98] In short, Gourou's medical misfortune hastened his configuration of the tropics as a discrete zone and object of study, and we should not minimise the part that World War II, and Gourou's travels towards the end of the war, played in pulling him away from exorbitant claims about the fecundity of the tropics and advancing a more pessimistic outlook.

Indochina was also part of Gourou's "viscous peace" and Moutet sought his expertise again as France sought to restore its foothold in Indochina in the wake of the Japanese occupation, and in the event of the Chinese and British holding the peace in the north and south respectively. From September 1945 onwards tensions between France (in the process of forming the Fourth Republic) and the Democratic Republic of Vietnam (DRV) mounted. Stein Tønnesson persuasively argues (against the prevailing French historiographical view) that a chain of events between March 1945 and the summer of 1947 made war inevitable.[99] Gourou was invited to accompany a French delegation led by Max André (a Christian Democratic party member of the French National Assembly), and overseen by Admiral Georges Thierry d'Argenlieu (French high commissioner in Indochina), that was sent to Dalat in April 1946 to negotiate with a communist Vietnamese team led by Võ Nguyên Giáp (Gourou's former school pupil in Hanoi) and overseen by Vietnam's foreign minister Nguyễn Tường Tam. Dalat was a colonial hill station in central Vietnam which had become the capital of the Indochinese Federation and which French diplomats viewed as a "calm, serene atmosphere away from the pressure of mass demonstrations."[100] D'Argenlieu had chosen the location and the two teams were to engage in preparatory talks over the future of the Indochinese Union and status of the DRV that would be taken up at a conference at Fontainebleau in July.

Gourou's remit was cultural and educational and was seen as integral to the rebuilding of robust federal institutions in Indochina. He made various proposals about the retention of French as a second language of instruction in Vietnamese schools, and about France's right to administer its own educational establishments.[101] He was greeted with stern opposition: from Văn Huyên, who, while greatly admiring Gourou's scholarship, thought the geographer was meddling unnecessarily in politics and argued that "Vietnamese is already a language of civilization; we have translated French high school manuals into Vietnamese. And we will do better"; and from Giáp, who likened the new Vietnamese nation to the shape of the yoke used by Vietnamese peasants to carry their rice baskets (in stark contrast to the apolitical wisdom of Gourou's Tonkin peasants), and who demanded Vietnamese partnership in the EFEO and other French research institutes and educational establishments.[102]

By all accounts the tone of the meeting was not good, and d'Argenlieu left thinking that the diplomatic focus needed to be switched away from the north, where matters were still complicated by the presence of the Chinese as well as the Viet Minh's growing internal control of the Tonkin territory, to Cochinchina, where the prospects of refashioning a French Union seemed more plausible. Gourou left aghast at the deceit that crossed the negotiating table. He recalled his former pupil's rude challenges to d'Argenlieu's authority.[103] Indeed, on the eve of the conference, Gourou's friend Joseph Inguimberty, who was about to leave Indochina – his family had been threatened by the Japanese and their future in communist Vietnam seemed uncertain – wrote to Gourou, "You have been given a very difficult mission. You will be dealing with deceitful people who will try to lie to you about their past, and are looking for a nice little civil war." As for Giáp, "He is intelligent, but vain and devoid of scruples. All these people hate us for mystical, but also realistic reasons: they have tasted the most absolute power and they want to keep it."[104] Gourou partly concurred, later recalling how Giáp came across as "a highly intelligent man but a ferocious communist."[105] For all their differences, however, he admired Giáp's courage, intellect and adoration of his country.[106] Nor did Giáp make a secret of his adoration of his former teacher, and to whom, in 1994, he paid an informal visit whilst on a private trip to Europe for medical treatment.[107]

Impasse at Fontainebleau prompted Gourou to write the most directly political essay of his career, *L'avenir de l'Indochine*, an English translation of which ("The future of Indo-China") was submitted by the Comité d'études des problèmes du Pacifique as a document for the IPR's Tenth Conference held at Stratford-upon-Avon in September 1947, and a revised version of which appeared in the IPR's journal *Pacific Affairs* as "For a French Indo-Chinese Federation" later that year. Gourou's chaired the conference at Stratford, and his basic argument in *L'avenir de l'Indochine* was that

> France must assume its federal and federating role without any hidden colonialist agenda, but it must enjoy, for the sake of its nationals, its cultural mission, and its economic enterprises, and a loyally consented freedom. Beyond such principles, we can see only disorder and ruin that will benefit no one.[108]

In the shorter English-language version of Gourou's synopsis, he reasoned:

> Out of various mutually alien and hostile elements France moulded a peaceful whole from which domestic wars were excluded. Irrespective of France's right to intervene in Indo-China, the fact is that she brought about a state of affairs which, viewed in terms of the peaceful relations established among the peoples of the Federation, was certainly not undesirable. Accordingly, the Federation deserves to survive. To do so, however, it must be watched over by an arbiter possessing some authority; otherwise the stronger members will be sure to oppress the weaker ones, who in turn will seek protectors, and an era of international complications will result which will profit no one.[109]

In his view, France was the right arbiter, and building on arguments he made in the 1934 geography school textbook on the region that he wrote with Jean Loubet, where Indochina is described as a geographically "rational creation of France," he goes on to explain how "facts" of geography and history infused Vietnamese support for the perpetuation of the French Union, and with only the "Annamites" to the north "less enthusiastic" about it.[110]

As we said at the start of Chapter 3, the idea of Indochina made sense to Gourou not as an exotic peasant land but as a French geo-political construct: an entity based on economic, political and intellectual ties, investments and infrastructure that made the country into a coherent whole surmounting its pre-colonial diversity and disunity.

Gourou was troubled not only by the friends he had lost in Europe in two world wars but also by the way many of his former pupils and Vietnamese colleagues in Hanoi had got behind the revolution and become vehemently anti-French. For all his misgivings about the benefits of French colonialism, intellectuals like him and the humanist-reformist assemblage to which they were attached, struggled to get past the question of French ethnocentrism. In general, albeit in different ways, Paul Sorum argues, French intellectuals, including many anti-colonial intellectuals, deemed French culture to be distinguished from other cultures by its universalism, or special accessibility to all people.[111] Vietnam's rejection of France amounted to the rejection of the gift of a superior civilisation.

But this imperial thread in Gourou's view about the future of Indochina was tempered by the experience of war and problems of reconstruction. The politics and emotion in *L'avenir de l'Indochine* are interwoven with a more dispassionate appraisal of "the unusual geographical configuration and population distribution" of the region, and how it shaped regional differences and inequalities in standards of living. He hoped that the presence of a French arbiter would facilitate the search for "a formula which will recognise the existence of a large Annamite majority in the Federation without compromising the liberties of the other peoples."[112]

However, Gourou was unable to control the ends to which his own scholarship was geared. At Dalat, Giáp and Nguyễn Văn Huyên began to translate his insights about peasant life, and texts on the geography of Vietnam, into the language of Vietnamese nationalism. As Văn Huyên later reflected, the intellectual apparatus of French imperialism and scholarly works like Gourou's could be made to speak to revolutionary ambitions.[113] Gourou's image of an age-old peasant society in delicate balance with an unruly tropical nature began to be reconfigured as part of a Vietnamese revolutionary tradition. As we saw in Chapter 2, the peasant masses of an ethnically diverse and elongated country also started to be incorporated into Giáp's 'people's war' against the French, and later the Americans.

Pélissier thought that 1946–1947 was a vital tipping point for Gourou: one that "turned him off politics for good."[114] He had striven for a negotiated solution to the Indochina War, but it had not worked, and he was critical of French Communist Party, then in government in France, for its inertia over the colony. For Georges Condiminas, this made Gourou less of a colonialist than radical geographers subsequently claimed.[115] Gourou's life and work had become caught up in the exigencies of war, and he had used his academic reputation to leverage for peace rather than conflict (albeit an arbitration that would have kept Indochina within France's imperial yoke). His reaction to and disappointment over how things panned out helps to explain the focus and aesthetic tenor of his contribution to the 1949 *France Illustration* special issue (Chapter 2). He was well placed to talk about the geo-political evolution of the conflict, but chose to return to the bucolic beauties of a delta then under arms and under threat of destruction.

My tropicalist orientation

In a short speech, titled "My Tropicalist Orientation," to a 1988 colloquium in his honour, an elderly Gourou remarked: "I devoted myself to the geography of the tropical world after a long evolution and a few accidents which served my aspirations."[116] We have been concerned in this and the last chapter with the personal and political expediencies and "accidents" that were involved in the "evolution" of his work from 1936 to 1950, and have sought to show that the 1930s build-up to war in Europe and the Far East, and the 1940s experience of occupation, played a much more significant role in the trajectory of his work than has hitherto been disclosed. And as we will now show, this experience left a lasting impression on how Gourou's post-war project of tropical geography would develop and became aligned – in fact, constructively and critically misaligned – with the dominant forces of development and decolonisation that traversed the tropical world after World War II.

Notes

1 Daniel Gorman, *International Cooperation in the Early Twentieth Century* (London: Bloomsbury, 2017), 185.
2 AULB: IP-713 Fonds Pierre Gourou (correspondence 1936–1945).
3 AULB: IP-713 Fonds Pierre Gourou – Gourou to Marius Moutet 6 July 1936.
4 See Geoffrey Parker, "French geopolitical thought in the interwar years and the emergence of the European idea," *Political Geography Quarterly* 6, no. 2 (1987): 145–150.
5 He soon developed a strong desire to do fieldwork in Brazil and the Amazon, following in the footsteps of de Martonne (who had a short six-month stay in Brazil) and Pierre Monbeig (at Saò Paulo). Gourou's personal dossier at AULB states that he undertook 60 hours of teaching per year on human geography, and a further 90 hours of practical classes; 45 hours on regional geography, with a further 60 hours of practical activity, and 15 hours of methodology teaching. Effectively, he taught everything singlehandedly for many years. AULB: IP-713 Gourou Papers (correspondence 1935–1945)
6 ACDF: Pierre Gourou – personal dossier.
7 Moutet appointed Robert Delavignette Director of the École in 1937 – a position he held through the war (harbouring students involved in the Resistance) and until 1946, when he became (for a year) High Commissioner of Cameroun, before becoming a key figure, from 1947 to 1951, in Moutet's Ministry for Overseas France. This is when he got to know Gourou. Delavignette's 1946 *Service africain* became the standard work on French colonial administration in tropical Africa. Rivet was also a professor at the École.
8 N.T. Long, "Some features on Indochina institute for human studies (1937–1944)," *Social Science Information Review* 3, no. 2 (2009): 46–52.

200 Gourou en guerre

9   The Institut published a newsletter between 1938 and 1943, and classes in history, archaeology and ethnology were held at the University of Hanoi under its auspices. Gourou played a long-distance advisory role. In a 1948 letter to Gourou, Paul Lévy (then director of the EFEO) addresses him as his "friend and mentor," Paul Lévy to Pierre Gourou 18 December 1948 AEFEO: P74 Dossier du P. Lévy.
10 Febvre and Demangeon collaborated on a short book, *Le Rhin* (1936), and argued that the river was not only a break, but above all a link, between north and south, east and west. They renounced the Europe of nations for that of union. Similarly, Jacques Ancel (who would perish in 1943), in his 1938 *Géographie des frontieres*, criticised German geopoliticians for introducing the racial dimension and using geostrategic analyses that aimed solely to fuel nationalist sentiment. Complicating German geographers' quest for a 'just and natural' frontier, Ancel argued that mountains and rivers were not natural frontiers but places of contact. As Mike Heffernan surmises: "the enthusiasm for imagining new European solutions to national problems tends to flourish at precisely the moments when Europe is most divided and its future most uncertain." Michael Heffernan, *The European Geographical Imagination* Hettner-Lecture 2006 (Stuttgart: Franz Steiner Verlag, 2007), 44.
11 See M-C. Robic, A.M. Briend and M. Rössler *Géographes face au monde: L'Union géographique internationale et les congrès internationaux de géographie* (Paris: L'Harmattan, 1996).
12 Neil Smith, *American Empire: Roosevelt's Geographer and the Prelude to Globalization* (Berkeley: University of California Press, 2003), 277–282.
13 Cited in M. Rössler, "La géographie aux congrès internationaux: échanges scientifiques et conflits politiques," *Relations internationales* 62 (1990): 183–199, at 189.
14 Cited in Rössler, "La géographie aux congrès internationaux," 189.
15 Dany Bréelle, Interview with Pierre Gourou, Bruxelles 29 August 1995, Appendix H in Dany Bréelle, "The regional discourse of French geography: The theses of Charles Robequain and Pierre Gourou in the context of Indochina," Unpublished PhD dissertation, Department of Geography, Flinders University, 2003, 336–337.
16 Gourou, Preface, Jacques Weulerrse, *Noirs et blancs: A travers l'Afrique nouvelle: de Dakar au Cap* (Paris: CTHS, 1995), ix.
17 Pierre Gourou, *Terres de bonne espérance, le monde tropical* (Paris: Plon, 1982), 396.
18 Gourou, *Terres de bonne espérance*, 396.
19 Albert Demangeon, "Géographie politique à propos de l'Allemagne," *Annales de Géographie* 48 (1939): 113–119, at 114.
20 AULB: IP-713 Pierre: Gourou to Paul Hyman 7 December 1939.
21 AULB: IP-713 Pierre Gourou to ULB Administration 13 May 1940.
22 John Kleinen, Interview with Pierre Gourou, 24 August 1994, n.p., unpublished typescript, cited with the permission of the author. And see Jean Gottmann and Pierre Gourou, "Albert Demangeon (1872–1940)," *Bulletin de la Société languedocienne de Géographie* 12, no. 1 (1941): 1–15.
23 N. Nicolaï and A. Jaumotte, "Notices – Pierre Gourou," *Belgian Royal Academy* 34 (2001), www.academieroyale.be.
24 In French state universities, the position of *chargé de cours* (or *chargé d'enseignement*) is equivalent to a teaching fellow or sessional instructor in British and North American universities. The position of *maître de conférences* is roughly equivalent to the rank of assistant professor in North America, and lecturer or senior lecturer in the UK.
25 J-P. Chevalier, "Éducation géographique et Révolution nationale: La géographie scolaire au temps de Vichy," *Histoire de l'éducation* 113 (2007): 69–101.
26 Hugh Clout and Peter Hall, "Jean Gottmann," in *Biographical Memoirs of Fellows*, vol. 2 (British Academy, London: Oxford: Oxford University Press): 201–218, at 205.
    Gottmann was left bereft by the deaths of Demangeon and Sion in the same month.
27 The new 'network' to which Gourou became connected in Montpellier crossed the English Channel. Patrick Geddes established the Collège des Ecossais (Scots College)

in Montpellier in 1924 to facilitate exchanges between British and French geography students and professionals. See Hugh Clout and Ian Stevenson, "Jules Sion, Alan Grant Ogilvie and the Collège des Ecossais in Montpellier: A network of geographers," *Scottish Geographical Journal* 120, no. 3 (2004): 181–198.
28 Cited in Rössler, "La géographie aux congrès internationaux," 194.
29 Gourou, *Terres de bonne espérance*, 26.
30 Fernand Braudel, *On History* trans. Sian Matthews (Chicago: University of Chicago Press, 1980), 7–10, 33.
31 Aimé Césaire and Suzanne Roussi eds., *Tropiques* [1941], 2 vols. (Paris: Jean-Michel Place, 1978), I, x–xiii.
32 Pierre Mabille, "La jungle," [1945] *Tropiques*, II, 187.
33 Suzanne Roussi, "Le grand camouflage," [1945] *Tropiques*, II, 267–272. Also see Kara M. Rabbitt, "Suzanne Césaire's significance for the forging of a new Caribbean literature," *The French Review* 79, no. 3 (2006): 538–548.
34 Aimé Césaire n.t. [1941] *Tropiques*, I, 3.
35 J. Michael Dash, *The Other America: Caribbean Literature in a New World Context* (Charlottesville: University Press of Virginia, 1998), 21–35. Also see Gary Wilder, *Freedom Time: Negritude, Decolonization, and the Future of the World* (Durham, NC and London: Duke University Press, 2015).
36 Laurent Beauguitte, "Publier en temps de guerre: les revues de géographie française de 1939 à 1945," *Cybergeo: European Journal of Geography* [Online] 16 September 2008, http://journals.openedition.org/cybergeo/19853; DOI: 10.4000/cybergeo.19853.
37 Aimé Perpillou, "Geography and geographical studies in France during the war and the occupation," *The Geographical Journal* 107, no. 1 (1946): 50–57, at 50.
38 Perpillou, "Geography and geographical studies in France," 51–54.
39 See Chris Pearson, *Scarred Landscapes: War and Nature in Vichy France* (London: Palgrave Macmillan, 2008).
40 Cited in Olivia Harris, "Braudel: Historical time and the horror of discontinuity," *History Workshop Journal* 57 (2004): 161–174, at 174. For a broader survey of the different ways in which stealth and survival became geographers' watchwords, see Hugh Clout, "French geographers during wartime and German occupation, 1939–1945," *Journal of Historical Geography* 47, no. 1 (2015): 16–28.
41 Paul Pélissier, Interview with the authors. But Pélissier added that resistance was also a matter of individual conscience and method, and some of Gourou's students had their own minds. As Gourou's colleague Louis Papy wrote of one of them: "Jean Borde just disappeared one summer day after passing brilliantly his diploma. In fact, he had joined the maquis. He was captured by the Germans and subjected to a harsh interrogation. He managed to escape, and, via the Pyreneees, reached Spain, from which he reached the Free French in North Africa." Louis Papy, "Jean Borde (1921–1977)," *Annales de Géographie* 87 (1978): 78–81, at 78–79.
42 Pierre Gourou, *L'Afrique tropicale: Nain ou géant agricole* (Paris: Flammarion, 1991), 140. And see François Rouquet, "L'épuration des universités: défense et justification," in Marc Olivier Baruch ed., *Une poignée de misérables: L'épuration de la société française après la seconde guerre mondiale* (Paris: Fayard, 2003), 515–529.
43 Cited in Olivier Dumoulin, "La Langue d'Ésope: les revues historiques entre science et engagement," *La Revue des revues* no. 24 (1997): 45–71, at 52.
44 Philippe Burrin, *La France à l'heure allemande 1940–1944* (Paris: Seuil, 1995), 322–325.
45 BNF: Fonds Gottmann: Pierre Gourou to Jean Gottmann 19 September 1941. Also see Luca Muscarà, "The long road to Megalopolis," *Ekistics* 70, no. 418/419 (2003): 23–35.
46 BNF: Fonds Gottmann: Gourou to Gottmann 30 March 1942.
47 BNF: Fonds Gottmann: Gourou to Gottmann 13 April 1942.
48 Raoul Blanchard and André Cholley were both students of Vidal de la Blache who specialised in regional geography. Cholley became director of l'Institut de géographie in 1944 and Professor of Geography at the Sorbonne in 1945.

49 BNF: Fonds Gottmann: Gourou to Gottmann 30 August 1942.
50 See Jacques Leclerc, "Amsterdam 1938: un tropique bien blanc, sinon rien," in Michel Bruneau and Daniel Dory, éds., *Les Enjeux de la tropicalité* (Paris: Masson, 1989), 91–97.
51 See e.g. Ian Ousby, *Occupation – The Ordeal of France 1940-1944* (New York: St Martin's Press, 1998); Lynn Taylor, *Between Resistance and Collaboration: Popular Protest in Northern France, 1940-1945* (New York: St Martin's Press, 2000); Richard Vinen, *The Unfree French* (New Haven: Yale University Press, 2006).
52 Pierre Gourou to Marinette 2 May 1944. Letter in private possession of Michel Bruneau, quoted with permission.
53 Details in the following few pages are based on the information in: AMB: 7880 H 1: Comité départemental de la libération, Gironde, Reports and correspondence, 1944–1945 – here 29 August 1944; and ADG: 57 W 24: Commission du Ravitaillement, 1944.
54 AMB: 7880 H 1: 1 September 1944. France-Liberté was a resistance movement established in Lyon in 1940. It changed its name to Franc-Tireur in December 1941. By 1944 it had merged with Libération-Sud and Combat, forming the Mouvements unis de la Résistance, although Gourou continued to identify with the original group. It has been estimated that there were 45 resistance 'movements' in France during the war, over 250 'networks,' and over 1,000 clandestine publications, and with the movements having a more integrated system of recruitment, organisation and knowledge dissemination than the networks, which were often looser assemblages. See Christopher Lloyd, *Collaboration and Resistance in Occupied France: Representing Treason and Sacrifice* (London: Palgrave Macmillan, 2003), 27.
55 AMB: 7880 H 1: 1 September 1944.
56 Pétain, October 1940, quoted in Kenneth Mouré, "Food rationing and the black market in France (1940–1944)," *French History* 24, no. 2 (2010): 262–282.
57 Eric Alary, Bénédicte Vergez-Chaignon and Gilles Gauvin eds., *Les français au quotidien, 1939–1949* (Paris: Perrin, 2006).
58 Pearson, *Scarred Landscapes*.
59 Michel Cépède, *Agriculture et alimentation en France durant la IIe guerre mondiale* (Paris: Genin, 1961).
60 A Bremen wine merchant, Heinz Boemers was appointed Bordeaux's *Weinführer*, and by 1944 over 15 hectolitres of wine had been confiscated the Third Reich, including the first growth wines of the most distinguished chateaux. Boemers insisted that he bought and traded wines in an orderly fashion, and the region's wine producers and merchants realised they could profit by the German occupation (providing much grist for allegations of collaboration down the line). However, Germany's devaluation of the Franc (1940) and imposition of a 20 per cent tax on a bottle of wine soon encouraged black market activity and the requisitioning rather than just sale of wine. See Donald Kladstrup and Petie Kladstrup, *Wine and War: The French, the Nazi, and the Battle for France's Greatest Treasure* (New York: Broadway Books, 2001), 57–89.
61 Lloyd, *Collaboration and Resistance*, 32–41.
62 AMB: 7880 H 1: Reports and correspondence 12 September 1944.
63 The same day Caillier resigned as President amidst a row about how suspicions of collaboration should be dealt with, Gourou voted with the communist bloc on the committee, to the effect that all suspicions should be investigated and settled by means external to the committee.
64 The British praised Grange for the detailed and accurate information he had provided on the movement of cargo ships and submarines, and his protestations to German Commandant Kühnemann in August 1944 were instrumental in the German navy not acting on orders to destroy the port. See Max Vignes, "In memoriam – Alphonse Grange (25) 1906–2001," *La Jaune et la Rouge* online, www.lajauneetlarouge.com/article/memoriam-alphonse-grange-25-1906-2001. Also see Louis Papy, *Aunis et Saintonge* (Collection *Les Beaux Pays*) (Grenoble: B. Arthaud, 1937).

65 Paul Pélissier told us that Gourou was close to Delaunay, who went on to become the regional prefect of Aquitaine after to the war. Paul Pélissier, Interview with the authors.
66 AMB: 7880 H 1: 26 September 1944.
67 AMB: 7880 H 1: 7, 21 September 1944. He also oversaw the ousting of a committee member for holding pro-Vichy views, and the reprimand of one for socialising with a collaborator.
68 AMB: 7880 H 1: 21, 28 November 1944.
69 AMB: 7880 H 1: 7 December 1944.
70 AMB: 7880 H 1: 7 December 1944.
71 AMB: 7880 H 1: 7 December 1944.
72 Gilberte Bray. Interview with the authors.
73 Antonio Gramsci, *Selections from the Prison Notebooks* Quentin Hoare and Geoffrey N. Smith eds. and trans. (New York: International Publishers, 1971), 137.
74 Gourou, *Terres de bonne espérance*, 182.
75 Pierre Gourou, *Terres de bonne espérance*, 183.
76 AULB: IP-713: Gourou to Rector 5 May 1945.
77 ANF: F/17/16701–16705 Conseil académique d'enquête de l'Académie de Bordeaux.
78 ADG: 57 W 25: "Note pour M. Gourou,"
79 ANF: F/17/16705 Gourou dossier, 1945.
80 ACDF: Gourou – CV.
81 Megan Koreman, *The Expectation of Justice: France 1944–1946* (Durham, NC: Duke University Press, 1999), 10.
82 Tony Judt notes that collaboration "was not a pre-existing crime with legal definitions and stated penalties." It raised many "conundrums" as to what constituted harm and who was to take responsibility. Prosecution and punishment varied greatly in time and space, and "there were multiple procedural irregularities and ironies, and the motives of governments, prosecutors and juries were far from unsullied – by self-interest, political calculation or emotion." But it was remarkable in 1945 that the rule of law and justice was efficacious at all, for a whole continent had to define a new range of crimes on a continental (and not just French) scale. Between 1944 and 1951 official courts in France sentenced 6,763 to death for treason, although only 791 executions were ever carried out. 'National degradation' (introduced in August 1944) was the (loosely defined) crime most often punished, usually by being denied position in society: 49,723 received this punishment, and 1.3 per cent all state employees were removed from their jobs. Overall, *épuration* affected 350,000. Tony Judt, *Postwar: A History of Europe Since 1945* (London: Heinemann, 2010), 42–47.
83 The provinces were more severe in their punishments – sackings, demotions and fines, and sometime imprisonments – than was the capital, although Paris provided the largest number of cases of university academic punished (Bordeaux came fifth). Such disparities reflected the vigour of local reactions rather than the real extent of collaboration. A 1950 amnesty bill pardoned many, but such was the sensitivity surrounding the subject of the purges in France that Novick's important 1968 study was not translated into French until 1985. See Peter Novick, *The Resistance Versus Vichy: The Purge of Collaborators in Liberated France* (London: Chatto and Windus, 1968); Koreman, *The Expectation of Justice*, 94.
84 ANF: F/17/16705 Gourou dossier, 1945.
85 Judt, *Postwar*, 45.
86 In fact all sides – Vichy, the communists, and the liberal coalitions that drafted the constitution of the fifth republic – saw empire as a compensation for the trials and tribulations of war and occupation.
87 Details from: UBCA: Institute of Pacific Relations Fonds, 51–59 Hot Springs.
88 Paul F. Hooper ed., *Remembering the Institute of Pacific Relations: The Memoirs of William L. Holland* (Tokyo: Ryukei Press, 1995), 37.

89 What is known about the conference has been pieced together from conference agendas, published summaries, delegate submissions (such as Gourou's paper) and newspaper coverage: Institute of Pacific Relations, *Security in the Pacific: Proceedings of the Ninth IPR Conference Hot Springs 1945* (New York: IPR, 1945); Yutaka Sasaki, "Foreign policy experts as service intellectuals: The American Institute of Pacific Relations, the Council of Foreign Relations, and planning the occupation of Japan during World War II," in G. Kurt Piehler and Sidney Pash eds., *The United States and the Second World War: New Perspectives on Diplomacy, War, and the Home Front* (New York: Fordham University Press, 2010), 293–332, at 297–299.

90 Gourou, *Terres de bonne espérance*, 368–369.

91 Summary of the American position in William C. Johnstone, "The Hot Springs conference," *Far Eastern Review* 14, no. 2 (1945): 16–22, at 20; draft memoranda, UBCA: Institute of Pacific Relations Fonds, 54–54; Gourou, *Terres de bonne espérance*, 369. Johnstone was Dean of the School of Government as George Washington University and was particularly interested in the fate of the Japanese emperor.

92 Gourou, *Terres de bonne espérance*, 369.

93 Hooper, *Remembering the Institute of Pacific Relations,*39; IPR, *Security in the Pacific*, 5–9.

94 Gourou, *Terres de bonne espérance*, 345.

95 ACDF: Gourou 1947 "Leçon inaugurale du cours de Pierre Gourou, 4 décembre 1947" ms.

96 Gourou, *Les terres de bonne espérance*, 43–45.

97 It was common practice for geographers from French state universities to examine baccalaureate students taught in overseas institutes. Under the direction of Théodore Monod, IFAN and its satellites attracted radical African and French intellectuals, including the Marxist geographer Jean Suret-Canale (see Chapter 1).

98 German visions of an edenic tropics were fuelled not only by Nazi expansionist ideology, but also by German cinema – for instance F.W. Murneau's 1931 film *Tabu*, set in the South Seas. See Anthony Coulson, "Paradise islands: Van Dyke's *White Shadows in the South Seas* and F.W. Murneau's *Tabu*," in Wendy Ellen Everett and Axel Goodbody, eds., *Revisiting Space: Space and Place in European Cinema* (Bern: Peter Lang, 2005), 133–156.

99 Stein Tønnesson S., *Vietnam 1946: How the War Began* (Berkeley: University of California Press, 2010).

100 Martin Shipway, *The Road to War: France and Vietnam, 1944–1947* (Oxford: Berghahn Books, 1996), 180.

101 ANOM: FP 56 PA/6P Dalat [April 1946]

102 ANOM: FP 56 PA/6Monguillot 1946, Papiers, annex 5, "Instructions du ministre des affaires étrangères pour la négociation d'un accord avec le gouvernement provisoire du Vietnam,"; and " Note sur le conférence de Dalat, 18 Avril–11 Mai 1946."

103 Gourou, *Terres de bonne espérance*, 32.

104 Joseph Inguimberty to Pierre Gourou 29 April 1946, Papiers Gilberte Bray. The Inguimberty family had to avoid Marseille on their return home to avoid communist dockers who had seized the harbour.

105 Gourou, "Itinéraire," 4–5.

106 Gourou, "Itinéraire," 3.

107 Kleinen, Interview with Gourou, n.p.

108 Pierre Gourou, *L'avenir de l'Indochine* (Paris: Centre d'études de politique étrangère, 1947), 54.

109 Pierre Gourou, "For a French Indo-Chinese federation," *Pacific Affairs* 20, no. 1 (1947): 18–29, at 19.

110 Pierre Gourou and Jean Loubet, *Cours de géographie: Enseignement primaire supérieur franco-indigène, 4e année: L'Asie moin, l'Asie russe, l'Indochine* (Hanoi: Imprimerie tonkinoise, 1934), 42; Gourou, "For a French Indo-Chinese federation," 24; In *L'Asie*, and with the Indochina War still in full swing, Gourou maintained

that "the main problem facing Indochina is a political one. If France ceases to exert a moderating and conciliating influence there is little reason why the political structure it built should survive." The following year he noted, "Immense in space and time, Asia will be the place for fermentations and evolutions that, by themselves, will be of the greatest interest and that will strongly influence the evolution of the rest of the planet." Pierre Gourou, *L'Asie* (Paris: Hachette, 1953), 336; Pierre Gourou, "Introduction" Roger Lévy, *Situations en Extréme-Orient* (Paris: Centre d'études de politique étrangère, 1954), 3.

111 Paul Sorum, *Intellectuals and Decolonization in France* (Chapel Hill, NC: University of North Carolina Press, 1977), 212.

112 Gourou, "For a French Indo-Chinese federation," 24, 29. The IPR was reorganised after World War II to reflect the shift in global politics: communist China and the Soviet Union withdrew, and Japan had only a shadow presence, while new national councils representing newly independent countries (e.g., Burma, India, Indonesia, Pakistan) were created. The IPR later suffered at the hands of Senator McCarthy's Congressional hearings.

113 Nguyen Van Huyen, *Seize ans d'essor de l'education nationale au Viet Nam (1945–1961)* (Hanoi: Éditions en Langues Étrangères, 1961).

114 Paul Pélissier, Interview with the authors.

115 Georges Condominas, Interview with the authors.

116 Pierre Gourou, "Mon orientation tropicaliste," in Henri Nicolaï ed., *Lecons de geographie tropicale et subtropicale: colloque organise en l'honneur de Pierre Gourou, Nimègue, 27 octobre 1988* (Faculteit der Beleidswetenschappen: Katholieke Universiteit Nijmegen, 1989), 4–7, at 4.

6   Affecting the tropics

'The world has changed completely'

Reporting to the Assembly of Professors at the Collège de France on 24 November 1946 and again on 17 February 1947 about the need for a chair devoted to the study of the tropical world, and proposing geographer Pierre Gourou for the position, the *Annales* historian Lucien Febvre underlined "the growing moral and scientific authority of a scholar who does not think you should close your eyes to the contemporary world," and explained:

> The world has changed completely in the last ten years. A chasm has opened up between 1936 and 1946. Surprising means of destruction and creation, and an uninterrupted series of 'progressions,' have transformed the old world and deeply modified the Universe. Twenty years ago, the spotlight shone only on the temperate lands of Europe and North America. Today the spotlight now shines on lands previously neglected. The satellite or annex lands. Some of them frozen, like the Arctic and Antarctic worlds. Others burning: I refer to the inter-tropical lands. . . . The tropical world is not a label, but a reality. It constitutes, at the centre of the globe, a powerfully individualised zone; it has its own physical geography; relations between man and milieu are not posed in the same terms there as they are in temperate lands.

In his second testimonial Febvre continued:

> One man has thought about the problems I have expressed . . . [and] has spent twenty years working on them . . . with a singular breadth of vision and insight. Qualities of a man. Qualities of a scholar. I learned to recognise and honour the scholar by reading his remarkable books. As for the man, I recall a memory that is very dear. It was one of the last meetings I had in Paris with Marc Bloch, who then led the hard and heroically simple life of the Resistance. He evoked his time with Gourou in Montpellier – and all the ugliness and nastiness . . . they had witnessed. But also all the nobility and generosity, the passionate devotion to France. . . . And Bloch told me: Pierre Gourou is not just a great geographer. He is also a man. Don't lose sight of that; nor that he does not think that one should close one's eyes to the contemporary world.[1]

The spotlight Febvre placed on the tropics and other "lands previously neglected" illuminated a rupture in the order of things. Febvre had written a great deal about the idea of civilisation as a narrative of the development of Western society from a primitive to an advanced state, and how, by the 1930s, civilisation was no longer a historic term denoting the collective life of a human group, or culture, but had become a contentious ethnographic concept used to measure and rank different cultures against European norms.[2] Over the previous 100 years the idea had acquired sinister colonialist, racist and nationalist overtones. Potent theses about the determining influence of climate on culture and civilisation were in the air too and Febvre had debunked them in his 1922 *La terre et l'évolution humaine*, which was subtitled "a geographical introduction to history" and was published in English with that title in 1925.[3] Both Febvre and Gourou acknowledged the impress of physical conditions on human life, particularly in the tropics, but strenuously rejected environmental determinism. Thanking Febvre for his support at the start of his inaugural lecture at the Collège de France nine months later, Gourou argued that the overriding virtue of the *Annales* historian's work was that it "keeps us from the traps of determinism."[4]

However, Febvre and Gourou struggled to escape a residual problem: namely, that due to a long history of colonial racism and violence, Western research on the non-western world was viewed from that world – and it soon was by Césaire – as laden with prejudice and white-man supremacy. This formed part of the "chasm" to which Febvre alluded. At the same time, Febvre was adamant that this rupture in the order of things heightened the need for thinkers and scholars to keep their eyes open to the world. 1947 was a moment of profound uncertainty and breakage, but also of hope and opportunity. As with Henri Lefebvre's *Critique of Everyday Life*, which was published in February 1947, Febvre's remarks perhaps reflect "the optimism and new found freedom of the Liberation, but appeared only a few weeks before the big freeze of the Cold War set in. . . . The year 1947 was pivotal, Janus-faced. Euphoria gave way to caution and disillusionment."[5]

The dawning of a brighter world was mediated by both belief and doubt regarding the ability of Western thinkers and scholars to provide some clarity and leadership. The history of civilisation needed to be rethought in relation to what Febvre termed "the dark side of modernity," which had issued from interwar Europe and Nazi Germany and enveloped the world.[6] His image of "surprising means of destruction and creation, and an uninterrupted series of progressions" chimes with the famous 'all that is solid melts into air' passage in Marx and Engels' *Communist Manifesto*, which conjures with the contradictory (creative and destructive) nature of capitalist modernity. The "progressions" to which he was referring were of a new and perplexing order: nationalist and ethnic assertion and independence struggles around the world; emerging Cold War bipolarity and the spectre of nuclear annihilation; racial as well as class struggle (including widespread strikes in the French empire); widespread rationing and food shortages; and a specious development discourse (empire's new clothes) that was starting to come out of the U.S.[7]

World War II further undermined the ability of colonial powers to claim they could offer leadership in either a moral or practical sense. In 1947, the French empire in Africa, Southeast Asia and the Caribbean was unravelling, and the incongruities between the politics of metropole and colony had reached crisis point. Many French colonial troops, three quarters of a million of whom hailed from North and West Africa, were bewildered about why and for whom they had gone to war. The French Communist Party had left the coalition government in Paris, and with fears that Moscow was seeking to foment communist revolt in French Africa and Polynesia there were urgent calls for trustworthy information about hitherto little known and understood places, many of them in the tropics. When French school textbooks started to be printed again after World War II, the tropical world attained a heightened visibility.[8]

This was also a time of great privation and interest in means of subsistence and reconstruction. France and other Western countries looked to the tropical world anew, as a material lifeline for food, raw materials and markets. In 1945 United Nations called for geography to be a core subject in school curricula around the world, believing the subject well placed to promote international collaboration and understanding.[9] We have also seen how the French strove for a revivified Union française, partly by giving assurances that the *niveaux de vie* (living standards) of colonial subjects would be raised and there would be a formal end to the subordinate (*indigène*) status of French colonial subjects. Gourou embraced this idea.

Febvre was by no means the only European commentator to think about decolonisation as a 'progression' in these vexed terms. He was no apologist for empire and not alone in noticing that most of Europe's colonies were in the tropical belt, and that the tropics were rich in natural resources. Accordingly, the 'progression' of decolonisation would generate acute demands for knowledge about how tropical nature might be exploited, and again geography was deemed to be ideally placed to provide it. Once a byword for foreign adventure and the exotic, the tropical world was now a "powerfully individualised zone" and education and expertise were the keys to grasping its problems, and thus averting disorder and strife. Similarly, the American geographer Owen Lattimore observed that victory over Japan would spur American interest in the Far East and generate an immediate need for factual knowledge of soil, climate, land use, people, village life, natural resources and industries. "A fresh integration is demanded of us because of the historical juncture," he wrote in 1945. "Of all the phases of history, the phase that immediately follows a great crisis is the one that demands the most subtle analysis of the way in which change works, and the most delicate handling of the process of change."[10] In the American case, this was expedited through the creation of area studies programmes, albeit none of tropical studies, and, perhaps ironically, with geography and geographers largely shut out of the ones that were established (partly since more than a quarter of American universities with such programmes did not have a geography department).[11]

By 1947 Gourou was widely acknowledged as a leading authority on the tropics, and Febvre's tribute to him hinged on the kind of epistemological outlook touted by Lattimore. But the epistemological tenor of Febvre's first testimonial cannot be fully grasped unless the emotional undercurrents of his second testimonial linking

Gourou 'the scholar' to Gourou 'the man' are unpacked. This second testimonial points to what Febvre saw as important ethical and affective dimensions to Gourou's work. For Febvre, as for Marc Bloch (Gourou's friend and Febvre's ally), intellectual stoicism was an important element of wartime survival and postwar wisdom. Writing to Febvre on 22 January 1942 from Montpellier, Bloch said that he had met Gourou and was deeply impressed by his 1940 survey *La terre et l'homme* (see Chapter 4) Gourou was "one of the three or four geographers who count," Bloch remarked, and by the end of the war the geographer's reputation had been cemented in the eyes of these two leading *Annales* historians.[12] To Bloch and Febvre, it was Gourou's fortitude and poise in the face of crisis that linked scholar to man, and he regarded such stoicism as of continuing importance after World War II, when, as Frantz Fanon portended, impartiality and solemnity in the face of violent emotion was neither appropriate nor possible in emerging situations of decolonisation.[13]

In 1934 Bloch secured a contract with the publishing house Gallimard to edit a book series titled *Les paysans et la terre* (Peasants and the earth), which would be a comparative and global project, and involve historians, anthropologists and sociologists (in the Durkheim stable) as well as geographers. Bloch recommended Gourou as a possible contributor. As Susan Friedman relates, the moral project at the heart of Bloch's concern with the structures of peasant life lay in his "belief in 'the attraction and sacred character of any human effort under whatever skies or by whichever branch of the great family of men that one accomplishes it.' . . . Bloch suggested that human life had proved more powerful than the events which challenged it. The message here was one of hope and resilience during what were exceedingly disrupted times."[14]

Febvre grappled with the emotional stakes in how intellectuals responded to violence and vehemence in a 1941 essay on the affective dimensions of history which was written from his censored wartime bunker at the Collège de France.[15] He conjured there with a distinction between "emotional life" and "intellectual life," describing the former as a life of fear, enmity, despair, unbridled passion and blind devotion that was spurred by experiences of occupation, conflict, repression and exploitation, and the latter as a stoical life of reason, dignity and compassion, and a capacity for toleration and adaptation in the face of adversity and suffering.[16] He worried that between 1900 and 1945 emotional life had overflowed into intellectual life and undermined the grounds upon which destructive and malicious attitudes (as well as constructive sentiments of affection, hope and empathy) could be construed as frameworks of inquiry in the first place.

He argued that it was only by finding moments in history when extreme and destructive emotion had been curbed or channelled – when "the emotions became a sort of institution . . . controlled in the same way as a ritual" – that the historian could start to tackle the "dark side of modernity" (its crippling violence and wanton desire) and know how to identify and contest dark forces if they came along again.[17] A history of "civilised life," he continued, should not extol the West, because the West was the fount of colonisation as well as civilisation. Rather, he called for an "affective history" that would track, in global and comparative terms, when and where eruptions of emotional life had been channelled back into intellectual life, and where they had not.

How would scholars and intellectuals respond to the welling up of emotional life and the fear and hatred it unleashed? On what basis would they take sides? How did they remain composed and dignified when they encountered malice and horror? Would they become collaborators, passive witnesses, or armed adversaries? Was it possible for them to keep a distance from scenes of violence, excess and evil? Febvre lived these questions and knew that his friends Gourou and Bloch had too. He rationalised intellectual life as the life of the French Resistance and knew that taking sides did not follow a single or simple script. Gourou had served the Resistance as a planner, Bloch as a combatant (who was captured and shot by the Nazis), and Febvre as a subversive operating from his university quarters. Each of them knew that stoicism, fear and insecurity did not come in distinct packages of class, gender, race, nationality or region. Febvre urged that the scholar and intellectual had to be on the side of intellectual life but acknowledged that the relations between emotional and intellectual life were complex and fraught. What some counted as positive, bright and progressive emotion, others counted as negative, dark and repressive.

Michel Bruneau describes Febvre, Bloch and Braudel as Gourou's "intellectual family," and it is with Febvre's testimonial in mind that we return, in this chapter, to the scene of difference with which we began the book (but left in abeyance) and explore Gourou's post-war entanglement with tropicality.[18] *Les pays tropicaux* was published just a few months after Febvre gave Gourou this ringing endorsement, and the gulf in opinion between Césaire's admonition of the book and the laudatory reviews it received within the academic press at the time did not melt into air in ensuing decades. We shall consider the range of (often conflicting) motivations, experiences and influences that shaped understanding of the tropics after World War II, with reference to *Les pays tropicaux*, a considerably larger body of work by Gourou, and broader and shifting geographical and political mind-sets. Again, our aim is to relate the particulars of Gourou's project to wider processes – here of development and decolonisation – while underscoring what was unique about his interests and arguments.

As Febvre intimated, Gourou's election to the Collège de France happened on the cusp of fast-changing and uncertain connections between temperate and tropical regions. Gourou's comparative and zonal interest in the tropics became geared to a peacetime order which was imbued with fraught distinctions between dream and reality, and reason and emotion. His attempt to grapple with these distinctions spawned a tropicality that, yes, cast a geographical spell over the tropics, as Césaire averred; and yes, was pessimistic and backward-looking, as Arnold subsequently observed. In our view, however, it was more supple and ambivalent than this, and more interesting for it. Deeply affected by World War II, Gourou was also traumatised by the fact that Vietnamese communists had assassinated many of his former Hanoi pupils.[19]

The adjective 'affecting' is deployed in our chapter title to capture the burden of our argument, which is that the tropical world was both an affected and affective domain: impinged upon and transformed in material terms, and moved by tangled webs of emotion, many of which were bound up with the ends of empire and question of how order, harmony and rationality would be recuperated from brutality, and with new – post-colonial – forms of covetousness. Remarkably little attention has been paid to the affective dimensions of post-war tropicality, and in our account of them here we consider Gourou's concern with geographical truth, order and harmony, and concomitant emotions of scepticism, nostalgia and compassion.

'Illusory riches'

As Febvre surmised, Gourou's post-war project can be read as an emotional as well as epistemological response to change, at the heart of which was a drive to channel fantasy and exuberance regarding the tropics back into factual realism, and with what the American historian Palmer Throop, with respect to Febvre, perceptively identified as an outlook of "ironic compassion" that had many affinities with Bloch's injunction to the historian "Not to judge, but understand" and as Febvre added, "to reveal and demonstrate as well as explain . . . and look and see in order to comprehend."[20]

Gourou questioned German representations of the tropics as bounteous, and there was a patriotic as well as disciplinary edge to his efforts. Reflecting in 1948 on his encounter with the Nazi geographers 10 years before (see Chapter 4), he noted: "It is now well established that the old dreams of wealth and exuberance inspired by legendary notions have no relation to tropical reality."[21] While Germany's colonial empire had been confiscated after World War I, ideas of tropical fecundity and excess had lived on and taken root in some surprising places. In his Preface to *Les pays tropicaux*, Paul Rivet remarked that Gourou had provided a wake-up call to those (especially the young) whose "need for escape" after World War II might lead them to tropical regions thinking that they were "the promised land, where life is easy, the earth is fertile, and effort fruitful."[22] Gourou "shows forcefully, and with precision, that there is a great distance between dream and reality." The fertility and wealth of the tropics had been wildly exaggerated, and the promise of riches had spawned ill-conceived colonial schemes and yielded ill-gotten gains.

This 'wake-up call' was part and parcel of Febvre's plea to draw intellectual life back from the excesses of emotional life. At the same time, Febvre recognised that the distinction he had drawn revolved around deeply troubled matters of location and perspective. How might one distinguish between a progressive and insidious science? When did devotion to a cause or a discipline or nation become blind devotion that closed one's eyes to the world? Césaire opined that the West had arrogated to itself the power to decide on how such distinctions were drawn, and Fanon began *Les damnés de la terre* with the assertion: "Decolonization, which sets out to change the order of the world, is, obviously, a program of complete disorder."[23] There was not a level playing field from which a bright new post-war or post-colonial future might emanate from a dark and destructive history of European warfare and imperial aggrandisement.

While the ostensible focus of Febvre's essay on affective history was the Middle Ages and Johan Huizinga's account of its 'childlike' emotional state, he was also thinking (in code) about how to explain Vichy France, and in the second testimonial he alludes to the ongoing relevance of his concern with affective history for understanding France's fast-changing post-war relationship with its empire. When it came to the tropics, Febvre saw Gourou as 'the man' to address the chasm between 'emotional' and 'intellectual' life that had spilled out of fascism and war into the 'progression' of decolonisation. Their response to this chasm and progression was to disclose the delicate harmony and balance – in other words, order – between people and land, and underscore how easily it was upset. It was on this material ground that the scholar and intellectual would be able to discern not only how people would deal with the dislocation and suffering inflicted by occupation, but also the rapid change portended by ambitious Western aid and modernisation projects, and decolonisation agendas in the tropics.

In his reflections on Gourou's work, Febvre traces the process by which individual experiences *of* emotion gain meaning in relation to broader discourses *about* emotion – at this juncture with feelings of hope, trepidation, resilience and uncertainty, and their bearing on ideas of order and civility, being the affective watchwords.[24] In a review of Gourou's *La terre et l'homme*, Febvre noted:

> I would feel guilty if I did not quote here the sentences of the Introduction [to the book], which show the peasants 'suffering terribly from the current disorder in the Far East' – a disorder that they did not determine and from which they will draw no benefit – innocent and pitiful victims of conflicts caused by those who, having lost the peasant sense of the harmony of relationships, fall from violence into violence, through weakness of character and poverty of spirit – 'But,' concludes M. Gourou: 'it is on the peasants that a Far Eastern order will be rebuilt. The present storms will not break the union of man and land which has made Far Eastern civilisation endure and which will save it.'[25]

By the time this review appeared Japan had occupied much of the peasant territory surveyed by Gourou, and Braudel's *longue durée* provided a refuge from the storms of the present and the fall of the peasant "from violence into violence," as Febvre put it.[26]

In 1917 Febvre had predicted that "when peace returns . . . a painful and tormented era of social conflicts and extraordinary class upheaval will begin."[27] As many other witnesses observed, World War I fractured the moral and political authority of Europe's ruling classes, and redrew the social as well as political map of Europe. The war forced many to think about the concept of Europe for the first time. European civilisation and superiority – indeed the very idea of Europe – were no longer taken for granted, and communism vied with liberal democracy as a bulwark of resistance against fascism.[28] While the nation-state assumed a preeminent position with politics, and worked to sideline the interests of minorities, Europe became exposed to new pressures from within and beyond its borders. Yet Febvre and colleagues such as Demangeon and Bloch sought to argue that, as

André Burguière puts it, "Behind the apparent force of nation-states, their governments and their institutions, it was in actuality social structures and their capacity for resistance that had the last word."[29] It was this capacity as it pertained to the Far East that Gourou had in mind when he spoke in interviews late in life about how World War I had crippled "the idea Europe." In *La terre et l'homme* he was not denying that the Far East was being rocked by war and revolution so much as trying to think past political winds of change, and about the durability and adaptability of social structures in the face of crisis and catastrophe.

This is what Febvre saw in Gourou's book on the Far East too; but only perhaps in part; for one senses that Febvre was also talking in code about the suffering and resilience of the French people in the face of German occupation.[30] Indeed, some years later, in a textbook on geography aimed at senior lycée pupils that was published at the height of the Algerian War, Gourou and his wartime colleague at Bordeaux, Louis Papy, wrote about France in similar terms:

> Each generation contends with new problems and yet in each landscape of the France of yesterday, like that of today, each human construction carries the imprint of permanent characteristics. . . . Revolutions, like those occurring today, impose an extraordinary agitation on the human soul; they put in movement a crowd of desires and new ambitions; they inspire regrets in some and chimeras in others. But this trouble must not divert us from the heart of things. . . . The attentive study of what is fixed and permanent in the geographical conditions of France must be or become more than ever our guide.[31]

By 'revolutions' they meant the agricultural modernisation plans that unfurled in France immediately after World War II, which transformed French peasant farming, as well as Communist revolutions and independence struggles farther afield.[32] Akin to Febvre, Gourou and Papy were not saying that politics, war, revolution or modernisation were unimportant. Rather, such "agitations" have a spectral – *fantômatique* – presence in *La terre et l'homme*, in Febvre's review, and in this textbook, as forces that geographical and historical inquiry can declare worrying, and perhaps an aberration (fostering regrets and chimeras), and that should spark concern over "what is fixed and permanent."

There are also strong affinities between the comparative and zonal project Gourou started to undertake in wartime Montpellier and Bordeaux, and Febvre's remarkable essay on affective history. Febvre explored the social and institutional (or ritual) structures of emotion, and the way human consciousness could be split so as to allow those caught up in violence and uncertainty to cope and remain intellectually vigilant. The carnage of World War I led Febvre to the view that history could not be based on a linear model of progress and ideal of emancipation. History was not an ineluctable march to freedom, and with world history following French and European history. Mass slaughter on European soil had trounced this model, and Febvre and Bloch objected to the way it turned the historian into a puppet. They emphasised the multifarious nature of historical processes, and sought to substitute analysis and comprehension for prophesy and teleology. The ability of societies to sustain and transform themselves in the face of adversity constituted a core element of this historical vision, and Gourou's analytical exaltation of the distinct nature of human-environment relations in the tropics fitted this vision well. Geography, or more particularly the issue of the adaptive capacity of tropical civilisations in the face of adversity, could be used to break-up universal history.

So it was that a project which Césaire and later Arnold rightly chastised as purveying a Eurocentric discourse that hinged on the distinction between a superior temperate and inferior tropical world was judged by Febvre to be one which displaced Eurocentric and Gallocentric reasoning. Indeed, as Arnold observes, Febvre's 1922 *La terre et l'évolution humaine* "anticipated at least some of Gourou's latter findings. . . . [He] denounced the 'illusory riches' and supposed natural abundance of the tropics as the product of an 'over-idealistic geography.'"[33] In a 1949 article on the relations between geography and civilisation, Febvre also paid homage to how Gourou had developed an "original" view of "civilisation as environmental response" and as "the result of centuries of civilising the plant."[34] Gourou's "geography lesson," Febvre concluded, centred on "the determinism of civilisation" rather than that of environment or race, which they both deemed specious and dangerous.[35] Gourou had shown how civilisations in different parts of the world had responded differently to similar physical conditions, and the "comparative method" upon which this insight was based did not necessarily privilege Europe as the pre-eminent or ideal civilisation.

Even so, talk of "weakness of character and poverty of spirit," and of 'something tragic' in Gourou's claim about the difficulties involved in raising the living standards of the peasants of the Far East was patronising. Let us reminds ourselves of where we started, with what Aimé Césaire took umbrage at: Gourou wrote in *Les pays tropicaux,* "We who live in temperate lands find it difficult to realise how baneful nature can be to humanity there, or to grasp that in many tropical regions water may swarm with dangerous germs, numerous blood-sucking insects may inject deadly microbes into the body, and the soil may be very harmful to touch."[36] Tropical soils "are less favourable to humanity than temperate ones," he continued, and this "basic problem of salubrity" was aggravated by "pathogenetic complexes" unique to the region, by agricultural practices "which in the last analysis provide an insufficient economic basis for a bright civilisation," and by colonial practices that had been "ill adapted to tropical environments and their indigenous societies."[37] He concluded that the tropical world "suffers from a certain number of inferiorities,"[38] and while geographers did not immediately pick up on Césaire's indictment of such claims, by the late 1960s denunciation of Gourou's project, and particularly his presentation of all of this as a matter of fact, had started to arise from within the ranks of the field of study – tropical geography – he had shaped.

Tropical emotion

We shall return to this criticism later. At this point we want to note that it draws our attention to a potent and deeply flawed – ethnocentric and colonialist – distinction between reason and emotion, with the former deemed the putative preserve and magnanimous aim of the objective and high-minded Western scholar and expert striving to know the tropical world for what it is (as a "powerfully individualised zone") while professing compassion towards its plight, and the latter viewed from the West as the sentient outcry of the colonised and 'tropicalised' aggrieved at how they have been treated by the West and striving to shape their own tropical geographies and destinies.

Dominique Moïsi argues that following World War II fear became "the dominant emotion of the West," chiefly as "a reaction to the events and feelings taking place elsewhere" and revolving around the perception that the West was "no longer calling the tune."[39] For Fanon, bitterness was an overriding emotion in the colonised world, and Moïsi notes that it compounded an "identity crisis" within the West which stemmed from the "self-destruction" of World War I and the "suicide/murder" of World War II. Anti-colonialism brought a new urgency to Western attempts to study and know colonial peoples and places, in order to keep them within the yoke of empire, and brought a mixture of fascination, calculation and loathing in their train. There was a starkly instrumental side to how, in the post-war era, this knowing was geared to the exploitation of the tropical world by Western capital, science and technology, most palpably in the forms of agricultural and industrial modernisation and export schemes, and natural resource extraction. However, Gourou's "viscous peace" (see Chapter 5) was about more than this; it was thick with a range of emotions and glutinous matters of freedom and fear.

Of course tropicality's modalities of adventure, othering and appropriation are intrinsically affective. The Spanish in the New World felt lost and exposed as the prospect of finding El Dorado and the Garden of Eden in the jungles of the Amazon receded from view, and stifling heat, blood-sucking insects and the spectre of cannibalism (read as signs of the devil and origin sin) took their place. And as Arnold writes, "One can understand the preoccupation of the British in early nineteenth-century India [or the French in interwar Indochina] with death and 'death-like-separation' as constituting a major strand within a wider empire of affect." Moreover, "Employing this phrase is not meant to imply a stark opposition between an Empire of Emotion and an Empire of Reason – indeed, emotion and reason seemed to many at the time to complement, not contradict, each other and to be equally important in representing personal feelings and responses."[40] If little has been made of the question of emotion in the genealogy post-war tropicality, then we might also note that only recently have "mind-numbing abstractions like 'reason' and 'rationality'" been seen as "having histories" (and geographies) "rather than [as] eternal glassy essences."[41] To grasp what is at stake here for a critical understanding of post-war tropicality, questions of affect need to be hooked up to ones of epistemology and ethics. Let us say a little about each, and in so doing give the story that follows a postcolonial framing.

First, Febvre's "chasm" was marked by shifting categories and knowledges, and with the critical onus swinging away from the agendas of colonisers and the metropole, and towards the plight of colonised peoples and their future in independence. The anti-colonial and independence movements of the post-war era paved the way for latter-day critical examinations of Orientalism (Said), tropicality (Arnold), and other colonial discourses within the disciplinary project of postcolonialism. Césaire's critique of Gourou anticipates the postcolonial claim that the very idea of 'the tropics' was bound up with the question of how this part of the world had been represented from the vantage point of the colonising West, and Césaire was not alone in pushing this claim. The American zoologist Marston Bates mused about such matters in his 1952 semi-autobiographical *Where Winter Never Comes*, which he characterised as an inquiry into "understanding, by northern peoples, of the nature and problems of man and civilization in the tropics," and where he homes in on the idea of "burden" (weight, responsibility, affliction, guilt) and its colonial genealogy.[42] The "white man's burden in the tropics" is not simply "the burden of educating, improving, or governing the poor benighted natives," he suggested, but more fully "the burden of his own culture which he has carried into an alien environment."[43] Was it possible, he asked, for the tropics to be known and experienced in anything other than a Western fashion, especially given that tropical nature had been produced – cultivated, transformed and abused – by the West? The burden of which Bates wrote was at once spiritual and quotidian. How should Westerners now relate to an alien environment they had ransacked as well exoticised as a zone of adventure? By ploughing on regardless? By conserving or salvaging its exoticism? By arriving at a more accurate or compassionate understanding of it?

Texts like Césaire's *Discourse on Colonialism* had an important influence on how the project of postcolonialism was subsequently launched, among other things as a critical endeavour that would restore understanding of the long and tangled histories of resistance and critique that imbued empire, including Western self-critique going back to figures such as Montaigne (whom Gourou admired), and with colonisation deemed the enlightened West's main and lasting sin.[44] This framing permeates much of what we have been talking about to this point.

Second, as Leela Gandhi impresses, and Gourou knew full well (quoting frequently from European philosophers), on the flip side of this sin were the hopes and dreams of the "modern rationalist" and "egalitarian subject" who makes otherness "the privileged figure of ethics . . . [and] should be able to identify equally with all being and things."[45] Gourou imbibed this egalitarian outlook from a long line of thinkers from Montaigne and up to his contemporary Raymond Aron, quoting the latter's maxim "Man alienates his identity if he imagines he has said the last word," but also frequently quoting Asian proverbs.[46] But try as this subject might to foster a peaceable sense of itself in relation to the Other, it frequently comes up short and is short-circuited by what Gandhi and other postcolonial thinkers describe as the 'double-time' of modernity: the temporal contradiction between promising freedom, equality and mutual respect on the one hand, and constantly suspending the prospect of their attainment to the colonised on the other. This suspension of the full promise of modernity – liberation to the colonised – was made all the more visible and risible, Fanon proclaimed, by the way the colonial world worked as a starkly "compartmentalized world," with a zone for colonisers and a zone for the colonised.[47]

By extension, a developed temperate world, which "in three centuries has gone through the millenia-old cycle of tropical humanity and found rational means of mastering nature," as Gourou put it, was cut off from a backward tropical world that was unable to break out of its cycle of poverty.[48] One of the chief ways in which the deferral of freedom took place was in the hands of Western experts, like Gourou, who monopolised the terms of representation.

But third, and now looking to rehabilitate the humble and compassionate egalitarian subject who gets into this epistemological and ethical trouble, good might still be found in what Gandhi terms a "spirit of renunciation."[49] She means a determination to break ties; to renounce dominant ideas, or at least treat them with scepticism; to refuse to be become powerful, and be prepared to stand alone; and to take refuge in what is overlooked and deemed marginal – there and then "to save Europe from its worst self," or save 'development' from the West's worst developmental self, and so forth. During the first half of the twentieth century, she suggests, this spirit fostered an "affective community" of thinkers and activists that surmounted boundaries and divisions of class, gender, ethnicity, nationality, ideology and empire. We now continue our story with the feeling that something of this spirit of renunciation can be found in Gourou too.

Gourou's post-war missions and 'system'

Between 1947 and his retirement in 1970 Gourou produced seven books and atlases, including two major syntheses, *L'Asie* (1953) and *L'Afrique* (1970), and over 200 essays, reports, reviews, textbooks and edited collections. He was prolific by any standard. While he published in a wide range of journals, many of his essays were published in the *Annales de Géographie*, the *Bulletin de la Société belge d'Etudes Géographique*, *Les Cahiers d'Outre-Mer*, and *L'Homme* (the interdisciplinary humanities journal, established in 1961, which he co-founded and co-edited with his Collège de France colleagues Lévi-Strauss and Émile Benveniste).[50] He undertook "missions" (fieldwork and advisory trips) to Angola, the Congo, Kenya, Madagascar, Mozambique, Réunion, Rwanda and Burundi, Senegal and Tanganyika. He also visited Brazil, Britain, Canada, India and Japan to give talks and participate in international initiatives. He commuted between Brussels and Paris to teach, and advised the Belgian and French governments over some important tropical research initiatives, such as the Centre d'Etudes de Géographie Tropicale (CEGET), established at the Université de Bordeaux III in the mid-1960s, and the 'geography' section of the Centre médical de l'Université libre de Bruxelles en Afrique centrale (CEMUBAC).[51] It was with a good deal of anticipation and some degree of trepidation, Michel Bruneau told us, that CEGET staff awaited their "master's" arrival from Paris for his yearly inspection of its work and facilities on behalf of its state funder, the Centre national de la recherche scientifique (CNRS).[52]

A 1948 visiting professorship at the University of São Paulo gave Gourou an opportunity to venture up the Amazon River and left a deep impression on his understanding of tropical soils and forests.[53] He was also a driving force behind the creation, in 1949, of a "world land use survey" under the auspices of the International Geographical Union (IGU), and in that year began collecting the population data on Belgium's African possessions that he would use to compile various treatises and atlases published by the Institut Royal Colonial Belge during the 1950s and 1960s (and that would form a cornerstone of his 1970 survey *L'Afrique*). In 1950 he attended the Institute of Pacific Relations (IPR) conference in Lucknow, India, forming impressions about south Asian tropical conditions that he incorporated into *L'Asie* (1953) and a later work, *Riz et Civilisation* (1984). He played a key role in the "population commission" of the IGU and from the late 1940s had a long-standing advisory relationship with the United Nations Educational, Scientific and Cultural Organization (UNESCO) (see Chapter 5).

As in Indochina, he liked to work alone but could not have undertaken his research without the assistance of what he often simply referred to as 'the authorities.' He rarely worked with anything more than a skeletal research entourage, and worked most closely in the field with his students. But while he cut the figure of the single-minded lone scholar, he attended myriad international conferences and was a copious report writer and reviewer of books. The Collège de France did not admit students, but Gourou supervised researchers at the Université libre de Bruxelles (ULB), and during the war years had worked with and taught a number of geographers (Gilles Sautter at Montpellier, and Paul Péllissier, Guy Laserre and Louis Papy at Bordeaux) who went on to hold key positions within tropical research laboratories funded by the CNRS and the Office de la recherche scientifique et technique outre-mer (ORSTOM).

Les pays tropicaux was "still *the* book that first comes to mind if we are asked by a student or colleague to suggest a readable, succinct introduction to the geography of the humid tropics," American geographer Eugene Kirchherr wrote in 1967, and Gourou's international reputation rested mainly on his books and articles.[54] He eschewed the label 'tropical geographer,' preferring to describe himself as a geographer who specialised in the study of tropical lands, and some question whether he founded or presided a school of thought or field of study, at least in a direct manner. It did not start so brilliantly for Gourou. He confided to his friend Jean Gottmann that his first lecture course at the Collège de France in the winter of 1947–1948 was a "rather depressing" affair: "For the inaugural lesson I had a large audience, made up of friends. But since then, the audience is limited to five or six people, of whom three or four are old imbeciles asleep. It's all rather grotesque and useless."[55] But by the 1950s he had become the doyen of a research field that was quintessentially associated with him, although interest in the tropics was also spurred by scholars and works from others fields, such as Bates, and economist Douglas Lee (especially his 1957 *Climate and the Economic Development of the Tropics*), and there were schools of tropical geography in Britain (with Robert Steel at Liverpool a driving force) and Portugal (led by Orlando Ribeiro, whom Gourou befriended), and a larger group of geographers from around the world who specialised in tropical questions of climate, disease and population.[56]

In Chapter 1 we noted how commentators thought Gourou's diverse interests were framed by a 'system' of thought. This is the opportune moment to say more about this system, which has four elements, and about the drawbacks of thinking in this way.

First, Gouou's overriding concern was with the organisation of space and the humanised landscapes of rural areas. "People, these makers of landscapes, only exist because they are members of a group which is itself a tissue of techniques," he proclaimed in his main work of theoretical synthesis *Pour une géographie humaine* (1973), upon which he started to work the day he retired, and as if to say that he would never stop working: "Every human group is supported by techniques which make its members 'civilised.' There is no such thing as a 'savage'. . . . Our consciousness of the landscape in its historical and physical depth is a source of joy, a school for progress, and certainty of an inexhaustible activity" (also see Chapter 3).[57] He used the expression *techniques d'encadrement* (production and landscape moulding techniques) more frequently as time went by to capture this concern.

Second, while the physical environment played an important role in shaping tropical spaces and landscapes, for Gourou it did not play a determining role. He was adamant about this. It was an argument he made consistently from the beginning until the end of his career. "Geography teaches people that they made themselves and are not the playthings of natural conditions. Certainly, people have too often made bad use of their agency, but that results from a deficiency in their *techniques d'encadrement*," he wrote in 1966.[58] In *Les pays tropicaux* he took this claim back in time and across the tropical world, and instructively to the Maya of Central America, which he deemed a 'high' tropical civilisation. And contrary to Césaire, Gourou deemed this civilisation original to this part of the world, as "having not sprung from seeds imported from elsewhere."[59] 'Old' Mayan civilisation collapsed (and seemingly quite suddenly) during the sixth century, and Gourou argued that neither purely physical explanations revolving around how climate change spawned drought (a view propounded by his two *bêtes noires*, German geographers, and environmental determinists such as Elsworth Huntington), nor purely human ones, focusing on disease and invasion from the north, sufficed. He considered the "interdependence of people and environment" crucial and concluded that "soil exhaustion borne of the *Milpa* system is probably the best explanation." Mayan subsistence was based on a system of shifting cultivation (principally of maize), and it collapsed from within due to the shortening of fallow periods and the growing distance between Mayan settlements. A sophisticated civilisation became unhinged by "a rudimentary method of exploiting nature," from simplistic *techniques d'encadrement*, he argued.[60]

Here was an example of how "Geography demystifies the relations between human groups and their natural base, and contributes to giving people a more acute sense of their freedom and responsibility," Gourou declared in a 1961 conference paper on "geography and our times" that was delivered before an international group of policy advisors in Switzerland.[61] "A civilisation is the result of a combination of internal evolutions and external influences where the human dominates the physical," he observed in an earlier essay in Febvre's *Annales*.[62] And in an impassioned letter to the Association of American Geographers about Karl Wittfogel's influential 1957 book *Oriental Despotism*, he complained that the term 'Orient' was used much too vaguely and insidiously, "making us believe that the term is something other than a geographical expression," and that Wittfogel's study was "worrying because it reintroduces physical determinism to history and geography."[63] Finally, in a survey of "tropical geography and problems of underdevelopment" published in 1969, he argued that "Geography demystifies 'underdevelopment' and returns to it its character as a traditional, historical and normal condition of humankind, showing that 'under-development' is not a fatal effect of natural conditions and can be dominated only by a revolution of techniques, and especially those of encadrement."[64] That these essays, spanning 20 years, were republished in a collection marking his retirement underscores how seriously he felt about the matter.[65] Environmental determinism had tarnished the reputation of the discipline of geography, and "*déterminisme physique pas mort*," he retorted at a sociology conference in 1955.[66] "The paddyfield does not determine a civilisation and a civilisation does not determine the paddyfield," he wrote to Gottmann in the month that *Riz et civilisation* was published. "And so I continue along my merry way, which consists of trying to show that human geography eludes determinisms and models, and is more interesting for that."[67]

Third, Gourou maintained that "people and landscapes are conditioned by the civilisation to which they belong."[68] He was concerned with *civilisations* and preferred this concept to the anthropologist's 'culture,' which he thought was used in ways that swung too vehemently and erroneously between either trivialising the influence of nature on culture or asserting a much too deterministic connection between environment and race. The civilisations of the tropics, as Gourou understood them, could be compared in terms of their population densities, and in general those with temperate origins were seen as the most complex and harmonious. "It seems that the birthplaces of Asia's superior civilisations are extra-tropical," he proclaimed with gusto in *L'Asie*, although southern India was a further example (along with the Maya) of a superior civilisation that had arisen from within the tropics itself.[69] He also often used the term "administration" not only as a synonym for organisation, but also to explain how civilisations both worked internally and were affected by external conquests and forms of economic and political exploitation. "The Chinese masses are destined to remain poor," he noted (in typically sweeping fashion) in 1957, "but an orderly government will spare them misery."[70] He geared such generalisations to different tropical settings – delta, forest, mountain and plain – and argued that the civilisations of rice-growing deltas with high population densities were organisationally richer (albeit poor in material terms) than those in the sparsely populated regions of tropical Africa and the Amazon.

On the whole, Gourou's penchant was to think about tropical spaces as localised, and intensively organised and self-contained, especially in contrast to the extensive – seafaring, imperial and far-flung – spaces that were instrumental to empire and the dominance of the temperate world over the tropics. Western civilisation operated at this larger scale. But he was also committed to revealing the extraordinary environmental and cultural diversity of tropical Africa, Asia and America, and especially to dispelling the stereotyped portrayal of Africa as a place almost entirely given over to an itinerant slash and burn agriculture. There was a tension within what Jean Gallais described as Gourou's "Africanist period" of the 1950s and 1960s between respecting this diversity and seeking to elucidate the essence of tropical 'backwardness.' Gourou began *L'Afrique* with the declaration: "This geography of the African continent directs our attention to the relations between the inhabitants and the territory they occupy. If the inhabitants do not appear to be very happy, there is comfort in the observation that their unhappiness is not determined by natural conditions."[71] Small comfort perhaps.

And fourth, while he endeavoured to address questions of overpopulation and living standards, and the imperial and development discourses with which they were bound up, his inclination was to see progress – as it was propelled through mechanisation, export agriculture, industrialisation and urbanisation – as an intrusion, and as both exacerbating environmental and social calamity (breaking down traditional *encadrements*) and putting a strain on the adaptive capacity of people in the tropics to deal with tropical nature in their own ways.

If this amounted to a 'system' of thought, then it was both an elegant and elusive system: stylish and innovative in the way it sought to carry forward the geographer's post-enlightenment quest to bring society and environment into a unitary analytical frame; dogged over some issues; but also hard to get hold of because terms like *civilisation* and *techniques d'encadrement* were used in elastic and sometimes vague ways.[72] Gourou was fond of saying that geography "demystifies" understanding of human societies and landscapes, but often smuggled his own mystifications into the scene.

We now want to press on with the other suggestion we made in Chapter 1, that however this 'system' is construed, it should not be seen as hermetically sealed. It was not simply the product of some internal evolution or maturation in Gourou's thought. Rather, it gained meaning and direction in relation to changing personal, disciplinary and global circumstances and dynamics. We have seen that wartime experience was crucial. But so too were the post-war dynamics of development and decolonisation, and particularly the question of whether the latter was meaningful without the former. Less noticeable in Gourou's outlook and writing is the refraction of the Cold War strife through tropical Africa and Asia.[73]

Tropical conditions and Western recipes

Rather, Gourou trailed the ends of empire. His research was undertaken mainly in late colonial times, and his last major stint of fieldwork was in Portuguese Angola in 1962 with his friend Ribeiro as part of a Mission funded by the Junta de Investigações do Ultramar (Portuguese Council of Overseas Research). By then he was under no illusions about imperial dissolution. Decolonisation was by then an irreversible process, and "due to numerous factors," he and Papy informed school pupils:

> Europe built colonial empires at a time when its world hegemony was uncontested. . . . But ravaged by two world wars, weakened, dominated by two new economic powerhouses (the US and USSR), Europe was no longer mistress of her distant lands. Encouraged by both the US and USSR, and aggravated by international tension between East and West, the current of decolonisation spread throughout the world. The progress of means of communication and transmission of thought (notably by radio) allowed erstwhile 'dependent' countries to become conscious of the political and economic problems of the globe and aspire to independence. The unfortunate Indochinese experience showed that it was vain to oppose this new current of history. . . . The 4th and 5th Republics resolutely engaged in a policy of decolonisation whilst ensuring technical and financial aid to the newly independent states. The problem of aid to the 'Third World' became the major preoccupation of our time.[74]

This trajectory influenced how Gourou imagined and represented the tropics. Akin to the work of other geographers (many of whom had served their countries overseas during World War II), Gourou's responded to this new age of imperial retreat, and aspirations for freedom and development, by insisting that "the absence of land use surveys [and basic geographic knowledge] in underdeveloped countries makes it impossible to formulate [anything] other than very generalized schemes of development."[75] His work found meaning and significance within a framework of knowledge-production that was geared to both the extension and dissolution of empire. As Corey Ross relates, and as Gourou had witnessed first-hand at the IPR conferences he attended in 1944 (Hot Springs), 1947 (Stratford-upon-Avon) and 1950 (Lucknow),

> For most states in Western Europe, the enormous task of economic reconstruction at home, coupled with the strategic reliance on a U.S. government that had rather different ideas about the post-war international order, made a far-flung empire seem like a luxury that they could ill afford to maintain. On the other hand, the urgent demand for raw materials and the pressing need to earn US dollars highlighted the potential of colonial territories as a source of national strength. Amidst the bottlenecks and acute financial pressures of post-war recovery, overseas colonies seemed like necessities that Europe could ill afford to relinquish.[76]

Western governments and colonial ministries renewed their missions (detailed surveys, with multi-disciplinary research teams) to tropical Africa, which had been a hallmark of the link between science and empire during the 1930s, and targeted perceived troublespots: the Congo and Rwanda for Belgium; Ghana, Nyasaland and Tanganyika for Britain; and Côte d'Ivoire for France. Indeed, Éric Jolly argues that the period 1930–1960 might be seen as distinct era of colonial study within which the social sciences became professionalised.[77] Geographers – including Gourou, Sautter, Steel, Frank Debenham and Jean Dresch – played pivotal roles in these late colonial surveys, and the work they undertook left a profound impression on how they saw the world.[78]

In many of Gourou's books and articles an ostensibly dispassionate academic interest in the nature and evolution of human landscapes is punctuated by commentary and reflection on burning political issues of the day: whether the coming of the temperate world to the tropics should be regarded as a matter of contamination; whether tropical lands could be developed and tropical peoples civilised; on the resistance of the 'tropicalised' to being developed; whether the multi-ethnic societies of many nations could foster equality and mutual toleration; and on how direct an impact the tropical environment had on all of this. He saw and felt these questions most acutely, perhaps, in relation to Rwanda. And there, we suggest in the next chapter, as elsewhere, we note here, he was both out of step and in tune with his times.

Arnold points us in the former direction, and we shall say more soon about the idea of a 'phantom tropics' and blinkered tropicality. For now, suffice it to say in the latter regard that Gourou's contribution to the first issue of *L'Homme* was a detailed and prophetic discussion of how the post-war anthropological study of human origins and dispersals through blood types and genetics was playing dangerously with racial determinism; and that he followed this, in 1963, with a withering critique of the long shadow that Montesquieu's climatic determinism in his 1748 *De l'esprit des lois* cast over Western thinking about the tropics.[79] Neither climatic nor racial determinism were throwbacks to the imperial past, he insisted, but very much alive in the present, and their mobilisation to justify inequality, violence and misery needed to be contested. France's leading colonial administrator, Robert Delavignette, noted that Gourou

> elaborated his work at a time when totalitarian racism was rife. And it was in the course of unheard-of upheavals, provoked by two world wars, that he advanced our knowledge of Asia and Africa in the discipline of geography. He showed that the study of tropical regions dispelled the determinism which, it was claimed, oppressed their inhabitants by depriving them of the ability to modify and improve their organisation of space.[80]

The tropical geographer needed to be vigilant about lazy and perilous deterministic thinking, Delavignette implored, and he applauded Gourou's portrayal of American and German geography as arch protagonists in this respect.

Gourou himself was aware that the pessimism that many associated with *Les pays tropicaux* was political and expended great energy on trying to qualify his message in subsequent writing and editions of the book. "My tropics are attractive," he was still at pains to aver in 1982; "when they exude sadness, it is not for 'tropical' reasons but because of the unfortunate condition of the inhabitants, born of insufficient techniques and an unhappy history."[81] His pessimism was the product of a particular moment and mood: of "post-war," as Tony Judt terms it; of how the world might pick itself up from the ravages of war and make a new start. Gourou envisioned a world that he deemed neither stuck in the past nor frozen in the present, but as a challenging and vulnerable development environment. Echoing Bloch and Febvre, he envisioned geography as a potent means of "putting on trial everything we see, and of ruining what is apparently obvious."[82] Gourou's language of trial, misfortune and ruination was geared to a particular time. It was neither simply emblematic of a 'watchdog of colonialism' stance, as Césaire had it, nor just a negative spin on tropicality which could be traced back into colonial history, as Arnold sees matters. It was also a profoundly emotional reaction to the violence and disorder of the 1930s and 1940s, and Gourou's engagement with the dissolution of empire after World War II was in turns partial, progressive, radical, reactionary and passé.

Ross writes of the contradictory "mixture of imperial retreat and imperial reinforcement" in the post-war "political ecology of late colonial development" between 1945 and the 1960s.[83] *Les pays tropicaux* heralded a change in the direction and scale of Gourou's concerns, away from Indochina and monsoon Asia and towards not only a zonal characterisation of the tropical zone as a whole – or what Gourou more often called "the hot, wet lands" – but also, and as Césaire intimated, a late colonial diagnosis of this zone's prospects under the pretence of science and objectivity. Gourou addressed questions of colonisation only fleetingly in his 1947 primer, and preferred to use the more innocuous sounding term "European intervention."[84] His concern was with what he compactly, if elusively, termed "geographical conditions," and he argued that the 'backwardness' of the tropics was not wholly – and in some respects was only partially – attributable to the evils of empire and inroads of what he described to Gottmann as a "motorised world."[85] In the years that followed, tropical Africa and the last pangs of empire there became central to his work, even if he rarely talked directly about politics. However, he initially advocated the injection of Western capital and technical expertise, where and when appropriate, to boost living standards and consumption levels. We shall see that the caveat was important, and that Gourou perhaps did not believe some of things he was saying about foreign investment and agriculture modernisation. The general point, however, is that political questions of development and decolonisation were in the air wherever Gourou travelled and lectured. He could not avoid them. They shaped how funding for his research was solicited, how much of his research was designed and presented, and how much of his work was read. These big questions had a thoroughgoing influence on his tropicality.

The tropical world had not benefitted from eight interlocking "features of development" which had put Europe on a path to superiority in the sixteenth century, he asserted in his 1956–1957 Collège de France lecture course on "Underdevelopment in the tropical world": "demographic growth, a decline in mortality, improvement in material conditions of existence, medical progress, progress in individual productivity, an expansion in machine technology, and the centralised organisation of space, with States rather than small and dispersed communities, and low rural population density." Interestingly, he ignored the growth of towns, trade and empires, which are also routinely included in accounts of the so-called 'European miracle.' As these "basic" (political, social and demographic) conditions were not there in the tropics, he continued, "when Western recipes were applied to tropical conditions, without adaptation, they were failures. Furthermore, the variety of situations of 'under-development' indicate that no single model of development should be followed."[86]

The problem of geographical order

At a 1952 French intercolonial symposium in Bordeaux, Gourou noted that in the aftermath of World War II France had struggled to pull its colonial populations out of poverty but had largely reached three other objectives which would continue to make its empire legitimate: "(1) maintain order and peace; (2) develop a network of convenient and fast communication channels; and (3) control epidemic and endemic diseases."[87] For Gourou post-war dreams of economic development and imperial rejuvenation could not simply be packaged in grandiose models, promises and targets and projected seamlessly on to the tropics. He grasped that the citizens of newly independent countries hankered after 'development,' and that there was great pressure on their leaders to deliver it quickly, and not least with models imported from the West and the communist bloc. But he was among the first to press the point that, as Nick Cullather explains matters, "Models are a form of selective forgetting . . . [in which] history is displaced by a capsule narrative, part fact, part fiction, designed to convey a formula for development. Context and motives are stripped away, leaving an image of outside experts isolating problems and then introducing sharply focused solutions."[88] Césaire, of course, placed Gourou in this category – of surreptitious meddler – whereas others focused on the former – selective forgetting – element of Cullather's formulation to portray Gourou's fervently anti-model outlook as a positive thing.

Les pays tropicaux was published at a time when new ways of "assembling the planet" and understanding "human-Earth relations" – a new "politics of globality" as Rens van Munster and Casper Sylvest put it – was coming to the fore.[89] "The post-war planetary perspective was driven primarily by large-scale technological innovations," they suggest, spawning the 'nuclear age,' 'consumer age' and 'space race,' and adding new sensibilities of "verticality" to extant understanding of global relations revolving around horizontal processes of expansion and encirclement.[90] The Cold War became the pivot of competing "statist conception[s] of national security" and a "vision of the globe as an integrated ecological system" that was at once a rich and a vulnerable home to humanity. Tropicality does not make an explicit appearance in this discussion, and perhaps should, for it played a key role in another important facet of these politics: a new form of temporality in which "the present is viewed through the prism of urgency and impending disaster," with the threat and consequences of nuclear apocalypse its most intense expression, and with the linear time of progress, premised on the "expectation that science and intellectual activity would ultimately work to improve human life," placed in doubt.[91]

226 *Affecting the tropics*

By the time the third edition of *Les pays tropicaux* – which was the highest-selling English edition – appeared in 1961 Gourou had a good deal of international travel and African fieldwork under his belt, and Georges Courade notes that there were different views about whether his work amounted to a "conformism of the moment," and chiefly because there was no consensus, at least within French geography, about what constituted 'conformism.'[92] To our minds, however, by the 1960s Gourou had settled on two views. The first was that in many parts of the tropics traditional tropical *encadrements* needed to be protected from rapid change:

> The typical agriculture of hot, wet lands is remarkably in harmony with health, well-being and pedological conditions. It takes great effort to respect the natural equilibrium and to interfere as little as possible with the slow, delicate processes by which the soil succeeds in maintaining itself, and in keeping a certain degree of fertility in the special circumstances imposed by the tropical climate. This agricultural system is perfectly rational . . . [and the] disasters brought on by agricultural methods which have taken no account of the treasures of wisdom and experience accumulated in the old tropical system are sufficient proof of the latter's value.[93]

He expounded on this view in a 1963 report for the UN on "Africans and their milieu," and here, as in much else he wrote around this time, he placed the word development (*développement*) in inverted commas in order to register his reticence towards to the idea.[94] African societies "have perfectly respectable" communal ritual land-holding system, he insisted, and "are not naturally suited to 'development,'" by which he meant regimes of private property, intensive cultivation, and the attempt of colonial and post-independence states to levy taxes on land and basic commodities. "Peasants smoke their own tobacco – or hashish – and drink the beers they have made themselves," he opined; and in his mind the introduction of maize and manioc into African agricultural regimes were among the few "beautiful examples of 'development' [in Africa]: spontaneous and without heralding significant change in traditional farming techniques." And any advice that was imparted by "foreign experts and agencies" needed to be borne by Africans themselves.[95]

Gourou wrote this and much else about agriculture in the throes of African decolonisation, and struck a gloomy tone on the whole. "African agriculture provides insufficient food, both in quantity and quality, for the rural population; provides inadequately for the urban population; and provides inadequate exports," he wrote in a 1965 essay that was published in a notable volume with essays by many leading Western and African thinkers. "This calls for explanation," he continued, and it should not be attributed to climate, which "would seem to be one of the most favourable in the world for agriculture," but rather to a slew of human "deficiencies" (a lack of animal protein in the diet and "prejudices and harmful customs" surrounding hygiene; a "mere idling of the soil" during the dry season; and the spurning of "capital equipment").[96] All of this worked to limit the area and intensity of cultivation, and as a consequence the average population density of tropical Africa was around 65 people per square kilometre of the total area, and around 1,900 people per square kilometre on cultivated land.

Outside the delta areas of irrigated rice production in Asia, tropical cultivators had arrived at a similar solution to the challenges set by climate and soil: the so-called ray system of cultivation, whereby patches of forest, bush and savanna were cleared of vegetation by fire; hoed, manured and cultivated (chiefly for carbohydrates) for a few years, and subsequently abandoned; and with all agricultural work "usually accompanied by a grand display of magic."[97] Gourou referenced an international literature which reported on the diverse ways in which this basic regime operated in tropical lands, reiterated the view he had elaborated in his 1940 survey of Indochina, that it allowed for a meagre existence and averted excessive soil erosion, and thus questioned how it was generally seen by colonial officials, which was as wasteful of time and labour, and as environmentally deleterious (and particularly as a cause of deforestation).[98] He was dismissive of the colonial view that shifting cultivators were 'primitives' in need of development.

He was less sanguine than others writing around this time who were arguing that population growth would stimulate economic development by transforming traditional agricultural systems (e.g. Colin Clark and Margaret Haswell), or that population pressure would presage the modernisation and intensification of peasant agriculture in situations where it was otherwise unlikely to take place (e.g. Ester Boserup). Gourou questioned the viability of rapid population growth.[99] In turn, Boserup noted that while he had successfully explained why long-fallow systems of cultivation "represent an adaptation to the special conditions of soil and climate in the tropical zone," he had assumed too quickly "that the soil in most of the tropical zone cannot be used for other systems of cultivation." Boserup disputed Gourou's conclusion that "most of the tropics is sparsely populated because the land is unable to support cultivation for more than one year out of twenty and therefore unable to support a numerous population," and found his claim that "additional population must largely be accommodated by means of industrialisation and reliance on foreign trade" too pessimistic.[100]

This leads us to his second view, which was that while many of the 'problems' of the tropics were shaped by external forces (especially colonialism), many were inherent to hot, wet lands. As the civilisations of the tropics stemmed from a lengthy evolution, rapid change was often not only unfeasible, but might also be calamitous. "The case of Brazil clearly shows the vicissitudes to which white colonisation [there by Portuguese settlers] is subjected if it does not respect the limitations imposed by the tropical environment," he wrote; this was a counterproductive effect of the belief that tropical nature could be mastered.[101] The colony's plantations, "giving excellent results at first, but ending in the ruin of the land they occupy" were the main source of this difficulty.[102] And looking east, he declared that "The superior civilisations of monsoon Asia have led the populations into a dead-end from which it will be difficult for them to escape. It is difficult for a new civilisation to transform what a preceding civilisation has made."[103]

228 *Affecting the tropics*

Three well-known surveys – Daryll Forde's 1939 *Habitat, Economy and Society*, Grenfell Price's 1939 *White Settlers in the Tropics,* and Auguste Chevalier's 1946 *Révolution en agriculture* – loom large in Gourou's chapter in *Les pays tropicaux* on "European intervention," and are read by Gourou as examples of the view that there is "a direct relation between the physical environment and human activities," and thus as texts that should be approached "with interest but also caution"[104] Gourou saw it as his job to root out and reproach environmental (and especially climatic) determinism wherever he found it – and even in mild forms.[105] Seeking, in a non-deterministic vein, to explain the marked difference in the density of population between two vast tropical regions with similar equatorial forest environments, the Amazon Valley and Congo Basin, and with that of the latter 7.5 times greater than that of the former, Gourou observes that is not questions of terrain that provide the answer, for access to the Congo is impeded by impassable rapids whereas the Amazon is an easily navigable waterway, but ones of imported disease. The Indian population of the Amazon had been decimated (we now think to the tune of 90 per cent) by diseases brought by Europeans (especially smallpox, measles, and influenza) from which Native people had no natural immunity.[106]

Nevertheless, while "tropical soils can certainly give better yields than they usually do today," Gourou observed in 1961, close attention needed to be paid to their inherent "defects" and "drawbacks":

> while difficulties can be overcome . . . advance has depended on the great progress made during the nineteenth and twentieth centuries in zootechny [sic] and scientific agriculture. But here again we find a further drawback, a slowing down, due to the physical environment, for progress in the nineteenth and twentieth centuries in science and technology was achieved in temperate lands by research workers in those countries, and the transference to the tropical world of the progress so made is achieved slowly and gropingly and with many setbacks, since conditions in the tropics differ so greatly from those in the temperate regions.[107]

"It would be mere delusion to seek the economic safety of tropical lands in an agricultural system imitated from temperate latitudes" (mixed farming), he continued in *Les pays tropicaux*.[108]

Gourou's project was thus in some ways open-minded and reformist, and in others (as Arnold stressed), pessimistic and out-of-date. And because of the boldness and clarity with which he put his formulations across, it is easy to find in *Les pays tropicaux* what one wants to find. To be sure, the book's central concern with the organisation of space and the diverse humanised landscapes of the tropics does not resonate with the outlook of anti-colonial thinkers such as Fanon, for whom disorder and transgression through violence were held to be progressive forces that would bring Western dominance to an end. "Decolonization, which sets out to change the order of the world, is, obviously, a program of complete disorder," Fanon proclaimed at the start of *Les damnés de la terre*.[109] Such a programme would come about not "as a result of magical practices, nor of a natural shock, nor of a friendly understanding," he continued, but as "a historical process. It cannot be understood, it cannot become intelligible nor clear to itself, except in the exact measure that we can discern the movements which give it historical form and content."[110] Fanon's dialectical conception of the 'movements' of the era – national liberation movements and struggles; the break-up of colonial empires and attainment of independence; the creation of a plethora of new nation-states; and the violent percussion of imperial crisis and dissolution – underscore the idea that the historical process of decolonisation was messy and protracted rather than a tidy event or easily measurable transformation.

This process was about emotion – feelings, sensations, longings, passions and resentments – as well as about epistemology (a re-ordering of knowledge and truth), and geographers felt it. Gourou talked about Westerners' passionate desire for escape and expectations of luxuriance, a 'get-rich-quick' colonial mentality, and their visceral reaction to disease and how it had made the tropics a space of danger, illusion and degeneration. Emotion was pivotal to decolonisation. As historians such as Martin Evans and Jordanna Beilkin have shown, in the postwar decades of imperial retreat the emotions of French and British settlers were often strident and were deemed in need of 'management,' and there were trenchant and tetchy debates about how new bonds between centre and periphery, and citizen and subject, would be formed.[111] Social scientists understood emotion in diverse and conflicting ways: as a desire for order, to be sure, but also a search for peace and a denial of violence, as good faith, and with welfare and well-being becoming standards by which the quality and ethics of academic scholarship and advice needed to be judged. "Discipline is needed," Gourou cautioned in a 1948 survey of the problems of the tropical world in *Les Cahiers d'Outre Mer*, "but who will impose it": "At a moment when the authority of the colonial powers is collapsing, who will impose upon the natives and the European colonisers the indispensable groupings and the total obedience which are the conditions for the correct starting up of the tropical economy? . . . On the other hand, democratic methods generally lead to compromises which are incompatible with the demands of scientific knowledge."[112]

Gourou venerated order and organisation, and the sense that World War II had culminated in uncertainty and the prospect of further chaos, rather than a lasting peace, prompted him to get on the bandwagon of imperial federalism as a solution (however temporary). But the colonial development and welfare programmes initiated in the 1930s and 1940s had fallen short of the rhetoric and targets to which Gourou referred in 1952, and Điện Biên Phủ sounded the death knell of peace and order in the Far East. His remark about the "dead-end" population situation in monsoon Asia was written in the wake of France's ignominious retreat from Indochina – at a moment of doom and uncertainty.

This concern with order and discipline was by no means restricted to Gourou or French geography. For example, in a 1945 assessment of how rural Burma might recover from the wartime ransacking of forest and field, village and peasant, and crops and animals, by the Japanese the British geographer Oskar Spate observed

230 *Affecting the tropics*

with respect to "The Burmese village," which was as sacrosanct to him as the villages of the Tonkin Delta were to Gourou:

> The devastated towns were after all only grafted onto Burmese life; the fields and the forests lay at the very heart of it. The reclamation of fields ruined by silting and salt-water flooding and the rehabilitation of gutted forests may well prove a much sterner task than the provision of schools and police stations and railroads. Nevertheless, there are not wanting signs that behind the advance of the Allied armies some of the most urgent needs – cloth, seed, medicines, such humble essentials as needles – are returning; some sort of settled economic life, beyond a mere hand-to-mouth existence, can begin again. And there is still that unspectacular peasant pertinacity and courage which, not less than the demands of the world market and European capital, in three generations turned the vast desolation of the Delta swamps and jungles into the greatest export granary of the world.[113]

On the other hand, Gilles Sautter lamented that while the advance of "capitalist companies" and "monetised economies" into tropical Africa after World War II was the source of many "sins," the "obduracy" with which rural communities and leaders defended their traditional systems of peasant production against all efforts at "improvement" was a delusional as the blindness of Western planners to the virtues of peasant organisation. There was a battle, he suggested, between "two rationalities" of state "operational planning" and "peasant strategy," and with each freighted by defensive and specious thinking about the other.[114]

Sautter hinted that underlying this battle was the question of whether order could be found or recovered, and not just geographical order. Benoît L'Estoile writes of a drive by French social scientists working in tropical Africa from 1945, and frequently in the face of crisis (in the Congo, Gabon, Guinea and elsewhere) and a fragile French presence, to "put in order – both cognitively and politically – African reality" by fashioning new objects of investigation (village life, population, and migration structures), ideas (of modernisation, dependency, motivation and adjustment), and disciplines (applied sociology and tropical geography) that would foster new forms of colonial governmentality and ways of understanding (and for some supporting) anti-colonial resistance.[115]

In a 1963 review of post-war geographical writing on Africa, the American (Cal Tech) geographer Edwin Munger suggested that surveys, such as Gourou's *The Tropical World* (3rd edition 1961), R.J. Harrison Church's *West Africa* (3rd edition 1961), George Kimble's two-volume, thousand-page, *Tropical Africa* (1960), and Dudley Stamp's *Africa: A Study in Tropical Development* (1953), served as a "basis for action."[116] Munger noted that he had experienced the "anarchy" of imperial breakdown as "a prisoner of rampaging Congolese troops" during the 1950s while reporting on Africa for the American Universities Field Service. For him too, disclosing the underpinning and enduring geographic factors shaping an "emergent Africa" was a basic, if fraught, task.[117] In a lengthy review of *Geographers and the Tropics: Liverpool Essays*, edited Robert Steel and Mansell Prothero, Gourou himself noted that while, for British geographers, "there is no such thing as 'tropical geography,'" by a charming inconsistency, they have published a collection of fourteen essays [over half of them on tropical Africa] devoted specifically to 'tropical' subjects," and that these studies of diverse tropical regions were linked by a consistent desire to furnish basic and comparative geographical knowledge about questions of land use, agricultural practice, irrigation, resource extraction, and economic and port development.[118]

Take Kimble's *Tropical Africa*, for example. He was a former director of the American Geographical Society and his project was funded by The Twentieth Century Fund, a lynchpin of America's area studies programmes, which, in turn, were geared to America's business interests in Africa and Asia. His tome was widely reviewed, and the Fund's director, August Heckscher, remarked that the type of work undertaken by Kimble was "a startling kind of knowledge which may let Americans see what is really there and not merely what their imagination and unanalyzed assumptions have told them is there."[119] Seven years in the making, Kimble's synthesis was aided by 46 collaborators and 6 consultants, and highlighted the impress of Africa's "physical frame" on the history and development of the continent. In volume two he argued at length that the tribal unit and village was the centrepiece of African society, that it was the culturally and politically diverse nature of this centrepiece, and that the diverse physical environments in which it was situated impeded the formation and functioning of new nation-states.

The local scale was paramount to Kimble, and two of Gourou's wartime students, Pélissier and Sautter, concurred. During the 1950s and 1960s they became concerned with small-scale rural development projects and agricultural practices in tropical Africa and popularised Jacques Richard-Molard's tropical use of the French term "terroir" to capture their concern with the lands farmed by a given village. Pélissier reflected that he used the term to focus attention on "an agricultural landscape selected or built by the work of people."[120] It was a similar idea in some ways to Gourou's notion of *encadrement*, but as Pélissier continued, "I also saw the use of the word terroir as a way of reacting against the dominant view at the time that all African agriculture was similar to shifting slash-and-burn agriculture." His project brushed against the grain of his tutor's more zonal framing. Locality was Pélissier and Sautter's watchword in their 1964 "Pour un atlas des terroirs africains" in *L'Homme* (a piece commissioned by Gourou), and Kimble's project was similarly an ambitious (if unwieldy) attempt to dissuade Americans from generalising about tropical Africa.[121]

Like other work in this mould, reviewers lauded Kimble's treatment of Africa's "physical frame" as a way into questions of locality, but got tetchier when it came to his treatment of African politics and states.[122] Physical and political geography, and the different scales and temporalities through which they were often made to work, were uncomfortable bedfellows in Western geographical scholarship during this period.

Yet Gourou's and Kimble's invocation of geographical conditions had a broad, albeit ambivalent, provenance. For example, the British geographer Robert Steel observed in 1962: "It is a commonplace to say that tropical Africa . . . makes almost daily head-lines in our newspapers and that Prime Minister Macmillan's winds of change has stirred up a whole continent of emerging nations. But just because these things are so true, it is vital that there should be an informed opinion about African affairs, and any books that help to provide the background necessary for the study of current African developments are to be welcomed."[123] Steel rued the lack of geography in Lord Hailey's monumental 1938 *African Survey* (and Gourou intimated the same in *Les pays tropicaux*); and, like Kimble, Steel deemed geography an arch purveyor of basic information about the continent.[124] Yet he left it to others to interpret what he might have meant by "informed." In an appraisal of Hailey's endeavours, the geographers Alan Ogilvie and Karl Pelzer suggested that "Both in the reconnaissance stages of field work and in later stages of research when phenomena have been mapped, the geographer's habit of searching for cause and effect should prove to be a valuable asset to any responsible authority. No more effective liaison officer of the sciences could be imagined."[125] Such a liaison did not simply pertain to the 'physical frames' of climate, relief, drainage, soils and vegetation. As the geographer Anthony Young noted in 1968, an understanding of how the "physical environment interacts with all other factors in development planning, economic, social and political" was central to both conceptual issues of spatial pattern and variation, habitat and subsistence, and population and migration, and to applied projects of development.[126]

There was a significant difference between Munger's 1963 review and that of "African books" undertaken in *The Geographical Review* in 1926 by the American botanist Homer Shantz. Describing Africa as a "problem," Shantz's analysis was based on the reality of European colonial dominion over pretty much the entire continent, and for him the geographical questions connecting the works under review were about how "to adapt colonial practice and administrative schemes to the zonal arrangements of nature" and the challenges thrown up by "the fact that the territories in question are inhabited by diverse hunting, pastoral, and agricultural races whose social and political organizations are imperilled or overthrown by the white man."[127] If this was Shantz's 'problem,' then it was not exactly Gourou's, or Munger's. The delineation of the tropics as a problem in *Les pays tropicaux* was not *passé* in this sense. But while the times and messages were different, the invocation of geographical conditions was broadly the same. In other words, a concern with what Munger characterised as "background geography" was not born with decolonisation, but it had a pronounced resonance in that situation.[128]

The idea that geography is a discipline innately concerned with finding and analysing geographical order in the world has a long history, and one that came to the fore with particular zest during the so-called 'quantitative revolution' in 1950s and 1960s geography, especially in Britain and North America. Pioneers of this 'new geography,' as it was called, and especially those in North America, devised visionary and sophisticated models of spatial pattern, process and change in tropical Africa and Asia which moved in a fundamentally different direction to the local, and place- and detail-based orientations of the likes of Gourou, Kimble and

Steel. This new geography used the tropics as a laboratory within which to test the ambitions of spatial science. The tropical world was at a far enough remove from North American geography to damage the reputation of this metropolitan spatial science should its analytics or forecasts fall foul of that part of the world.[129]

But there was strategic underside to the geography affecting retreating colonial powers and the arrangements that were taking their place. What resources, grounds and investments would they seek to hold on to, or take over, or discover anew? What bounties and danger awaited them? In this respect, Munger noted, the provision of 'background geographies' was not an innocent exercise, for the generation of basic geographical knowledge was geared to development planning and Western economic interests, and Gourou remarked on the Liverpool geographers' particular interest in tropical port development on account of the city's long-standing colonial trade with the tropics.[130]

In the wake of the escalation of the Vietnam War in 1965 (the arrival of hundreds of thousands of American ground troops), the British political geographer Charles Fisher likewise described Vietnam as a "geographical problem," and again with a different outlook than Shantz. Fisher argued that "the Vietnamese problem assumes fundamentally different characteristics according to the basis on which its geographical context is defined," and that Americans had underappreciated how the Vietnam War came into view at interlocking scales: as a local civil war between northern and southern areas; as a "continental gateway" between China and Southeast Asia; and as a "menacing flashpoint" of a global Cold War formed in the wake of Japan's defeat and ensuing anti-colonial struggle and nationalist assertion in the Far East. Of these three geographical contexts, "the outstanding geographical effect of French rule" on widening differences between the north and south was of the utmost significance but was the one least well grasped by the Americans.[131] "Indochina formed a much better balanced geographical unit than the old attenuated Vietnam," he argued, even if French colonial policies had stacked up problems for the future.

Furthermore, Kimble lauded the idea that "African nationalism is what African nationalists like, and that African nationalists are those, of all racial origins, who believe in the right to govern and be governed in the way of their own choosing, and to resist, by whatever means in their power, those who would withhold that right."[132] A few years before, and on the back of his attendance at the 75th anniversary celebrations of the Royal Geographical Society of Egypt held in Cairo, and his subsequent "tour of several colleges and other research institutions of Central and West Africa" in his capacity as Secretary of the IGU, Kimble questioned the West's penchant to stereotype the continent, and noted that the only appropriate generalisation was that Africa was "changing":

> The present-day traveler in tropical Africa is likely to return home with a feeling of frustration, especially if he went there in search of those pregnant generalizations which it is at once the geographer's joy and pain to bring to birth. For the fact is that tropical Africa admits of few generalizations. It cannot any longer be characterized as "dark," if indeed it was ever as dark as

our ignorance of it; nor can it now be dubbed "dangerous," for it is possible to live a lifetime in the Belgian Congo without so much as seeing a snake or a mosquito; it cannot even be described as "hot," for many thousand square miles of it stand more than a mile above sea level and are cool enough for a blanket to feel good on the bed almost any night in the year. But this much is true of the whole of it – it is changing.[133]

Fisher was not miles away from such sentiments either. "Unfashionable though it may be to say so," he ventured with reference to the problems of decolonisation in Malaysia, "the Malays' gracious andante tempo of living is as appropriate to the uncrowded humid tropics in which it evolved – and to which it is still restricted – as are the dynamism and resilience of the Chinese to that of the more northerly and intensely congested areas in which they developed." Yet as crude as this "tempo of living" tropicality undoubtedly was, he maintained that

> until these two peoples, who during a century and a half of British rule have come to live side by side in Malaysia, can begin to see themselves in some such terms as these, which give no warrant for the contempt with which each is nowadays increasingly prone to display towards the other, and hence can get away from both the parrot cries of their respective extremists and pernicious Western half-truth that the Malays must somehow 'catch up' with the Chinese, neither political arithmetic, political economy nor political science will be able to do much to solve their problems.[134]

In sum, the world was in motion, and geographers were in motion with it, twisting and turning with their background geographies and the categories (of centre and periphery, geographical determinism, and so forth) that made and anchored them.

One of British geography's arch post-war purveyors of tropicality, Dudley Stamp, similarly gives an impression of the geographer thinking out aloud, and of twisting and turning with dominant ideas and assumptions in a fast-changing context. One of the most well-connected geographers of this generation, and prone to bombast, Stamp gives this impression in his 1964 *The Geography of Life and Death*. In the chapter, on "Living in the tropics," he starts with the hoary stereotype that "Peoples native to the tropics suffering … from a lowered efficiency and resistance have neither the physical nor mental ability to tame their environment," and goes on to regurgitate Ellsworth Huntington's determinist thesis about the climatic reasons for temperate vigour and ingenuity and tropical sloth (with the natives of the tropics "dull in thought and slow in action," as Huntington put it).[135] But he then goes on to declare "It is time there was some new thinking about life and living in the tropics," and to dispute the idea of northern Europe as an ideal climate and thus superior civilisation. For one thing, he noted, Huntington did not anticipate air-conditioning and central heating, which could well reverse some of the climatic measures of "health and efficiency."[136]

Some of Clarence Mills' wartime experiments on the corporeal effects of heat and humidity with mice and rats also "refuted" some of what was believed about human life in the tropics. Mills proposed a "vegetative" human existence (heightened sensitivity to climatic variation) in the tropics, and Stamp observed that such arguments helped him to understand his "experience during five years' residence and work in Burma," where he had worked as a geological advisor to the Indo-Burma Petroleum Company and frequently travelled "From the damp sticky heat of Rangoon in the Rain . . . [to] the Dry Belt" using "the overnight Mandalay express and stop[ping] off early in the morning at some wayside village to enjoy sunshine and a drying air." "It was World War II," he continued, "which finally broke . . . old traditions of tropical clothing and introduced the basic concept of leaving the body as free as possible," and the tenor of much of his discussion in this pot boiler about the sweltering tropics is that it was World War II, and the experience of Westerners like himself, that changed old habits of thought.[137] And it was with Stamp's care for detail rather than assumption, and his love of the "living word" rather than the "dead letter" in mind that Kimble, in his obituary to his former teacher, recalled that Stamp's "manner was mild, his grades generous, and his sympathies clearly with the underdog, whether in Asia or Aldwych."[138] Having opened some doors, however, Stamp soon closed them again in *The Geography of Life and Death*, returning to the long-standing idea of disease as a "the ruler of Africa" and barrier to any meaningful development there.[139]

In these ways and more, geographers' consideration of the post-war tropics as a 'problem' of geographical order was sometimes retrograde but also sometimes alive to change. Many of the geographers concerned had served overseas during World War II, and the spectre of war left a lasting impression on how they viewed the world.[140] No more so, perhaps, than in the case of the pioneer of behavioural geography, William Kirk, whose 1978 reflection on the nature of geography as an interpretative enterprise was titled "The road from Mandalay" and was shot through with his personal reflections on how war was a potent cipher for and conduit of the very idea of geography. "We constantly bemoan the fact that in contrast to other sciences we do not have a laboratory to study the spatial behaviour of human groups within prescribed conditions," he wrote;

> yet the majority of military campaigns afford such an opportunity. A host of problems of geographical interest arise: the adjustment and acclimatization of men and machines to various theatres of war; the influence of terrain on the technical levels at which war can be waged (the American operations in Vietnam present a classic illustration); strategic mobility in the media of land, sea and air; regional hierarchies of theatres of war from world strategic systems to the smallest tactical areas; cartography and the scale of war; territoriality as a function of conflict and conflict resolution; decision making in relation to perceived spatial constraints and objectives; game theory.[141]

Kirk had been a field reconnaissance officer with the British Army, behind enemy lines in Burma in 1944–1945, and many geographers' subsequent peacetime (and sometimes again wartime) engagements with decolonisation were rooted in a commitment to fieldwork and observation, and in both an idiographic immersion in foreign lands and communities, and a broader effort at regional synthesis. These were tried and tested (albeit to some tired) means of doing geography, and not just in the tropics.

Rarely did geographers working in or on the tropics wax lyrical about decolonisation or development, even if most of them also found significant fault with the colonial policies of the past. Gourou was concerned about the dictatorial inclinations of colonial and post-colonial elites. Many of them were self-interested and aggrandising, he thought, and out of touch with the people, and especially rural folk. The elites had their own interests at heart in the way administrative districts and ploys were hatched, and economies and infrastructures re-structured. The rural peasant was a casualty of change and political transformation.[142] Like the other geographers invoked here, Gourou was neither an apologist for colonialism nor an advocate of rapid imperial retreat.

Geographical work in tropical regions took much of the speed and optimism out of the desire for change. Munger noted the urgency with which African leaders set about nation-building once independence had been attained; but his review also points to how geographers tied the speed of decolonisation up in a knot of geographical problems. Luella Dambaugh, of example, characterised Jamaica's 1950s road to independence as a rocky road:

> Jamaicans brush aside their notorious past with a hearty laugh, but are quite sanguine about their future. The island has experienced throughout its history a number of violence upheavals, both natural and economic, that have left the people no better off than before, if anything a little worse. Poverty, disease, malnutrition, undernutrition, unemployment, underemployment are very much with them today. Also there are devastated and eroded hillslopes, impoverished soils, and above all, many landless peasants. While notorious buccaneers once claimed the limelight, notable bauxite interests hold the clue to her future.[143]

The French geographer Hildebert Isnard expanded on this in his 1971 primer *Géographie de la décolonisation*. He defined his topic as "the remodeling of all structures pertaining to the old [colonial] state of things . . . [and] the creation of a new geography that meets the requirements of an authentic national life." But he went on to argue that geographic conditions were an integral part of authentic national life, and that because imperial powers had shaped colonial space according to their needs and interests, decolonisation was destined to be a quintessentially "uncomfortable geographical experience." The watchword of decolonisation, he suggested, was incompletion (*inachèvement*).[144]

Similarly, in a 1968 survey of "the problem of Nagaland" the German geographer Ulrich Schweinfurth emphasised the idea that "Decolonization has presented the world with many new problems arising from the fact that many of the newly established nations, immediately after gaining their independence, have tended to resort to policies similar to those of which they accuse the former colonial powers."[145] Finally, the British geographers Brian Hoyle and David Hilling began their 1970 collection *Seaports and Development in Tropical Africa* by stressing how arbitrary colonial borders had been left largely intact by decolonisation and thus perpetuated a port system that was incapable of "supporting rational, co-ordinated economic growth." While exports from British West African territories (then decolonising) increased by 90 per cent between 1945 and 1970, and those from French West African territories trebled, this growth was based there and elsewhere in tropical Africa on "a number of major seaports at the coastal termini of railways penetrating in many cases far into the interior, but with no interconnection between railways, [and this] meant only that the vast areas of the continent were inadequately served but also the economic growth tended to be highly concentrated in favoured coastal and interior 'islands.'" They stressed the significance of this "geographical and economic background," and saw the search for "spatial generalization" as a wrong-headed export of American geography.[146]

For geographers, then, and by no means just Gourou, the problem of geographical order – the question of stability, harmony and organisation in people-land relations and the way development prospects were appraised – was seen as an elementary aspect of decolonisation, and fostered a common epistemological outlook. Many Western geographers associated geographical order with the maintenance of peace and stability, and many of them saw abstract and uniform projects imported from afar, and top-town methods, as counter-productive. Many regional, zonal and general (continental) geographies from this period – from Gourou's surveys of Asia and Africa, to Kimble's *Tropical Africa*, Spate's *India and Pakistan* (1954), E.H.G. Dobby's *Monsoon Asia* (1961) and Stamp's *Africa* (1953) – probed how the problems and energies of the post-war tropics were anchored in its geography.

Commentators pointed to the uneven quality of the usually sweeping coverage offered in such texts, and as Gourou himself noted with respect to Stamp's *Africa*, this genre of geography was best when broaching local details and circumstances that were known to the author, and was often shakier when operating at the level of grand generalisation.[147] *Les pays tropicaux* was credited with being one of the few texts that struck an insightful balance between detail and generalisation, and Gourou insisted that works of tropical geography should not necessarily be castigated for their "biases and omissions," for they furnished information that was vital to planners, developers, the state and international organisations.[148] But he and other geographers working on the tropics were better, more insightful, when they were writing about places they knew than ones they had only read about, and sometimes seemed quite easily persuaded by what they had been told. Just weeks after attending the IPR in Lucknow, for instance, Gourou observed: "The Indian government is wisely trying to hurry through the development of irrigation (in order to alleviate the food crisis, which remains serious) and the hydroelectric industry. It is inspired by the [American] Tennessee Valley Authority to realize three great [dam] projects."[149] To Gourou, this was not a good thing.

We have been suggesting that these engagements with the post-war tropics do not just come down to knowledge and epistemology. To return to Febvre, the problem – tropes – of geographical order sketched above also need to be brought into an emotional register. The inclination to think in terms of geographical order was in part an emotional response to the devastation of war, the brashness of development thinking, and the uncertainties of decolonisation. 1945 was not a blank slate. It raised the problem of how people from all walks of life would come to terms with horrific violence, mass destruction, wholesale displacement, and rapid change. Perhaps most of the geographers discussed here responded by trying to see something permanent and orderly in what Febvre had described as "surprising means of destruction and creation, and an uninterrupted series of 'progressions.'" Gourou was just one of these geographers.

An understanding of geography provided a ballast, a beacon of hope, a grounding against the welling up of emotional life and unravelling of empires. In his 1964 book *The Image of Africa*, a survey of British imperial imagery and colonial experience between the late eighteenth and mid-nineteenth centuries which was written at the height of African decolonisation, the American historian Philip Curtin extolled how Gourou had smashed the long-standing "myth of tropical exuberance," in which Britain as well as Germany had indulged.[150] Delavignette drew attention to this debunking and added that Gourou had impressed upon French planners and politicians the urgent need "to grasp the significance of geography better."[151] "Let's return to Gourou's *L'Afrique*," Delavignette suggested.

> The very cover of the book gives off the reddish atmosphere of bush fire. The frontispiece is a photograph of black women, children on their backs and hoe in hand, bent over to dig the earth. Already our attention is attracted by what is essential: how does one impose modern administration on an organisation of agricultural space where the peasants expend their time and strength working, far from the village, in extensive cultivation? . . . In this situation, political leaders and their administrative servants are tempted by easy solutions that adopt the mask of force.[152]

Both Gourou and Delavignette recognised that while many of the important post-war geographical works on tropical Asia and Africa were published during the 1960s, most of them had been started immediately after World War II and were borne in wartime experience. In the 1970s, one of Gourou's and Febvre's mutual friends, Georges Condominas, the pre-eminent post-war French anthropologist of Southeast Asia, reflected that many rural communities of the developing and decolonising world were victims rather than winners of the putatively 'progressive' and 'emancipatory' dynamics that had changed their lives, and lamented that all Western social scientists had to do to demonstrate their compassion for the plight of such communities was to "decolonize" their categories and disciplines: make themselves less Eurocentric and haughty. They did not have to "re-orient their whole way of life in the way that those they studied had to."[153]

Nevertheless, Condominas added that Western scholars' own experience of rupture and devastation had played a significance part in how they came to re-orient – defend, revise and overhaul – their categories in the light of change. One of the key manifestations of this rupture and re-orientation in the tropics, Ross suggests, was "an unprecedented escalation of state intervention into the lives of rural people, their use of land, and the environments they inhabited," and this move itself spawned a further range of epistemological and emotional reaction.[154] This was "the heyday of the scientific expert, the comprehensive plan, and the monumental mega-project," with late colonial states pressing to "convert . . . natural riches into economic growth" by boosting agricultural productivity, promoting commodity exports, prospecting for new land and resources, and expanding and improving infrastructure and health and educational provision.[155] The question of discipline extolled by Gourou – the need to order and regulate resources and opportunities, and bring discipline to indigenous agricultural systems and understanding of how, and how far, it could be changed – loomed large in all of this, as did awareness that the harder colonial powers pushed, the more perilous their efforts often became in economic, political and environmental terms.

Tropicality thus became a mode of state and corporate judgement about the ways and extent to which change was possible, and it was deeply implicated in the post-war rejuvenation and demise of empire. Late colonial development advisers, planners and practitioners developed a plethora of views about what was at stake, and such views were cross-cut by a range of political agendas and sympathies. Gourou took a particular line, the exposition of which points to wider debates among agronomists, ecologists, economists, geographers and politicians. He thought that while the drive to produce new and more export commodities (everything from tropical fruits and oils to tin and rubber) was unstoppable, and might bring a higher standard of living to colonial populations as a whole, it often came with severely counter-productive (he usually stopped short of saying 'unacceptable') economic and political consequences. Many development projects ruined tropical ecosystems and human-environment relations.

Gourou was particularly animated by the question of whether indigenous agricultural systems and landscapes needed to be 'developed.' Development practices that failed to get to grips with indigenous knowledge, ecological systems and agrarian practices would backfire, he argued. And he shunned the Marxist concept of 'under-development' because he thought it assumed to flippantly that there was some elevated state of development to which tropical peoples and places might naturally aspire. "Geography demystifies 'under-development,'" he declared in 1969, "by returning to its character as a traditional, historical and normal condition of mankind," and by "insisting that it is not a fatal effect of natural conditions."[156] The problem of development, he believed, was fundamentally distributional and cultural rather than quantitative or aspirational; it would come "only from a revolution of techniques, and especially those of *encadrement*."[157] The adoption of modern technology (machinery, breeding and cropping methods, and artificial fertilisers) did not present a quick or untrammelled means of boosting indigenous agricultural productivity or of easing problems created by population pressure, he thought, and peasants were less receptive to it than planners and officials thought and expected, even in situations of poverty and precarity.[158] At the same time, Gourou was largely silent about late colonial and post-independence projects of land and resource conservation, which also shaped development

240 *Affecting the tropics*

discourse during this era and included attempts to protect vulnerable indigenous groups from modern and foreign encroachment.[159] In short, his tropicality was cautionary and selective.

Les tropiques fantôme ou dialogiques?

Yet there were many difficulties in knowing how to find and evoke geographical order. Organisation and discipline lay at the heart of Gourou's quest, and he held Japan, on the northern fringe of monsoon Asia, up in this regard. "I will send you my reflections on Japan," he wrote to Gottmann in 1983: "How did the country pass from the rural stage to the present one? If we can't explain that, how do we explain human geography at all?"[160] Japan had a long and strong commitment to *encadrement* (organisation), and this commitment underpinned the country's unique fusion of traditional and modern ways.[161] "While remaining faithful to the customs of a Far Eastern civilization," he told the 1950 IPR meeting in Lucknow, "Japan has created a modern industry with modern tools."[162] The country had had an empire with both a rural peasant and modern industrial base. Its traditional rural civilisation was highly disciplined, with severe constraints imposed by the state, nobility and village authorities, and which the peasant "meekly accepts," he observed. But discipline had brought many benefits. He extolled "an authoritarian enlightened government" that had displayed "a lively awareness of the public good," had "abstained from prodigality," and had intervened "vigorously" and "progressively" in both agriculture and industry, perhaps showing "the path to be followed" by newly independent countries such as India and Indonesia.[163]

Shogun authorities had not allowed pastoralists and horse breeders to compete with rice cultivators for land, and in Gourou's estimation "a hectare of rice provides five times more calories if the rice is consumed directly by humans than if the rice is first used to fatten pigs or feed cows."[164] One in seven farms was "'officered' by agricultural technicians . . . who spread knowledge of technological progress," and peasant cooperatives managed "matter[s] of credit, insurance, purchase and sale." Between 1945 and 1950 the deficit between food production and consumption in Japan was lower than it was in Belgium and Great Britain. Since the late nineteenth century modern know-how had been deployed to tackle problems – particularly of poverty, allaying peasant unrest in densely populated and intensively cultivated (chiefly rice-growing) lowland areas – that had been less easily dealt with in other parts of the tropics and Far East. Gourou put the gradual increase in Japan's rice yield between 1940 and 1970 down to the "careful and precise organisation" of technology (irrigation, agricultural experimentation, disease control, and the use of "mechanical appliances").[165]

Over a 100-year period, Gourou noted in *La terre et l'homme*, "Relations between man and the land have been profoundly modified in Japan, not by any transformation of the physical environment but simply be technological revolution. Here is one more proof of the dominant role played by civilisation in creating the human landscapes of the Far East."[166] This still came with his tropicalist refrain: namely, that revolution "is not sufficient to raise the living standards of a Far Eastern population to those of a population in Western Europe or North America" (according to the World Development Index, by the end of the century they were about the same).[167] Gourou marvelled at a Japanese agrarian peasant civilisation that was based on a vegetarian dietary regime but was more receptive to change than its counterparts in other parts of the tropics. The Japanese landscape in 1972 was still "unintelligible without careful and continuous reference to its history," and is "still a superb example of the landscape-making quality of the civilisation."[168] One senses that he deemed the misery and subservience of the peasant population less important than the symbolism of organisation and discipline he beheld in the landscape.

Gourou found elements of this archetype elsewhere in the tropics, but only elements. More commonly, he wrote of what he saw as clashes between rural and urban, traditional and modern, and indigenous and foreign values.[169] During the 1950s he reflected on problems of rapid population growth in China and India, and the "failure of 'the great leap forward,'" which he attributed partly to the Chinese state's avoidance of upland and "Indianised" (Himalayan) areas. "Could the problems of Asia" – and not least "the failures of China's 'great leap forward'"– not be addressed in 'the Japanese way,' which emphasises scientific and technical progress," he pondered in his 1962–1963 Collège de France lectures.[170]

Gourou professed to Gottmann that "Economists are dangerous illusionists . . . [who] see the 'economy' as the queen of all disciplines."[171] But haughtiness was also a theme writ large in geographers' outlooks. Some were not shy about advancing bold generalisations about the fate of vast regions, and resorting to crude stereotypes. Many saw tropical regions through the prism what political scientists and areas studies moguls termed 'strategic interests.' Here, for instance, is the politically well-connected American geographer Isaiah Bowman offering the U.S. Department of State with "a mess of words about the Far East" in 1949.[172] Starting with the mantra that problems of geography were pivotal to "organization for a free world," he described China, Burma, Siam and Indochina as the "no-man's land of political venture" and "outposts of irresolution" that were "alien to democratic doctrines." He recommended a policy of "wait and see," and warned: "We can lose our shirt in the swamps and canyons of the hinterlands of Burma and Siam," and that "many of the problems of the world cannot be 'settled': they can only be wrestled with."[173] He was right about the shirt, but it is the scale at which he writes about the 'problems' of the Far East that interests us. Such an outlook emanated from Washington, of course, but it was also made possible by three decades of geographical scholarship that had rendered the Far East as a geographical unit.

Another prominent American geographer, Derwent Whittlesey, wrote in similar terms in 1956 about Britain's fledgling Central African Federation (the merger of its three colonies of Southern Rhodesia, Northern Rhodesia and Nyasaland – now Zimbabwe, Zambia and Malawi – into a single political structure). "Until now," he asserted,

> the Zambezi River has been the demarcation between contrasting European attitudes toward the question of the races. South of the river, Europeans are

in a minority (in Southern Rhodesia, about 1 to 15 Africans). Nevertheless, European interests always take precedence. It is assumed that the common roll for voting will grow, but that only those Africans who live in the European manner will be registered. The best that most Africans can hope for is guidance to improve their situation by working into the economic structure being developed by and for permanent European settlers. North of the Zambezi the Europeans constitute a tiny elite that exploits natural resources to satisfy a world market. Nearly all will return to the middle latitudes upon retiring (at an early age). Under favorable suzerainty, such as that of Britain, the Africans are treated considerately. To do so is humane, and it pays.[174]

This 'what is best for Britain, or America, or the West is best for the rest' view of the world was a potent part of the way problems of geographical order were posed as problems of paternalism, and views like Whittlesey's did not come from afar: he had undertaken two stints of fieldwork in the region. And as this reference suggests, haughtiness was not confined to geographical writing about the tropics, but to the decolonising world more generally, even if much of that world was in the tropical zone (see Figure 6.1).

However, there was a good deal more subtlety and sophistication in the writing of figures like Whittlesey or Bowman than meets the eye from passages like this. Whittlesey offered the rudiments of a historical geography of colonialism in Southern Rhodesia and stressed the importance of thinking about the varied character of the region's 'velds' (open rural landscapes) as channels of contact between African pastoralists, European farmers, and colonial land commissioners. It is important to register such tensions between facetious opinion and insightful commentary.

Febvre, we recall, noted that there was "something tragic" in Gourou's conclusion that "the peasants" lived happily but in misery, and the theme of 'geographical misfortune,' which might be deemed another form of conceit, was writ large in Gourou's work during the 1940s and 1950s.[175] Who was to say that his vision of tropical misfortune was not also a form of selective forgetting? Or to return to our epigraph from Said, where did compassion end and a more insidious way of knowing kick in? During the 1950s and 1960s an impressive young cohort of American anthropologists (Clifford Geertz, Marshall Sahlins and Eric Wolf among them) extolled Gourou as a scholar of traditional, pre-capitalist civilisations of the tropics, and thus, by implication, as a scholar who worked at a remove from the present.[176] On the other hand, they found in Gourou a valuable story about how, in the Far East, high agricultural over-specialisation in areas of high population density had fostered agricultural 'involution' rather than intensification.[177] Geertz imbibed Gourou's pessimism regarding this matter, as did leading tropical economists and agronomists such as Douglas Lee and John Phillips, respectively.[178]

At the same time, Brian Hodder and other commentators reflected on the "excessively pessimistic" tone of the first edition of *Les pays tropicaux*, and René Dumont rebutted some of same imagery that Césaire had chastised, stressing that

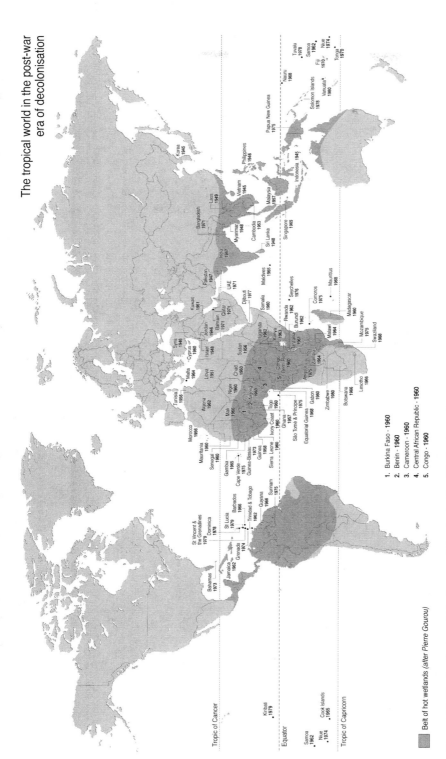

Figure 6.1 The tropical world in the era of decolonisation (G. Sandeman)

"there is no [geographical or ecological] curse on Black Africa [for] while the soil is fragile, without a doubt, and the climate unpredictable . . . the situation is by no means desperate."[179] One of Britain's pre-eminent scientists, John Russell, thought that Gourou passed much too glibly over the "devoted labour" of British and French scientists "who have done so much to develop tropical agriculture."[180] And foreshadowing Arnold, the New Zealand economic historian John Gould saw Gourou's primer as somewhat anachronistic as well as pessimistic: a more accurate description of the tropical world as it was in the 1920s than of the tropics of the 1940s or 1950s.[181]

Gourou frequently walked a fine line between compassion and conceit, and was by no means the only geographer to do so. In his 1954 *India and Pakistan*, for instance, Oskar Spate wrote about his idyllic Indian village thus:

> Though the substratum of life – the gruelling round of the seasons – remains and will ever remain the same, though a miserable livelihood extracts an exorbitant price in endless toil, there have been great changes, material and psychological, since Edwin Montagu, Secretary of State for India, spoke in 1918 of the 'pathetic contentment' of the Indian village. Pathetic it still too often is; contented, less and less; which is as it should be. . . . Now new motifs are changing the tempo of life in the large village. . . . There may be loss as well as gain in all this: but it is idle to bewail the break-up of integrated by religious, social, and economic sanctions which were a complete denial of human dignity.[182]

Spate was not simply describing a foreign reality to a Western audience. He was also moralising about "changes, material and psychological," and was weighing up different Western and Indian perceptions of human satisfaction, dignity and change. He was thinking about the categories, norms and distinctions that pertained to this situation. And in 1961 Gourou was still rehearsing nineteenth-century debates about the susceptibility of white colonists in the tropics to becoming physically and morally "degenerate" due to the impress of climate and disease, and whether it was possible to lift tropical peoples designated as being in a "primitive state" out of their backwardness.[183]

At the same time, the likes of Gourou and Spate were acutely aware that the systems of knowledge and inquiry that geographers had wielded were no longer acceptable to people who were now seeking to throw off the shackles of Western domination. We have seen how Césaire threw down a gauntlet by haranguing liberal-minded academics as "watchdogs of colonialism" (see Chapter 1). There was no room in his view of the world for assessments like the following, from Fisher, on the Vietnamese situation: "There is no denying the fact that most South-East peoples are addicted to an easy-going tempo of life and a related insouciance of behaviour, which . . . are probably related to the combination of an enervating climate with an absence, except in a few untypical places, of serious population pressure."[184] Fisher painted a political picture with the broad brushes of Orientalism and tropicality, the kind of picture that a latter-day postcolonial geography has rightly set about debunking.

Plenty more examples of such essentialist and patronising thinking come readily to hand. Gourou's gulf between temperate and tropical worlds can also be found in spades in Kimble's diagnosis of the legacies of colonialism. "In large parts of tropical Africa," he wrote,

> the helm is still a long way from the bridge. Much of the economic navigating is therefore done by remote control. Even such independent territories as Liberia and Ethiopia are not wholly free to chart their own courses; without American, and to a lesser extent, European help, their economies would not long remain on an even keel.[185]

The allusion here to the cover of the original French edition of *Les tropicaux* is palpable (see Chapter 1). According to Kimble, Europeans and Americans still needed to be masters and commanders of the ships (of capital, investment, governance and expertise) that sailed from temperate to tropical ports.

If we fast forward now to 1968, we find the firebrand New Zealand geographer Keith Buchanan giving a generational twist to how the spirit in which the likes of Gourou, Kimble and Spate had worked was being dismissed as *passé*. Those "born into, and living amid, a world in flux," he declared,

> cannot always realize that 'the earthquakes of change' to which you have grown accustomed, which indeed, for you represent the normal world condition, are symptomatic of the end of the world. And for those of us who are older, who grew up in a world whose major lineaments seemed fixed and unchanging (because we did not recognize that the Long March and the rioting in India and the shooting down of Africans were the twisting birth pains of a new world), it is no less difficult to adjust to the reality of an era in which most of the old and familiar landmarks – the Empire, the supremacy of Europe, the dependence of Africa, the inscrutable chaos of the East – have disappeared, along with Loretta Young and Laurel and Hardy: *Mais où sont les neiges d'antan*?[186]

Buchanan deemed Gourou's work reactionary and, as James Sidaway and Marcus Power reveal, sought to get it removed from the tropical geography curriculum at the National University of Singapore during his time there as external examiner there.[187]

Yet some disputed Buchanan's assessment that it was "inevitable that the methods and the techniques formulated [by geographers] to interpret a gradually evolving European scene should prove inadequate for the interpretation of a new, swiftly changing world scene."[188] Spate, for instance, named 45 Western geographers (including Gourou) who defied Buchanan's generalisation and had achieved high "intellectual standards" in their studies of the non-Western world.[189] Even so, when geographers did get to Buchanan's 'swiftly changing' world it was usually with the sentiment that rapid economic, social or political transformation would potentially unhinge centuries of intricate and often delicate human adaptation to environmental circumstances,

246 *Affecting the tropics*

and thus undermine what Bowman grandiosely described as "the meaning of geographical conditions" and the orderliness of such conditions.[190] Akin to Nicholas Spykman's plea that "An orderly world is a world not in which there is no conflict, but one in which strife and struggle are led into political and legal channels and away from the clash of arms," Gourou saw his geography, and many reviewers of *Les pays tropicaux* read it, as an 'orderly' endeavour in which the tropical world was led by the light of geographical reason to a faith assessment of its problems and prospects, and away from the perils and makeshift politics of a decolonising world.[191]

When Gourou suggested that war, capitalism, revolution and imperial dissolution were not the most important factors affecting the capacity of the tropics to embrace change, he was not saying that all attempts at change were doomed to fail. Nor was he or other geographers unequivocal believers, as Buchanan had it, in "the supremacy of Europe," "the dependence of Africa," or "the inscrutable chaos of the East." Rather, many geographers' appeal to what was "fixed and permanent" in the tropical areas and landscapes they studied was a response to what Gourou in his textbook (with Papy), written at the height of the Algerian War, had characterised as "extraordinary agitation" on the human soul."

This exchange between Buchanan and Spate raises the question of whether post-war tropicality, at least as it was purveyed through the discipline of geography, was dialogical (receptive to different views) or fostered a vision of *les tropiques fantôme* – a phantom and shady tropics that stemmed from a biased and retrograde, if acutely felt, outlook. These two – dialogical and phantom – sides of tropicality are encapsulated well in two mid-1950s conferences involving Gourou (and which also extend the 'networking' theme of Chapter 4), and in the way the promulgation of tropical geography in France has been narrated and critiqued. We will discuss the two conferences first, which enable us to place Gourou and tropicality in a wider arena, and then look at the more circumscribed question of French geography.

Césaire's *Discours sur le colonialisme* was re-issued by Présence Africaine in the year of the famous Bandung Conference, at which newly independent African and Asian nations gathered to take stock of the legacies of empire and discuss their future. Also in 1955, 70 scholars from around the world (albeit predominantly from Europe and North America) and a range of disciplines (if chiefly from geography) met at Princeton University for a symposium on "Man's Role in Changing the Face of the Earth" organised by the Swedish Wenner-Gren Foundation.[192] This gathering was sparked by growing concern over humanity's deleterious environmental impact on the planet, and a two-volume selection of the talks from the conference was published the following year, and at the moment when the Senegalese writer Abdoulaye Ly vividly characterised mounting revolts against colonial rule as "a ring of fire burning all along the Tropics."[193] At the symposium American geographer Carl Sauer lamented the destruction of "wise and durable native systems of living with the land. The modern industrial mood (I hesitate to add intellectual mood) is insensitive to other ways and values."[194] He also impressed on the organising committee the need for a historical and ecumenical focus on the relations between land and life, chiefly in his mind to counter the ahistorical and reductionist gaze of the post-war social sciences.[195] However, there was no sustained engagement at the symposium with either the post-war 'end' of empire or the emerging ideology of development. The terms 'colonization' and 'empire' appear only a handful of times, and then largely with reference to archaic Chinese and Indian empires. On can find no trace of Ly's world in the proceedings. Gourou's chapter, on the world of tropical cultivators, epitomises this blind spot.[196] He sketched the evolution of tropical cultivation over millennia and without any reference to contemporary dynamics.

On the other hand, in March 1954 some of the world's leading social scientists met in Paris under the auspices of UNESCO to discuss "the problems of black Africa." The UN rapporteur for the meeting was Georges Balandier, and he noted that the speakers had collectively agreed that issues of "motivation," "adjustment" and "conversion to modernity" were of utmost significance to Africans and came in a range of guises. Research into the nature of decolonisation – which Balandier defined as both the quest for independence and development, and the fraught process of attaining them – needed to be guided by a regard for both "the general situation" of opportunity, constraint and organisation, and "the differential variation" of this situation by locality, nationality, class, gender and ethnicity.[197] Balandier urged that research needed to be interdisciplinary, comparative and involve Africans. In other words, dialogue was needed. Only then, Lévi-Strauss, then a rising star of French anthropology, pleaded at the conference, would "the dangers of a unilateral outlook" be avoided. Balandier spoke a good deal about the "webs of necessity" and "adjustment" that connected the "traditional environment" and "modern environment" of the African, and as we have already seen, he lauded Gourou's concern with "adjustment to place." Balandier, Lévi-Strauss and Gourou were all ambivalent about the ideas of development that were being pushed at the conference by economists and planners. Balandier thought that development models lacked depth and questioned the presumption that economic growth necessarily entailed social progress or nurtured political stability. Questions of "under-development" were not the preserve of economists, he continued, just as psychologists did not have a monopoly on how colonialism and anti-colonialism were read in cultural terms. He thought that experts like Gourou offered a more "nuanced view" that avoided the distortions and limitations of grand theories and models, and remarked on the importance of Gourou's idea of *encadrements*, which Balandier interpreted as the colonial situation of people in urban and rural landscapes.[198]

This conference, which took place during the month that the siege of Điện Biên Phủ began, points to a shift in French colonial thought away from a Manichean concern with the cultural formations of colonisers and colonised, and towards a more holistic and dialogical appreciation of the diverse contexts and drivers connecting and separating different groups, and a concern with the inner conflicts of late colonial and post-colonial societies rather than just with how they looked to the imperial centre. At the same time, Balandier worried that knowledge of the colonial situation in tropical Africa continued to hinge too much on a range of Western experts (academics, administrators, jurists, doctors, aid workers and security services). In short, this conference was a space of contestation where Africanists and tropicalists with disparate agendas and orientations churned ideas around in a spirit of dialogue, and we might infer from these two conferences two styles of representation – the dialogic and *fantômatique* – that combined the epistemology, ethics and emotion of tropicality in different ways.

Let us now say a little more about *les tropiques fantôme* with reference to French tropical geography. The period 1945–1970 was a highpoint in the connection between tropicality and geography. The project that Gourou and a small group of other French geographers pioneered, chiefly in the 'traditional' societies of rural Africa (as from 1945 Indochina/Vietnam was cut off to French researchers), the French Antilles, Indian Ocean and Brazil, gained significant backing from the French state through the CNRS, created in 1939, Institut Français d'Afrique Noire (IFAN), which opened in Dakar in 1940 to orchestrate research in French Africa, and ORSTOM. These "big science" initiatives, as Paul Claval calls them, straddled the late colonial and post-independence eras.[199] They strove to provide the overseas knowledge and understanding that France thought it needed to rejuvenate empire, abate demands for independence, and facilitate post-colonial development. ORSTOM, for instance, originated in two interwar institutions that were charged with "coordinating the colonial sciences, creating a network of research centres in French overseas territories, and providing specialist scientific training for research on the tropical world"; and from 1960, one of its three missions was to "undertake basic research for the development of tropical countries."[200] In Belgium, where Gourou had taught since 1936, and Henri Nicolaï became his most notable student, geography did not attain a firm and equal standing in the social sciences until 1929 but grew in prominence after World War II.[201]

Between 1945 and 1980 tropical geography accounted for between 15 and 20 per cent of the entire output of French geography. However, as Sautter noted in a 1978 review of work in the field since 1945, the rural/regional focus on 'traditional' society that prevailed became increasingly out of step with forces of modernisation, and particularly urbanisation. Paul Claval adds that while the regional perspective and study of traditional rural worlds generated both fastidious knowledge and good administration, it struggled to perceive the development of the modern economy, and had little to say about how rural areas were becoming subordinate to big cities, extractive industries and export zones, and how urban and regional areas were becoming polarised.[202]

Big science paid relatively small dividends, Claval suggests, and by the 1970s tropical geography began to be chastised as *passé* and reactionary by a younger generation of geographers who had been radicalised as students during the 1960s. While Gourou's area of study "came under strong criticism for its pessimistic view of prospects for industrialisation and urbanisation in the tropics, it seduced French geographers because it matched the contemporary interest in zonality and relied on a [still prevalent] *genre de vie* analysis of, typically, rural areas," Claval observes.[203] It also seduced sections of British geography, with tropical geography of a more economic and urban (albeit still descriptive) ilk attaining a significant foothold in geography departments in London and Liverpool.

In his 1971 retrospective "Quarante ans de géographie tropicale: Bilan et perspectives," Gourou wrote as if nothing much about the orientation of the field that he and others had shaped and driven had changed over 40 years. Much, of course, had changed, and Gourou knew it, but he chose not to write about it. Critics noted that that while works of tropical geography were often evocative, and in Gourou's hands highly literate and even philosophical, many of them had a descriptive, bucolic and apolitical hue, and their inventories of physical and human facts were out of step with the radicalism of the era. In 1984 Michel Bruneau (who worked on Thailand, and whose father had been taught by Gourou in Hanoi), and Georges Courade (an Africanist who had been taught at Nanterre in the late 1960s by Pélissier) opined that "tropical geography, embodied by Gourou, with its naturalist and colonialist perfume, still has many days ahead of it. It is only quietly and discretely . . . transforming into a geography of dominated countries." For now, it remains "the hard core bastion of Vidalian geography," the continued, and has kept itself in tact "by being able to make some compromises with a few neighbouring disciplines, mainly ecology and ethnology."[204]

Tropical geography became caught up in the radical – generational – politics of the era. Geographers "who jumped on the bandwagon of tropical geography often held simplistic views," Claval notes: "they focused on medical problems, rural areas and farming techniques," and shied away from processes of urbanisation and industrialisation.[205] That many tropical geographers, including Gourou, did not fully grasp what radicals such as Walter Rodney, André Gunder Frank and Samir Amin meant by "underdevelopment" (*sous-développement*) compounded matters. When Gourou discussed this term in a 1969 essay which was aimed at a broad social science audience, he placed it (as he did the term development) in inverted commas and characterised it as a 'lack of development' rather than in the way that Marxist theorists and geographers conceived it, which was as a "consequence of the highly unequal terms of exchange between suppliers of raw materials and suppliers of capital an finished goods," as Ross puts it, and thus as producing and regimenting "economic disparities between the industrial powers and the rest of the world," and extending the influence of empire.[206]

However, "a constant attitude of doubt and modesty gives a great scientific and human strength and worth" to Gourou's work, François Bart maintained the year after our doyen of tropical geography died, and the one in which Arnold re-ignited debate about his tropicality.[207] And as Pélissier wrote in his obituary to his teacher, critics "who seldom leave the banks of the Seine and hardly read Gourou's work accused him of observing rather than transforming the world," but "his culture and courtesy kept him from indulging in loud polemic."[208]

More recently, and riffing on ethnographer Michel Leiris's 1934 *L'Afrique fantôme*, Catherine Fournier-Guérin suggests that French tropical geography had a "phantom" quality that did much to shape Western perceptions of tropical regions.[209] Like Leiris in Africa, Gourou in Indochina and later Africa, many French geographers "were seduced by exotic landscapes and populations that they deemed authentic" and saw them as apparitions of difference. They chased exotic ghosts. This phantom or ghost tropics, she continues, was bound up with both primitivism (as an artistic, literary and late colonial form of pining), and the *plain-pied dans le monde* style of representation we considered in Chapter 3, which removed the author from the scene of inquiry.

"Agricultural development is the first condition of tropical development," Gourou maintained; "everything else will work if agricultural development succeeds. It is necessary to build this agricultural development on a precise and detailed knowledge of local farming systems and from the opinions of local farmers about their own agriculture."[210] His focus was on rural conditions, and "for many post-war French geographers the cities of sub-Saharan Africa were for a long time only ghosts, in the sense of figures absent from their writings on the continent, and cities still suffer today from this distorted outlook. The cities of Africa are doubly 'ghost towns' in geographers' work on Africa, either because they have long been neglected or because appear in a biased way."[211] Only 1 of the 40 contributions to the 1972 *Études de géographie tropicale offertes à Pierre Gourou* deals with cities.[212]

Geography was part of a throng of writing and image-making – in art, journalism, film, exhibitions, scientific and textbook coverage, and charity work – in which African cities became portrayed "with images of violence (gangs, war), urban pathologies (AIDS), pollution, pervasive misery and the domination of black market" – imagery, we shall note in Chapter 8, that lingers.[213] Geographers like Gourou neglected cities primarily because they were concerned with 'traditional' practices, but also because much of their fieldwork took place during the 1950s, and the rapid urbanisation of African societies did not begin until the 1960s. They also largely ignored questions of gender and gender violence. Furthermore, during the 1950s colonial authorities limited Africans' access to the city. In many territories there were restrictions on the permanent settlement of black Africans in urban environments. One upshot of this was that many geographers considered the city to be a European rather than African phenomenon, and white and not black, prompting them to ignore it. Are there thus strong grounds upon which to suggest that French geography nurtured *les tropiques fantôme*?

Les pays tropicaux played an important role in arresting what Paul Sutter sees as the "grand climatic theorizing and patronizing racial and civilizational generalizations" associated with tropicality. In the light of the poor and perverse results that accrued from the so-called 'green revolution' (the West's developmental faith that imported seed, irrigation, pesticide and fertiliser technologies would boost agricultural productivity), Gourou's insistence that rising standards of living would only stem from a combination of indigenous technologies and good land management policies was not, Jean-Pierre Raison avers, so "strangely timeless and oblivious to the great social and economic conflicts" of the day.[214] "The great problem facing Africa," Gourou maintained, "is that agricultural renewal puts into question the old African civilisation. This renewal does not depend only on the supply of better seeds, fertilisers and motorised machines; it depends primarily on the development of *techniques d'encadrement*. And that is something that only Africans can do."[215] It was these techniques that ensured "fixity and calm," and "while the agents of colonisation were not inactive, they hardly touched old rural techniques" in many parts of tropical Africa.[216]

Finally in these pages about the twists and turns in geographers' engagement with the ends of empire and coming of independence, and the ways and extent to which they wielded a particular – and in some ways narrow – disciplinary view, Gourou came to rue his inattention to urban issues, partly upon being quietly chastised about the matter by his friend Gottmann. In June 1966 Gourou confided to Gottmann that "Cities threaten to overwhelm mankind," but 20 years later, he lamented that while northern "Agglomerations asphyxiated by the motor car" did not interest him, "the huge shanty towns of Mexico City, Lima, Rio, Sao Paolo, Lagos, Kinshasa" were surely worthy of his tropical attention.[217] Indeed, in 1954 Roger Lévy wrote to General Secretary of the IPR, in connection with its upcoming Kyoto Conference, that Gourou was hoping to present a paper on the growth of large cities in Asia.[218] To our knowledge, this paper never materialised, but it suggests that Gourou was at least aware of this significant omission in his project, and he did write about town-country relations (albeit very sparingly) in due course.[219] While he contemplated African rural civilisations that to his mind "have been closed upon themselves and not undergone the modifying effects stemming from a conquest," he insisted that what he beheld "were nonetheless true civilisations and ones worthy of respect."[220]

Bruneau adds that there was never a full or free flow of ideas between Gourou and other scholars and disciplines, including his own intellectual family (Mus and Lévi-Strauss as well as Febvre and Braudel). He rarely strayed beyond his focus on people and milieu, and civilisations. Nevertheless, his 'family' and many of the Western scholars and experts he met in his travels through post-war international research and policy-making networks shared a well-intentioned concern with 'other' cultures, regions and histories, albeit at a time when it was difficult to do so without incurring the suspicions and often wrath of those on the receiving end of it. In many articles – for example, a 1955 article about the Congo that he wrote for the Brussels socialist daily newspaper *Le Peuple* – Gourou advocated the development of indigenous economies without a heavy-handed state interest or Western influence.[221]

There was a clear political edge to what Dipesh Chakrabarty describes as the "dialogical side of decolonization": research and conversation that circumvented and translated dominant ideas – particularly of development, modernisation and nation-state building – into other (if often occluded and now forgotten) forms, and that points to Leela Gandhi's spirit of renunciation.[222] Nicolaï told us that while Gourou was more than aware of how problematic it was to undertaken research at the tails end of empire, and how easily one's research could be used for ulterior ends, he was "convinced and convincing about the role geographers could play in preliminary research for development projects."[223] There was a strong dialogical element to a meeting between European and American geographers in Brussels in 1956, under the auspices of the IGU, to discuss ways of extending and standardising the population mapping of Africa. The University of Wisconsin geographers Glenn Trewartha and Wilbur Zelinsky, and Robert Steel and Mansell Protherto

252 *Affecting the tropics*

from Liverpool, pressed this agenda and probed how differential fertility, mortality and migration patterns helped to explain population distributions and demographic change. All Prothero said of this research was that he hoped it would prove "useful" and "practical" (it was presumably up to development actors to say how?).[224] However, Delavignette went much further, viewing Gourou's work as "essential" because 1950s Africa was "subject to profound social and political movements which oblige . . . colonial administrations to know geography better," and because Gourou "makes us reflect on the confrontation of a modern administration, issued from western civilisation, with different civilisations, which had for a long time been shut off from the rest of the world." It was unwise, Delavignette judged, to contemplate decolonisation and its possible effects, "without some liaison between the leadership of the administration and geographical research."

Delavignette was in no doubt about on what Gourou's expertise rested:

> The geographer enjoys an independence that the administrator lacks: he is in the service of truth, not of the State. And it is good that civil servants who have politico-administrative jobs are enlightened by independent research into geographical facts. It is in this spirit that Pierre Gourou taught at France's school for colonial administrators between 1940 and 1944, when all sorts of falsifications were the rage.[225]

Experts like Gourou were also believed to be aloof from the structures of nepotism and informal 'old boys' networks through which lines between public service and private businesses interests often became blurred.[226] And as Susanna Hecht reflects on the slippery distinction between a science and politics of decolonisation in the tropics, "scientists have always had an important role in colonial and modernization policy, but especially in tropical environmental and land politics since the mid-twentieth century. The tropical development literature is replete with interventions from scientists with stakes in development controversies . . . [They] have been entangled in mythologies of development (the idea of progress) and conservation (lost Edens)."[227] It is important to note that Delavignette's plea for 'geography' is often side-lined in the recent literature on the repositioning of academic disciplines and expertise in the era of decolonisation.[228] Furthermore, as Barbara Yates observed in 1964, while "tropical Africa has surged to world attention, primary materials on education are just beginning to be collected and systematically collected and critically analysed," and geography was a basic but ill-codified subject.[229]

The syntheses offered by experts like Gourou did not filter down into African curricula quickly, or smoothly. Some British geographers made the transition from late colonial to post-colonial employment, staying on in Africa in new roles as teachers and advisers to newly independent governments. There was a more wholesale flight of French scholars based in Africa back to France upon independence, to the extent that some African leaders had to call on foreigners to return.[230] Those, like Gourou and his students, who came to Africa on research missions from metropolitan institutions faced new constraints. Their work was looked upon with caution, and sometimes disdain, by a new generation of African geographers, many of whom had been trained in the West but either strove themselves, or were instructed by their governments to create, new school and university geography curricula that supported nationalist agendas and were untainted by colonialist ideas. For example, the Ugandan geographer and anti-colonial writer Jakayo Ocitti wrote a "revolutionary syllabus" for East Africa in which the study of indigenous links between territory and nation, and the tropical wealth of the region, were extolled.[231] And in 1960 Ocitti published a novel entitled *Every Dog Has His Day* (*Lacan Mako pe ki nyero*), where he began to explore African (Bantu) folklore as an instrument of geographical education.

Postcolonial criticism has belatedly restored recognition that post-war geographical research undertaken in the service of truth and order came from a Western mind-set that was 'impure' in the ways Césaire and Ocitti described. We have sought in this chapter to account for how such impurity worked in Gourou's case and wonder whether anything more sanguine might be pulled from the critical wreckage. As Ocitti sought to convey in his novel, it was one thing to declare that the colonial dog had had its day, but quite another to feel sure that what was replacing that dog would not commit the same sins – and Gourou perhaps understood the sentiment more fully than one might think.

Notes

1 ACDF: Assemblée des professeurs du 24 novembre 1946: Rapport de Lucien Febvre: pourquoi l'étude du monde tropical? Assemblée des professeurs du 16 février 1947: Exposé de M. Lucien Febvre.
2 Lucien Febvre, "Civilisation: Évolution d'un mot et d'un groupe d'idées," in *Civilisation – le mot et l'idée, Exposés par L. Febvre, É. Tonnelat, M. Mauss, A. Nicfero et L. Weber* (Paris: la Renaissance de livre, 1930), 10–59.
3 Lucien Febvre, *La terre et l'évolution humaine: Introduction géographique à l'histoire* (Paris: Albin Michel, 1922).
4 ACDF: Pierre Gourou, "Leçon inaugurale du cours," 5. Interestingly, Gourou did not reflect on how the anarchist geographers Pëtr Kropotkin and Élisée Reclus had earlier disparaged the racist roots and tentacles of environmental determinism. On this issue, see Simon Springer, "Foreword: Anarchy is forever: The infinite and eternal moment of struggle," in Federico Ferretti, Gerónimo Barrera de la Torre, Anthony Ince and Francisco Toro eds., *Historical Geographies of Anarchism* (Abingdon: Routledge, 2018), x–xi.
5 Michel Trebitsch, Preface to Henri Lefebvre, *Critique of Everyday Life: The One-Volume Edition* (London: Verso, 2014), 7.
6 On Febvre's frequent use of this expression and its wider genealogy – going back to de Tocqueville – see Barbara H. Rosenwein, "Worrying about emotions in history," *American Historical Review* 107, no. 3 (2002): 821–845.
7 See Frederick Cooper, "French Africa, 1947–1948: Reform, violence, and uncertainty in a colonial situation," *Critical Inquiry* 40, no. 4 (2014): 466–478.
8 Particularly with the publishing houses Ecole/Magnard, De Gigord and Hatier. See Hervé Théry, "Les pays tropicaux dans les livres de géographie: manuels de l'enseignement secondaire entre 1925 et 1960," *L'Espace Géographique* 17, no. 4 (1988): 299–306.
9 United Nations, *Some Suggestions on the Teaching of Geography* (Paris: UNESCO, 1949). The volume was drafted by the French geographer Robert Ficheux.

10 Owen Lattimore, Foreword to Karl J. Pelzer, *Pioneer Settlement in the Asiatic Tropics: Studies in Land Utilization and Agricultural Colonization in Southeaster Asia* (New York: American Geographical Society, 1945), n.p.
11 Ishan Ashutosh, "The geography and area studies interface from the Second War to the Cold War," *Geographical Review* 107, no. 4 (2016): 705–721, at 714.
12 Marc Bloch, *Lucien Febvre et les Annales d'Histoire économique et sociale: Correspondance Tome III 1938–1943* (Paris: Fayard, 2003), 187.
13 Jean Gottmann noted such "fecundity in ordeal" in his "French geography in wartime," *Geographical Review* 36, no. 1 (1946): 80–91, at 80.
14 Susan Friedman, *Marc Bloch, Sociology and Geography: Encountering Changing Disciplines* (Cambridge: Cambridge University Press, 1996), 162–163. Only one of the many volumes Bloch planned were published before he was killed – Henri Labouret's *Paysans d'Afrique occidentale* (1941).
15 Lucien Febvre, "La sensibilité et l'histoire: Comment reconstituer la vie affective d'autrefois?" *Annales d'histoire sociale* 3, no. 1–2 (1941): 5–20.
16 Febvre, "La sensibilité et l'histoire," 7–10.
17 Febvre, "La sensibilité et l'histoire," 9.
18 Michel Bruneau, "Pierre Gourou (1900–1999): Géographie et civilisations," *L'Homme* 153 (2000): 7–26, at 9–10.
19 On this point, see Yves Lacoste, "Géographie coloniale et géographie académique: approche epistémologique," in Michel Bruneau and Daniel Dory, eds., *Géographies des colonisations XV-XX siècles* (Paris: L'Harmattan, 1994), 343–348.
20 Palmer Throop, "Review of Lucien Febvre *Combats our l'Histoire*," *Journal of Modern History* 35, no. 2 (1963): 162–163, at 162; Marc Bloch, *The Historian's Craft* trans. Peter Putnam (New York: Alfred A. Knopf, 1954), 140; Lucien Febvre, *Combats pour l'histoire*, Volume III: *Pour une histoire à entire* (Paris: S.E.V.P.E.N., 1962), 425.
21 Pierre Gourou, "Les problemes du monde tropical," *Les Cahiers d'Outre-Mer* 1, no. 1 (1948): 1–7, at 3.
22 Paul Rivet, Preface to Pierre Gourou, *Les pays tropicaux: Principes d'une géographie humaine et economique* (Paris: Presses Universitaires de France, 1947), vii.
23 Frantz Fanon, *The Wretched of the Earth* trans. Richard Philcox; orig. pub. 1963 (New York: Grove Press, 2004), 2.
24 On these distinctions, see Alison Twells, "'Went into raptures': Reading emotion in the ordinary wartime diary, 1941–1946," *Women's History Review* 25, no. 1 (2015): 143–160.
25 Lucien Febvre, "Paysanneries d'Extrême-Orient," *Annales d'histoire sociale* 3, no. 1 (1943): 71–73.
26 Fernand Braudel, "The situation of history in 1950," in Fernand Braudel ed. and Sian Matthews trans. *On History* (Chicago: University of Chicago Press, 1980), 6–24.
27 Cited in André Burguière, *The Annales School: An Intellectual History* trans. J. Todd (Ithaca, NY: Cornell University Press, 2009), 35.
28 See Menno Spiering and Michael Wintle eds., *Ideas of Europe Since 1914: The Legacy of the First World War* (London: Palgrave Macmillan, 2002), 2–5.
29 Burguière, *The Annales School*, 35.
30 How censorship and anti-Semitic legislation affected the intellectual life of wartime France, and affected journals such as the *Annales*, has been keenly debated.
31 Pierre Gourou and Louis Papy, *Géographie, classe de première* (Paris: Hachette, 1962), 5–6. Papy studied for his doctorate at Bordeaux, dealing with the Atlantic littoral of France, but became Gourou's 'tropical' successor there after the war.
32 On the first of these, see Philippe Madeline, "Les constructions agricoles dans les campagnes," *Histoire & Sociétés Rurales* 26, no. 2 (2006): 53–93.
33 David Arnold, "'Illusory riches': Representations of the tropical world, 1840–1950," *Singapore Journal of Tropical Geography* 21, no. 1 (2000): 6–18, at 16.

34 Lucien Febvre, "La géographie – géographie et civilisation," *Annales. Économies, Sociétés, Civilisations* 4, no. 1 (1949): 73–77, at 75.
35 Febvre, "La géographie," 77.
36 Gourou, *Les pays tropicaux*, 7.
37 Gourou, *Les pays tropicaux*, 28, 49–50, 157. Gourou drew the expression "pathogenetic complexes" from Maxamillian Sorre's, *Les fondements de la géographie humaine,* Vol. 1: *Les fondements biologiques. Essai d'une ecologie de l'homme* (Paris: Armand Colin, 1943), and was particularly drawn to Sorre's argument that disease was not determined solely or ultimately by physical conditions, but in a fuller way by human factors of population density, culture, subsistence, lifestyle, and the diffusion of animals, plants, insects and microbes.
38 Gourou, *Les pays tropicaux*, 156.
39 Dominique Moïsi, *The Geopolitics of Emotion: How Cultures of Fear, Humiliation and Hope Are Reshaping the World* (London: Bodley Head, 2009), 90–91.
40 David Arnold, *The Tropics and the Travelling Gaze: India, Landscape and Science* (New Delhi: Permanent Black, 2005), 54–55.
41 Paul Erickson, Judy L. Klein, Lorraine Daston, Rebecca Lemov, Thomas Sturm and Michael D. Gordin, *How Reason Almost Lost Its Mind: The Strange Career of Cold War Rationality* (Chicago and London: University of Chicago Press), 9–10. Also see Jan Plamper, "The history of emotions: An interview with William Reddy, Barbara Rosenwein, and Peter Stearns," *History and Theory* 49, no. 2 (2010): 237–265.
42 Marston Bates, *Where Winter Never Comes; A Study of Man and Nature in the Tropics* (New York: Charles Scribner's & Sons, 1952), 9.
43 Bates, *Where Winter Never Comes*, 118.
44 Leela Ganhi, *Postcolonial Theory: A Critical Introduction* (Edinburgh: Edinburgh University Press, 1998).
45 Leela Gandhi, "It's not grand theory, it's makeshift" 2017 International Conference on "Postcolonialism and Theology of Mission," *Frankfurt am Main*, Institut für Weltkirche un Mission, video, https://iwm.sankt-georgen.de/en/kontakt/
46 Aron, epigraph in Pierre Gourou, *Riz et civilisation* (Paris: Fayard, 1984), 276.
47 Fanon, *Wretched of the Earth*, 3
48 Pierre Gourou, "Qu'est-ce que le monde tropical?" *Annales. Économies, Sociétés, Civilisations* 4, no. 2 (1949): 140–148, at 144.
49 Gandhi, "It's not grand theory," video.
50 See Henri Nicolaï, "Bibliographie de Pierre Gourou," *Revue Belge de Géographie* 64 nouvelle série, no. 2 (1998): 115–130.
51 Compilation from ACDF: Gourou – missions; AULB: Fonds Pierre Gourou PP153 and IP-713.
52 Michel Bruneau, Interview with the authors.
53 São Paulo was a favoured destination of Parisian academics, including Gourou's colleagues Braudel and Lévi-Strauss.
54 Eugene Kircherr, "Review of *The Tropical World* 4th edition," *Journal of Geography* 66, no. 6 (1967): 342.
55 BNF: Fonds Gottmann: Gourou to Gottmann, 31 December 1947.
56 Bruneau, Interview with the authors; Jean-Pierre Raison, Interview with the authors; Marcus Power, *Rethinking Development Geographies* (New York and London: Routledge, 2003), Ch. 3.
57 Pierre Gourou, *Pour une géographie humaine* (Paris: Flammarion, 1973), 10–13.
58 Pierre Gourou, "Pour une géographie humaine," *Finisterra* 1, no. 1 (1966): 10–32, at 18.
59 Gourou, *Les pays tropicaux*, 44.
60 Gourou, *Les pays tropicaux*, 47–49.
61 Pierre Gourou, "La géographie et notre temps," *Bulletin de la Société neuchâteloise de Géographie* LIII, no. 1 (1963): 1–9, at 2.

256 *Affecting the tropics*

62 Pierre Gourou, "Civilisation et malchance géographique," *Annales: Economies. Sociétés. Civilisations* (1949): 445–450, at 448.
63 Pierre Gourou, "Letter to the editor: à propos de l'ouvrage de K.A. Wittfogel, Oriental Despotism," *Annals of the Association of American Geographers* 51 no. 4 (1962): 401–402.
64 Pierre Gourou, "Géographie tropicale et problems de 'sous-développement,'" *Information sur les sciences sociales* 8, no. 4 (1969): 9–18, at 17–18; Pierre Gourou, *Leçons de géographie tropicale* (Paris: Ecole Pratique des Hautes Études, Vie section, 1971).
65 Pierre Gourou, *Recueil d'articles* (Bruxelles: Société Royale Belge de Géographie, 1970). On this wider disciplinary history and perception, see William B. Meyer and Dylan M.T. Guss, *Neo-Environmental Determinism: Geographical Critiques* (London: Palgrave Macmillan, 2017).
66 Pierre Gourou, "Déterminisme physique pas mort," *Revue de l'Institut de Sociologie* 3 (1955): 421–431.
67 BNF: Fonds Gottmann: Gourou to Gottmann, 2 August 1984.
68 Pierre Gourou, "Des inédits de Pierre Gourou," *Revue Belge de Géographie* 64, no. 2 (1998): 135–196, at 135.
69 Pierre Gourou, *L'Asie* (Paris: Hachette, 1957 edition), 4, 79–80.
70 Gourou, *L'Asie*, 1957 edition, 188.
71 Pierre Gourou, *L'Afrique* (Paris: Hachette, 1970), 5.
72 This elusiveness is opened up instructively by Michel Bruneau: "Tropicalité, tropicalisme, géographie tropicale: Evolution d'un débat, des géographes francophones aux géographies anglophones," in Hélène Velasco-Graciet ed., *Les tropiques des géographes* (Pessac: Maison des Sciences de l'Homme d'Aquitaine, 2008): 171–188, at 177–180; Michel Bruneau, "Civilisation(s): Pertinence ou resilience d'un terme ou d'un concept en géographie?" *Annales de géographie* no. 674 (2010): 315–337. Bruneau notes that *encadrer* can mean supervise, train, contain, frame, surround, restrict, control; that it had this range of connotations in Gourou's work, and some distinct economic, philosophical or architectural meanings in different areas and contexts of Gourou's work.
73 For the beginnings of such a study, and with a focus on technology and tropical nature, see Peter Redfield, *Space in the Tropics: From Convicts to Rockets in French Guiana* (Berkeley: University of California Press, 2000).
74 Gourou and Papy, *Géographie, classe de première*, 228.
75 The Geographical Association, "A world land use survey," *The Geographical Journal* 115, no. 4/6 (1950): 223–226, at 224.
76 Corey Ross, *Ecology and Power in the Age of Empire: Europe and the Transformation of the Tropical World* (Oxford: Oxford University Press, 2017), 351.
77 Éric, Jolly, "Marcel Griaule, ethnologue: la construction d'une discipline (1925–1956)," *Journal des africanistes* 71, no. 1 (2001): 149–190.
78 See, e.g., H. J. Fleure, J. N. L. Baker, Raymond Firth and R. W. Steel, "Ashanti Survey, 1945–46: An experiment in social research: Discussion," *The Geographical Journal* 110, no. 4/6 (1947): 177–179; Frank Debenham, *Nyasaland: The Land of the Lake* (London: Colonial Office, HMSO, 1955); Jean Dresch, *Un géographe au déclin des Empires* (Paris: Maspero, 1979); Olivier Dollfus Olivier, "Le regard attentif, le regard sélectif: Jean Dresch et Pierre Gourou. Entre monde tropical et tiers monde." *Hérodote*, no. 33–34 (1984): 73–88.
79 Pierre Gourou, "Hématies en faucille et géographie humaine," *L'Homme* 1, no. 1 (1961): 90–94; Pierre Gourou, "Le déterminisme physique dans 'L'esprit des lois'," *L'Homme* 3, no. 1 (1963): 3–11.
80 Robert Delavignette, "L'Afrique de Pierre Gourou et le point de vue de l'administrateur," in *Etudes de géographie tropicale offertes à Pierre Gourou* (Paris: EHESS, 1972), 279–286, at 285.

81 Pierre Gourou, *Terres de bonne espérance, le monde tropical* (Paris: Plon, 1982), 395.
82 Gourou, *Terres de bonne espérance*, 395.
83 Ross, *Ecology and Power in the Age of Empire*, 351.
84 A point emphasised by Jean Suret-Canale, "Les géographes français face à la colonisation: l'example de Pierre Gourou," in Michel Bruneau and Daniel Dory, eds., *Géographies des colonisations* (Paris: L'Harmattan, 1994), 155–169.
85 BNF: Fonds Gottmann: Gourou to Gottmann 7 August 1986.
86 Gourou, *Leçons de géographie tropicale*, 195.
87 Pierre Gourou, "Remarques sur les plans de *mise en valeur* des regions économiquement attardées," *Symposium intercolonial* (Bordeaux: Delmas, 1952), 97–100, at 97.
88 Nick Cullather, *The Hungry World: America's Cold War Battle Against Poverty in Asia* (Cambridge, MA: Harvard University Press, 2010), 45.
89 Rens van Munster and Casper Sylvest, "Introduction," in Rens van Munster and Casper Sylvest eds., *The Politics of Globality Since 1945: Assembling the Planet* (Abingdon: Routledge, 2016), 1–20.
90 Munster and Sylvest, "Introduction," 2–4.
91 Munster and Sylvest, "Introduction," 3–9.
92 Christian Taillard, Georges Courade, Gilles Sautter, François Durand-Dastès, Alain Durand-Lasserve and Michel Bruneau, "La géographie tropicale de Pierre Gourou et le développement," *Espace géographique* 13, no. 4 (1984): 329–337, at 333. Courade quoted from Gallais's and Sautter's essays on Gourou's 'system' of thought, largely to register the claim that they treated questions of politics as peripheral to his concerns.
93 Pierre Gourou, *The Tropical World* 3rd English edition trans. E.D. Laborde (London: Longman, 1961), 32. We quote from the English translation of the third edition because it was (in our estimation) the most widely cited of the four English editions, and thus the one perhaps most familiar to those readers of this book who read Gourou or used his primer in their teaching.
94 Pierre Gourou, "L'homme africain et son milieu," *Les Cahiers de l'Enfance*, UNICEF Paris (1963): 30–37.
95 Pierre Gourou, "Géographie et development," [1964] in Jean-Louis Boutillier and Yves Goudineau eds., *Cahiers des sciences humaines (trente ans 1963–1992)* (Paris: Orstom, 1993), 49–50, at 50.
96 Pierre Gourou, "Agriculture in the African tropics," in Peter J.M. McEwan and Robert B. Sutcliffe eds., *The Study of Africa* (London: Methuen, 1965), 128–138, at 129, 131, 134. The essay originated as a conference paper presented at the Institute of Agricultural Economics, Oxford University.
97 Gourou, *The Tropical World*, 3rd edition, 28.
98 Pierre Gourou, *L'utilisation du sol en Indochine française* (New York and Paris: Institute of Pacific Relations; Centre d'études de politique étrangère, 1939), 179–180. On this wider debate about shifting cultivation, and its place in colonial discourse and officialdom, see Nancy Lee Peluso and Peter Vandergeest, "Genealogies of the political forest in customary rights in Indonesia, Malaysia and Thailand," *Journal of Asian Studies* 60, no. 3 (2001): 761–812; William E. O'Brien, "The nature of shifting cultivation: Stories of harmony, degradation, and redemption," *Human Ecology* 30, no. 4 (2002): 483–502; Ravi Rajan, *Modernizing Nature: Forestry and Imperial Eco-Development 1800–1950* (Oxford: Clarendon Press, 2006); Amy Ickowitz, "Shifting cultivation and deforestation in tropical Africa: Critical reflections," *Development and Change* 37, no. 3 (2006): 599–626; and Ross, *Ecology and Power in the Age of Empire*, 276–298.
99 See Pierre Gourou, "Aspect économiques de l'agriculture de subsistence, d'après Colin Clark et Margaret Haswell," *Annales de Géographie* 76, no. 414 (1967): 226–228. He was reviewing Clark and Haswell's *The Economics of Subsistence Agriculture* (London: Methuen, 1964).

258 *Affecting the tropics*

100 Ester Boserup, *The Conditions of Agricultural Growth: The Economics of Agrarian Change Under Population Pressure* (London: George Allen & Unwin, 1965), 11.
101 Gourou, *The Tropical World*, 3rd edition, 117.
102 Gourou, *The Tropical World*, 3rd edition, 120.
103 Pierre Gourou, "Civilisation et géographie humaine en Asie des Moussons," *Bulletin de l'EFEO* 54, no. 2 (1954): 467–476, at 475.
104 Gourou, *The Tropical World*, 3rd edition, 29. Also see Pierre Gourou, "La colonisation blanche dans les pays tropicaux au regard de la science," *Le Monde Colonial Illustré* 16, no. 1 (1938): 3–4; Albert Demangeon, "La colonisation blanche sous les tropiques," *Annales de Géographie* 49, no. 278–279 (1940): 98–105, which is a more wide-ranging assessment.
105 For example, Gourou's review of Barry and Chorley's physical geography primer *Atmosphere, Weather and Climate* alights on a figure on page 248 entitled "The influence of climate on man." Gourou asked: "Should we believe that the hundreds of millions of people who live in the conditions of climatic discomfort indicated by this figure are really uncomfortable? Should we, as geographers, rely on laboratory experiments or on the study of the actual life and distribution of the population to draw such conclusions?" Pierre Gourou, "Review of *Atmosphere, Weather and Climate*," *Revue Belge de la Géographie* 91 (1967): 249.
106 Gourou, *The Tropical World*, 3rd edition, 126.
107 Gourou, *The Tropical World*, 3rd edition, 52.
108 Gourou, *The Tropical World*, 3rd edition, 96–97.
109 Fanon, *Wretched of the Earth*, 2.
110 Fanon, *Wretched of the Earth*, 2
111 Martin Evans, "Towards an emotional history of settler decolonisation: De Gaulle, political masculinity and the end of French Algeria 1958–1962," *Settler Colonial Studies* 8, no. 2 (2018): 213–243; Jordanna Bailkin, "Decolonizing emotions: The management of feeling in the new world order," in Frank Biess and Daniel M. Gross eds., *Science and Emotions After 1945: A Transatlantic Perspective* (Chicago: University of Chicago Press, 2014), 278–298.
112 Pierre Gourou, "Les problemes du monde tropical," *Les Cahiers d'Outre-Mer* 1, no. 1 (1948): 4–13, at 5.
113 O.H.K. Spate, "The Burmese village," *Geographical Review* 35, no. 4 (1945): 23–43, at 43.
114 Gilles Sautter, "'Dirigisme opérationnel' et stratégie paysanne, ou l'aménageur aménagé," *L'Éspace Géographique* 7, no. 4 (1978): 233–243, at 235.
115 Benoît L'Estoile, "Politique de la population, gouvernmentalité modernisatrice et 'sociologie engagée' en Afrique équatoriale française," *Cahiers d'Études africaines* no. 228 (2017): 863–919, at 863. For an overview of crises and struggles in the region between 1944 and 1960, see Tony Chafer, "Decolonization in French West Africa," *Oxford Research Encyclopedia of African History* Online 2017, http://africanhistory.oxfordre.com/view/10.1093/acrefore/9780190277734.001.0001/acrefore-9780190277734-e-166.
116 Edwin Munger, "Some African geographies," *Annals of the Association of American Geographers* 53, no. 2 (1963): 235–247, at 241.
117 Munger, "Some African geographies," 236.
118 Pierre Gourou, "Review of *Geographers and the tropics: Liverpool essays* by R. W. Steel and R. Mansell Prothero" *Annales de Géographie* 74, no. 403 (1965): 326–330, at 326–327.
119 George H. T. Kimble, *Tropical Africa*, 2 vols. (New York: The Twentieth Century Fund, 1960) I, Foreword.
120 Paul Pélissier, "The terroir, a tool for the recognition of farming knowledge in Africa: An interview with François Verdeaux," in Laurence Bérard ed., *Biodiversity and Local Ecological Knowledge in France* (Paris: Inra-Quae, 2006), 43–46, at 43.

121 Gilles Sautter and Paul Péissier, "Pour un atlas des terroirs africains," *L'Homme* 4, no. 1 (1964): 56–72.
122 See Munger, "Some African geographies," 239–241.
123 Robert Steel, "Tropical Africa: Land in rebirth. A review," *Economic Geography* 38, no. 2 (1962): 176–180.
124 Steel, "Tropical Africa," 178. In an earlier review of Dudley's Stamp book on tropical Africa, Steel complained that in Hailey's survey of the continent geography did not "receive more than casual attention," and noted that the originality of Stamp's efforts lay in his attempt to fill this void. Robert Steel, "Review of Africa: A study in tropical development," *Geographical Review* 45, no. 2 (1955): 289–291.
125 Alan G. Ogilvie and Karl J. Pelzer, "African triad," *Geographical Review* 29, no. 4 (1939): 653–658, at 654.
126 Anthony Young, "Natural resource surveys for land development in the tropics," *Geography* 53, no. 240 (1968): 242–249, at 247.
127 H.L. Shantz, "The problem of tropical Africa. A review of recent books," *Geographical Review* 16, no. 4 (1926): 597–613.
128 Munger, "Some African geographies," 241.
129 These tropical connections – or fodder – in the promulgation of American spatial science are most conspicuous in E.J. Taaffe, R.L. Morrill and P.R. Gould, "Transport development in underdeveloped countries: A comparative analysis," *Geographical Review* 53, no. 4 (1963): 503–529; Edward Soja, *The Geography of Modernization in Kenya: A Spatial Analysis of Social, Economic and Political Change* (Syracuse: Syracuse University Press, 1968); Peter R. Gould, "Man against his environment: A game theoretic framework," *Annals of the Association of American Geographers* 53, no. 2 (1963): 290–297; and Peter R. Gould, "Tanzania 1920–1963: The spatial impress of the modernization process," *World Politics* 22, no. 2 (1970): 149–170.
130 The allusion here is to a latter-day geographical literature on critical geo-politics, which proceeds from the claim that geography and geographical knowledge are not innocent but work as technologies of power.
131 Charles Fisher, "The Vietnamese problem in its geographical context," *The Geographical Journal* 131, no. 4 (1965): 502–515, at 502.
132 Kimble, *Tropical Africa*, II, 271–272. In a lengthy critical review of Ellsworth Huntington's 1945 *Mainsprings of Civilization*, Kimble took the Yale geographer to task over his assertion that "whereas resistance movements are a perfectly natural response to tyranny in lands with a stimulating climate such as Norway and the Netherlands, non-resistance is just as 'perfectly natural' in a hot, enervating climate like that of India. . . . Are we then to infer that . . . the organized resistance courageously maintained by native guerrillas in Burma, New Guinea, the Solomon Islands, and the East Indies, which, climatically, are, if anything, worse off than most parts of India, are completely *un*natural?" George Kimble, "Review: Mainsprings of civilization," *Geographical Review* 36, no. 1 (1946): 144–147, at 145.
133 George H.T. Kimble, "Tropical Africa in transition," *Geographical Review* 42, no. 1 (1952): 7–15, at 7.
134 Charles A. Fisher, "Malaysia: A study in the political geography of decolonization," in Charles Fisher ed., *Essays in Political Geography* (London: Methuen, 1968), 75–146, at 141.
135 Dudley Stamp, *The Geography of Life and Death* (London: Collins, 1964) 70.
136 Stamp, *Geography of Life and Death*, 7.
137 Stamp, *Geography of Life and Death*, 74–78.
138 George H. T. Kimble, "Obituary: Laurence Dudley Stamp 1898–1966," *Geographical Review* 57, no. 2 (1967): 246–249, at 247.
139 Kimble, "Obituary," 247.
140 See Power, *Rethinking Development Geographies*, 47–49.

141 William Kirk, "The road from Mandalay: Towards a geographical philosophy," *Transactions of the Institute of British Geographers* 3, no. 4 (1978): 381–394, at 383–384.
142 Gourou, *Les pays tropicaux*, 124–128.
143 Luella N. Dambaugh, "Jamaica: An island in transition," *Journal of Geography* 52, no. 1 (1953): 45–57, at 45.
144 Hildebert Isnard, *Géographie de la décolonisation* (Paris: Presses Universitaires de France, 1971), 2–4, 37–39, 66.
145 Ulrich Schweinfurth, "The problem of Nagaland," in Charles Fisher ed., *Essays in Political Geography* (London: Methuen, 1968), 161–176, at 176.
146 D. Hilling and B.S. Hoyle, "Seaports and the economic development of tropical Africa," in B.S. Hoyle and D. Hilling eds., *Seaports and Development in Tropical Africa* (London: Palgrave Macmillan, 1970), 1–7. They were particularly critical of Taaffe, Morrill and Gould's "Transport development in underdeveloped countries."
147 John D. Eyre, "Review of *Monsoon Asia* by E.H.G. Dobby," *Journal of Asian Studies* 21, no. 3 (1962): 370–371; Pierre Gourou, "Comtes Rendus, Stamp, *Africa: A Study in Tropical Development*," *Revue Belge de la Géographie* 91 (1967): 263–264.
148 Gourou, "Comptes Rendus, Stamp," 263. Dobby made a virtually identical point: E.H.G. Dobby, *Monsoon Asia* (London: University of London Press, 1961), 21.
149 Pierre Gourou, "Asie et Océanie," in Jacqueline Beaujeu-Garnier, Pierre George, Jean Dresch, Pierre Gourou, André Viaut and Charles Robequain, "Chronique géographique," *Annales de Géographie* 59, no. 317 (1950): 376–392, at 385.
150 Philip Curtin, *The Image of Africa: British Ideas and Actions 1780–1850*, 2 vols. (Madison: University of Wisconsin Press), I, 60.
151 Delavignette, "L'Afrique de Pierre Gourou," 279.
152 Delavignette, "L'Afrique de Pierre Gourou," 281.
153 Georges Condominas, "Notes on the present state of anthropology in the third world," in Gerrit Huizer and Bruce Mannheim eds., *The Politics of Anthropology* (Paris: Mouton, 1979), 187–200, at 190.
154 Ross, *Ecology and Power in the Age of Empire*, 352.
155 Ross, *Ecology and Power in the Age of Empire*, 352–353.
156 Pierre Gourou, "Géographie tropicale et problems de 'sous-développement,'" *Information sur les sciences sociales* 8, no. 4 (1969): 9–18, at 9–10.
157 Gourou, "Géographie tropicale et problems de 'sous-développement,'" 10.
158 On this question of peasant receptivity, see Paul Richards, *Indigenous Agricultural Revolution: Ecology and Food Production in West Africa* (Cambridge: Cambridge University Press, 1985).
159 Although he pointed to this shortcoming in tropical geography and its much firmer anchor in American geography: Pierre Gourou, "L'homme et la tortue," *L'Homme* 4, no. 3 (1964): 110–117.
160 BNF: Fonds Gottmann: Gourou to Gottmann 28 January 1983.
161 Pierre Gourou, "Techniques d'encadrement et géographie humaine au Japon," *Revue Belge de Géographie* 106, no. 3 (1982): 171–181.
162 UBCA: Institute of Pacific Relations Fonds, 86–11: Pierre Gourou, "The economic problems of monsoon Asia and the example of Japan," 1.
163 Gourou, "The economic problems of monsoon Asia, Gourou," 5.
164 Gourou, "The economic problems of monsoon Asia, Gourou," 3.
165 Pierre Gourou, *Man and Land in the Far East* trans. S.H. Beaver (London: Longman, 1975), 231–237.
166 Gourou, *Man and Land in the Far East*, 232.
167 See Dudley Baines, Neil Cummins and Max-Stephan Schulze, "Population and living standards, 1945–2000," in Stephen Broadberry and Kevin O'Rourke eds., *The Cambridge Economic History of Modern Europe* Volume 2: *1870 to the Present* (Cambridge: Cambridge University Press, 2010), 391–420.
168 Gourou, *Man and Land in the Far East*, 230–231.

169 Pierre Gourou, "Le Japon," *Extrême-Asie, Revue Indochinoise illustrée* no. 40 (1929): 701–709.
170 Gourou, *Leçons de géographie tropicale*, 123–124.
171 BNF: Fonds Gottmann: Gourou to Gottmann 7 August 1986.
172 Isaiah Bowman, "The Far East," 1, appended to Isaiah Bowman to Hon. Philip C. Jessup 23 September 1949, Isaiah Bowman Papers, Box 17. We are grateful to the late Neil Smith for sharing this item from his Bowman collection, and which he described to us (personal correspondence) as a "flatulent gem" (there was no end to Smith's encouragement and generosity). Smith's *American Empire: Roosevelt's Geographer and the Prelude to Globalization* (Berkeley: University of California Press, 2003) is a magisterial account of both the crankiness of, and many nuances in, Bowman's sense of a changing post-war world and America's place in it.
173 Bowman, "The Far East," 6–7.
174 Derwent Whittlesey, "Southern Rhodesia – an African compage," *Annals of the Association of American Geographers* 46, no. 1 (1956): 1–97, at 23.
175 Gourou, "Civilisations et malchance géographique," 445–450.
176 To access this reading of Gourou, see Stephen K. Sanderson ed., *Sociological Worlds: Comparative and Historical Readings on Society* (London and New York: Routledge, 2000), Ch. 1, 3, 8.
177 Clifford Geertz, *Agricultural Involution: The Process of Ecological Change in Indonesia* (Berkeley: University of California Press, 1963), 66–68; 144–147. Geertz was particularly drawn to Gourou's "Notes on China's unused uplands," *Pacific Affairs* 21, no. 3 (1 948): 227–238.
178 Douglas H.K. Lee, *Climate and Economic Development in the Tropics* (New York: Harper & Row, 1957); John Phillips, *The Development of Agriculture and Forestry in the Tropics* (London: Longmans, 1961). For an overview of this literature, see Dan Luten, "Empty Land, Full Land, Poor Polk, Rich Folk," *Yearbook, Association of Pacific Coast Geographers* 31, no. 1 (1969): 79–90. On the sway of this pessimism in Southeast Asian research, see Rodolphe de Koninck, "Southeast Asian agriculture post-1960: Economic and territorial expansion," in Chia Lin Sien ed., *Southeast Asia Transformed: A Geography of Change* (Singapore: ISEAS, 2003), 191–230, at 192–195.
179 B.W. Hodder, "Review of *Leçons de géographie tropicale*," *Bulletin of the School of Oriental and African Studies* 37 (1974): 285; René Dumont, *False Start in Africa* (New York: Preager, 1969), 29–30.
180 E. John Russell, "Review of *The Tropical World,*" *International Affairs* 29, no. 4 (1953): 489. On Russell's important place in these debates, see Martin Morgan Hodge, *Triumph of the Expert: Agrarian Doctrines of Development and the Legacies of British Colonialism* (Athens, OH: Ohio University Press, 2007), 90–116.
181 J.D. Gould, *Economic Growth in History: Survey and Analysis* (London: Methuen, 1972), 85.
182 O.H.K. Spate, *India and Pakistan: A General and Regional Geography* (London: Methuen, 1954), 181. Also see J.K. Gibson-Graham, "Area studies after poststructuralism," *Environment and Planning A* 36, no. 3 (2004): 405–419, at 408.
183 Gourou, *Les pays tropicaux*, 114–117, 145–146.
184 Fisher, "The Vietnamese problem," 511
185 Kimble, *Tropical Africa* I, 450.
186 *Where are the snows of yesteryear?* Keith Buchanan, *Out of Asia: Essays on Asian Themes* (Sydney: Sydney University Press, 1968), 21–22. In the same year Buchanan lampooned regional and tropical geography in a satirical piece published under a pseudonym "an exiled Celt who now finds himself, to his surprise and gratification, teaching in, and carrying the news of the Celtic resurgence to, an Antipodean University": Llwynog Llwyd, "A preliminary contribution to analysis of a Pooh-scape," *IBG Newsletter* 6 (1968): 54–63.

187 Marcus Power and James Sidaway, "The degeneration of tropical geography," *Annals of the Association of American Geographers* 94, no. 3 (2004): 585–601.
188 Keith Buchanan, "West wind, east wind," *Geography*, 47, no. 4 (1962): 333–346, at 345. 'West wind, east wind' is the title of a 1930 novel by Pearl S. Buck about 'winds of change' (family change) in China and as viewed from the U.S.
189 O.H.K. Spate, "Correspondence," *Geography* 48, no. 2 (1963): 206–207, quoting from Buchanan's "West Wind, East Wind" essay. Spate moved in the circles of the communist spies Guy Burgess and Kim Philby in 1930s Cambridge, and in 1947 was employed by the Muslim League in connection with the Punjab Boundary Commission and ensuing Partition of India and Pakistan. See Hannah Fitzpatrick, "The parallel tracks of partition, India-Pakistan 1947: Histories, geographies, cartographies," PhD dissertation, Department of Geography and Sustainable Development, University of St Andrews, 2015.
190 Isaiah Bowman, "The geographical situation of the United States in relation to world politics," *Geographical Journal* 112, no. 2 (1940): 129–142, at 140. Such apprehension about the rapidity of change informs the essays on decolonization in W. Gordon East, O.H.K. Spate and Charles A. Fisher eds., *The Changing Map of Asia: A Political Geography* (London: Methuen and Co. Ltd, 1950), with five editions to 1971.
191 Nicholas J. Spykman, *America's Strategy in World Politics: The United States and the Balance of Power* (New York: Harcourt, Brace and Co., 1942), 12; a formulation with which Bowman, in his "planetary" assessment of America's mid-century place in the world, concurred: Isaiah Bowman, "Political geography of power," *Geographical Review* 32, no. 2 (1942): 349–352, at 350.
192 The symposium was organised by William Thomas (a geographer and assistant director of Wenner-Gren Foundation), and Marston Bates and Lewis Mumford, and was chaired by Carl Sauer (all of whom we have already encountered). Founded in 1941 by the Swedish business magnate Axel Wenner-Gren (estimated to be the richest man in the world during the 1930s), the Foundation supported research in anthropology, geography and allied disciplines.
193 Abdoulaye Ly, cited in Todd Shepherd, *The Invention of Decolonization: The Algerian War and the Remaking of France* (Ithaca, NY and London: Cornell University Press, 2006), 56. The resulting publication – William L. Thomas Jr. ed., *Man's Role in Changing the Face of the Earth* 2 vols. (Chicago and London: University of Chicago Press, 1956) – sold more than 10,000 copies.
194 Carl O. Sauer, "The agency of man on Earth," in William L. Thomas Jr. ed., *Man's Role in Changing the Face of the Earth* 2 vols. (Chicago and London: University of Chicago Press, 1956), I, 68.
195 See Michael Williams, "Sauer and 'Man's Role in Changing the Face of the Earth'," *Geographical Review* 77, no. 2 (1987): 218–231.
196 Pierre Gourou, "The quality of land use of tropical cultivators," in William L. Thomas Jr. ed., *Man's Role in Changing the Face of the Earth* 2 vols. (Chicago and London: University of Chicago Press, 1956), I, 336–349. Sauer also reflected on rising production as "the twin spirals of the new age which is to have no end if war can be eliminated." But this was in stark in contrast to Mumford, who lamented humanity's seeming inability to curb its destructive ways, declaring that "man's future seems black." Sauer, "The agency of man on the earth," 66; Lewis Mumford, "Prospect," in William L. Thomas Jr. ed., *Man's Role in Changing the Face of the Earth* 2 vols. (Chicago and London: University of Chicago Press, 1956), II, 1146.
197 Georges Balandier, "General report on the round table organized by the International Research Office of the Social Implications of Technological Change (Paris, March 1954)," *UNESCO International Social Science Bulletin* 6, no. 3 (1954): 372–387, at 373.
198 Balandier, "General report," 376, 379, 385.

199 Paul Claval, "Colonial experience and the development of tropical geography in France," *Singapore Journal of Tropical Geography* 26, no. 3 (2005): 289–303, at 292–295.
200 L'Institut de recherche pour le développement, France. Historique, www.ird.fr/l-ird/historique.
201 Henri Nicolaï, "Geography in Belgium," *Belgeo* 1 (2004): 33–44.
202 Claval, "Colonial experience," 292–295.
203 Claval, "Colonial experience," 292–295.
204 Michel Bruneau and Georges Courade, "A l'ombre de la 'pensée Gourou'," *Espaces Temps* 26–28 (1984): 67–78, 67, 77.
205 Iain Jackson, *The Architecture of Edwin Maxwell Fry and Jane Drew Twentieth Century Architecture, Pioneer Modernism and the Tropics* (Abingdon: Ashgate, 2014).
206 Ross, *Ecology and Power in the Age of Empire*, 397.
207 François Bart, "A propos de Pierre Gourou (1953), *La densité de la population au Ruanda-Urundi*," in Henri Nicolaï, Paul Pélissier and Jean-Pierre Raison eds., *Un geographe dans son siècle – Actualité de Pierre Gourou* (Paris: Karthala, 2000), 123–127, at 125.
208 Paul Pélissier, "Pierre Gourou, 1900–1999," *Annales de Géographie* 109, no. 612 (2000): 212–217, at 216.
209 Catherine Fournet-Guérin, "Les villes d'Afrique subsaharienne dans le champ de la géographie française et de la production documentaire: une géographie de villes 'fantômes'?" *L'Information géographique* 75, no. 2 (2011): 49–67. Leiris's ghosts are pursued in James Clifford, *The Predicament of Culture*: (Cambridge, MA: Harvard University Press, 1988); and more recently in Denis Retaillé, "Fantasmes et parcours africains," *L'information géographique* 62, no. 2 (1998): 51–65.
210 Pierre Gourou, *Terres de bonne espérance*, 415.
211 Fournet-Guérin, "Les villes d'Afrique," 50–51.
212 Published: Paris: Mouton, 1972, 600p.
213 Fournet-Guérin, "Les villes d'Afrique," 50–51.
214 Nick Cullather, *A Hungry World*; Jean-Pierre Raison, "'Tropicalism' in French geography: Reality, illusion or ideal?" *Singapore Journal of Tropical Geography* 26, no. 3 (2005): 323–338, at 325.
215 Pierre Gourou, *L'Afrique tropicale: Nain ou géant agricole* (Paris: Flammarion, 1991), 8.
216 Gourou, *L'Afrique tropicale*, 60, 125.
217 BNF Fonds Gottmann: Gourou Gottmann 20 June 1966.
218 UBCA: Institute of Pacific Relations Fonds, 42–11: Roger Lévy to William Holland, 24 June 1954.
219 For example, Pierre Gourou, "Villes et campagnes du Cameroun de l'ouest," *Le Cahiers d'Outre-Mer* 38, no. 2 (1985): 201–203.
220 Gourou, *L'Afrique tropicale*, 117.
221 *Le Peuple*, 23 juin 1955. Clipping AULB: Fonds Pierre Gourou PP153.
222 Dipesh Chakrabarty, "The legacies of Bandung: Decolonization and the politics of culture," in Christopher Lee ed., *Making a World After Empire: The Bandung Moment and Its Political Afterlives* (Athens, OH: Ohio University Press, 2010), 45–68, at 46.
223 Henri Nicolaï, Interview with the authors.
224 Mansell Prothero, "African population maps," *Geograpfiska Annaler* 45, no. 4 (1963): 272–277, a 275.
225 Delavignette, "L'Afrique de Pierre Gourou," 283–284.
226 For an intriguing recent study of the impact of British 'old boys' networks in Africa, see Peter Brooke, *Duncan Sandys and the Informal Politics of Britain's Late*

Decolonization (Cham: Palgrave Macmillan, 2018). Sandys was Winston Churchill's son-in-law, and Secretary of State for Commonwealth Relations in the early 1960s.
227 Susanna B. Hecht, "Domestication, domesticated landscapes, and tropical natures," in Ursula K. Heise, Jon Christensen and Michelle Niemann eds., *The Routledge Companion to the Environmental Humanities* (Abingdon: Routledge, 2017): 21–34, at 24.
228 For a recent example of this notable absence, in an otherwise impressive volume, see Miguel Bandiera Jerónimo and António Costo eds., *The Ends of European Colonial Empires: Cases and Comparisons* (London: Palgrave Macmillan, 2015).
229 Barbara A. Yates, "Educational policy and practice in tropical Africa: A general bibliography," *Comparative Education Review* 8, no. 2 (1964): 215–228, at 215.
230 For poignant biographical and disciplinary accounts of these twin dynamics from the vantage point of the French social sciences, see Pascal Bianchini, *Suret-Canale: de la Résistance a l'anticolonialisme* (Paris: L'Esprit Frappeur, 2011); and Marie-Albane de Suremain, "Faire du terrain en AOF dans les années cinquante," *Ethnologie française* 34, no. 4 (2004): 651–659.
231 J.P. Ocitti, "A revolutionary school geography syllabus in East Africa," *East African Geographical Review* 9, no. 1 (1971): 3–10.

7 Gourou's 'colonial situations'

Geography and 'the colonial situation'

In this penultimate chapter we take a closer look at the different 'colonial situations' in which Pierre Gourou worked and his work was read, and understand this expression in the way it was formulated by the radical French sociologist Georges Balandier. He coined the expression 'the colonial situation' in 1951 to capture the need to bring the "ensemble" of economic, geographic, political, psychological and sociological relationships between colonisers and the colonised into a unitary analytical frame in order to understand "the colonial problem in its entirety."[1] His formulation was at once holistic and particular in its sense of the direction in which the social sciences should move in response to the rising tide of anti-colonialism after World War II. Significant elements of the style of critical inquiry he pioneered later became incorporated into the disciplinary project of postcolonial studies (although some were also lost).[2] There are three basic components to Balandier's idea, and we find iterations of each of them in Gourou post-war work. First, Balandier argued that "Colonialism, in establishing itself, imposed on subject peoples a very special type of situation" – one of "subjugation" and "dependency," and one which was variously addressed by coloniser and the colonised through "reconquest," "liquidation," "conditional concessions," the "granting of independence," and a new "technical phase in colonialism" (or neo-colonialism). Second, he suggested that Western academic engagement with the colonial situation came in two broad forms: researchers were either "obsessed with the pursuit of the ethnologically pure, with the unaltered fact miraculously preserved in its primitive state," or with "numerous practical investigations of very limited scope, satisfied with a comfortable empiricism scarcely surpassing the level of using a technique." In Balandier's view, these two visions, the one concerned with essences and destinies, and the other with "certain concrete situations" in the present, needed to inform one another, and he saw geography as one discipline within which theory and technique, and the archaic and contemporary, usefully met. And third, he recognised that these separations and connections were profoundly ideological, and a matter of intense debate within metropolitan and colonial salons of academia and government.[3]

Balandier had a high opinion of French geographical scholarship and thought it complemented a growing sociological interest in the relational – local/global, centre/periphery – nature of post-war development and urbanisation. From 1948 to 1952 he and Gilles Sautter (one of Gourou's former students) were seconded to Brazzaville by the French government's Office de la Recherche Scientifique pour les Territoires d'Outre-Mer (ORSTOM, created in 1944) to devise applied research and rural development programmes, and the two of them worked closely together. Balandier lauded Gourou for avoiding the type of "dogmatic mapping" he associated with anthropology's essentialist fixation with tribe and ethnicity, with the psychologising categories of anti-colonial thinkers such as Frantz Fanon and Albert Memmi, and with American modernisation theorists' resort to abstract economic models that subordinated questions of politics and culture to material drives and aspirations. Gourou's work had historical depth, Balandier noted, but also bucked "the current disregard for the present that has characterized French ethnology."[4] In *Les pays tropicaux* Gourou had declared that schemes of modernisation and development placed too much emphasis on "means of living," and too little on "the art of living."[5] Gourou's expertise rested, Balandier thought, on his emphasis on "adjustment to place," and which was at once a temporal and spatial process.[6] While many French geographers working in colonial situations neither rejected empire *tout court*, nor were committed to decolonisation in a political or theoretical sense, Balandier thought that their interest in the agrarian strategies of African communities portended a decolonisation of the scientific gaze.

Both Balandier and Gourou engaged with issues that became central to colonial historiography and post-colonial debate: the differential impact that colonial systems of direct and indirect rule had across Africa and Asia, and how this political legacy might affect development projects and post-independence forms of governance.[7] Balandier observed that Gourou's work eschewed ideological closure and pressed the view that development schemes needed to be attuned to locale and circumstance. Gourou was sceptical of imperial rhetoric which preached that the creation of centralised systems of territorial administration and development would reap benefits for all Balandier deemed such a reading of the 'colonial situation' vital.

Yet only in a few places in his work did Gourou remark on how colonists creamed off the best land and resources for themselves, and displaced indigenous people to create a captive labour supply to work on their farms and plantations, and in their mines; and in much of his writing he depicted colonial 'organisation,' as he often simply put it, as essential to how tropical regions had been placed on a secure economic footing. He seemed more in favour of systems of indirect rule, such as those devised by the Belgians in the Congo and Ruanda-Urundi, implying that is was a kinder and gentler form of colonialism – one more respectful of tradition. Yet it slowly dawned on him that in Ruanda-Urundi and other parts of tropical Africa this form of rule distorted post-colonial development by failing to deal with, and in some places fomenting, ethnic and territorial rivalries, and because post-independence regimes inherited disjointed and some places nepotistic neo-colonial state infrastructures. Finally, it is not hard to see Gourou as what Edward

Said, with reference to the writer V.S. Naipul, characterises as a "witness for the Western prosecution," specialising in the view that "we 'non-whites' [or tropical peoples] are the cause of all our problems, not the overly maligned imperialists."[8] Gourou's arguments about the innate inferiority of the tropical zone made him such a witness.

We shall turn initially to how Gourou pursued the first element of Balandier's thesis: how, in an array of essays and reports on different parts of late colonial Africa, he articulated a deep scepticism towards high-minded models of development, the ongoing imprint of racial segregation, and the strong desire on the part of newly independent peoples and their leaders for rapid change. We then consider how the second and third elements of Balandier's conception of the colonial situation, and their links to postcolonial thought, take on acute and agonising meaning in relation to Gourou's work in late colonial (Belgian) Ruanda-Urundi, and to the understanding of the Rwandan genocide of 1994. Lastly, and extending the theme of *les tropiques fantôme* introduced in the last chapter, we consider how, during the 1970s and 1980s, critical appreciation of Gourou's work and tropicality moved in two directions, with some nauseated by its "colonialist perfume," as Michel Bruneau and Georges Courade put it in 1984, and others drawn to its ongoing relevance and utility as a tool of anti-imperial critique.[9] These three concerns point to a tropicality that was in turns strident, ambivalent and militant.

'Not at home in empire'

Gourou was critical of both indigenous land use practices, which in some places he thought exacerbated an already precarious insalubrity, and Western modernisation ventures, which, he argued, were premised on environmental miscomprehension and risked the well-being of peasant communities by promising too much too quickly.

While there are "many prospects for tropical development" through commodity export initiatives, and thanks mainly "to chemical and biological discoveries made in the temperate belt," he observed in *Les pays tropicaux*,

> increasing the living standards of tropical populations will still pose very great problems; perhaps the tropical milieu causes more problems than it can solve. Is there not at the root of all these difficulties the poverty of tropical soils, which prevents those who cultivate them from attaining the same living standard as farmers of the tropical zone?[10]

The "penetration of Europeans into hot and wet countries" had frequently met with failure and caused "serious damage to nature and people," he continued, because Europeans, and then Cold War (Soviet and American) interlopers, were too optimistic about their ability to use technology and new systems of administration to conquer tropical nature.[11] He was propounding a view he had started to develop in the late 1930s (see Chapter 4); in his inaugural lecture at the Collège de France in December 1947 he further mused that "while the tropical world is the biggest potential reserve of vegetal matter on the planet, the peoples of the tropics are, with a few exceptions, the poorest of all and might always remain so."[12]

Reflecting much later – in 1991 – on the colonial footprint in tropical Africa and the development challenges it had faced in the wake of independence, Gourou insisted that agricultural and rural reforms "must be carried out prudently."[13] "Wisdom counsels moderate ways," he had written in 1953 with regard to a mostly decolonised Asia:

> Free from European domination, and happy with this liberation, the countries of Asia can progressively improve their lot by effective administration and an indispensable reduction in the birth rate. It is for the peoples of Asia to show that their old and brilliant civilisations can survive and prosper, with the necessary adaptations to technical progress. Why should they value their values only to accept our own, be it a civilisation, a monetary hierarchy, or a Communist civilisation?[14]

There was a consistent line to Gourou's thinking about decolonisation: he questioned what counted as 'improvement' and thought carefully about from where 'reform' should come. He did not explicitly conceptualise the role of the state in either development or decolonisation, and this omission should be included in what, in the last chapter, we termed *les tropiques fantôme*. The state has a ghostly presence in much of Gourou's writing, as an alien and intrusive force haunting rural lives and practices. For reasons he never properly explained, he argued that the onus for change lay with the peasant cultures of the tropical and decolonising world, and viewed the modern state as an entity "which does not arise out of the society of the subject population but is imposed on it as an alien force," to borrow Ranajit Guha's iconic formulation about how and why the British in India were "not at home in empire."[15] If, as Guha continues, this "irreducible and history necessary otherness was what made imperialism so uncanny for its protagonists in South Asia," then for Gourou – albeit usually by implication – this otherness made the ends of empire and coming of independence uncanny too, and his disquiet about the state helps to explain why he often argued at cross purposes about what counted as a 'good' or 'bad' form of development or independence.[16]

If colonial empires were to survive, he declared in 1953 with reference to Belgian Africa, they "must raise levels of consumption"; and yet they "have come to a point where the colony costs the coloniser more than it gives in return."[17] He saw the colonial state as an alien force with its own calculative field of cost and benefit (as we saw in Chapter 3, he thought about the French in Indochina in similar ways); and in any event, state or no state, empire or no empire, there was only so far that tropical development could go. The Belgian colony of Ruanda-Urundi (now Rwanda and Burundi) had frequently been described as 'the Switzerland of Africa': land-locked; hilly and green; self-sufficient and with manicured landscapes; and relatively prosperous compared with its neighbours.[18] "It would be nice to turn Ruanda-Urundi into a sort of African Switzerland, exporting its products," he mused; but he went on: "how can this be done with an illiterate population that lacks technical know-how and capital," and with "demographic increase threatening to cancel out all measures to raise the low level of consumption?"[19] And "The problem is the same in the 'overpopulated' countries of Monsoon Asia."[20] In short, he viewed the colonial situations in which he worked with a mixture of optimism and pessimism, and was as critical of colonial powers as he was of the administrations of newly independent countries for misconstruing the challenges they faced.

The colonial tropics and the 'colour bar'

In a pair of essays published in *Les Cahiers d'Outre-Mer* in the mid-1950s Gourou addressed the Mau-Mau crisis in Kenya and the British Groundnut Scheme in Tanganyika (1945–1951). In the former, which was originally a talk given to Belgian colonial administrators in Léopoldville (perhaps as a warning), he saw the Mau-Mau movement in British Kenya as a modern nationalist ideology, but dwelt on how tensions between British settlers and Kikuyu pastoralists and agriculturalists were rooted in the land and a racial geography of segregation:

> Indigenous Kikuyu people have not entirely been stripped of their land by British colonists, but have felt the fear of being stripped of it; they have been forced to work on British concessions; they have been unable to respond to their growth in population by colonising new lands; and they have settled in the cities and have found little satisfaction there. . . . Meanwhile, the discontented colonists see the climate of the highlands as excellent and perfectly favourable to them. . . . But, carried away by their concerns, the Whites of Kenya go beyond the limit of what is necessary for their interests and maintain a colour bar.[21]

He saw the vast Groundnut (peanut plantation agriculture) Scheme – "the flagship development project in British Africa at the time," geared to rapidly "remedy the acute shortage of fats on world markets," and turning 1.3 million hectares of land into mechanised production with wage labour, as Corey Ross describes it – as a blundering tragedy of environmental miscomprehension that was ruining the adaptive skills of rural populations and hastening soil erosion.[22] Gourou offered his critique at a moment when the French had not long initiated a similar (if smaller-scale) undertaking in Senegal and were encountering similar problems. Ross notes that the Groundnut Scheme had many "spin-offs" in Africa, and Gourou can be added to his list of critics who, though never fully discarding "the post-war faith in scientific expertise as the key to managing the natural world," nonetheless took the sheen off "prestige projects" and "mega-model[s]" of development, and drew attention to the virtues of smaller-scale initiatives in which 'traditional' and 'modern' know-how might be combined.[23] In this regard, Gourou concurred with French botanist Auguste Chevalier that while indigenous agriculture in the tropics was primitive, it was nonetheless viable and "perfectible."[24]

In a 1968 essay on Portuguese Angola and Mozambique, written at the height of a decade long war of independence, Gourou prophesied a bloody end to the peasant resettlement plan then in the offing in Angola, with upwards of 3,000 peasant families scattered over a 25,000 hectare area in the east and northwest of the colony being resettled (concentrated) in villages in order to give Portuguese settlers and companies unfettered access to diamond, iron ore and oil reserves in these areas, and greater control over the colony's lucrative coffee and sugar plantations. The settlements were closely patrolled by Portuguese militia and a network of spies and informants to offset infiltration and mobilisation by the National Front for the Liberation of Angola. The Portuguese also sought to stimulate agriculture and industrial employment as a means of winning hearts and minds. But Gourou grasped that the peasants so resettled – up to 1 million of them by 1974 – looked disdainfully on the 200–300 Portuguese *colons* who were handed land and mineral concessions in their lands. The economic and social prospects of the vast majority of the population, he surmised, were being stunted by "the same racial segregation problems as in South Africa and Rhodesia."[25]

It was with such grand models of development, and divisive resettlement schemes like this, in sight that Gourou reflected in 1991: "geographers cannot help be aware of the partial character of a view which works from Western norms and priorities."[26] His admission brushed against the grain of Césaire's view of him as an apologist for empire, and his readiness to question dominant ideas and assumptions impressed France's leading colonial administrator, Robert Delavignette, who noted, with reference to Gourou's work, that "the training of new social classes of technicians, whose professional activity is comparable from one people to the other, should not happen without some liaison between the leadership of the administration and geographical research. These obvious facts are often neglected and geography seems to have been cast aside by Governments and Administrations imbued with the sense of their own infallibility."[27]

To the Switzerland of Africa

Gourou's most trenchant thoughts about development and decolonisation came in connection with the Belgian Congo and Ruanda-Urundi. The Congo gained independence in 1960 and Rwanda and Burundi became separate states in 1962. Belgium's connection with the Congo dated back to 1885, when the region was terrorised by King Leopold II's rubber regime; and Belgian colonial rule began in 1909. Rwanda and Burundi was initially colonised by Germany in 1884 and became Belgian League of Nations–mandated territory in 1918. After World War II, the region became a United Nations Trust Territory under Belgian stewardship, and a colony once more in all but name. Gourou first journeyed to the Congo, and then on to Ruanda-Urundi, for three months in 1949. His visit was facilitated by the Medical and Scientific Centre of the Free University of Brussels in Central Africa (CEMUBAC, founded in 1938), which had a mandate to tackle health problems in Belgian Africa and was funded by The Belgian Ministry of Colonies. In 1949 CEMUBAC created a geography section, which Gourou headed. He was commissioned to undertake studies of land use and population density, and from 1949 began to gather the data necessary for producing a series of 1:750,000 scale maps of Ruanda-Urundi (Figure 7.1, on a purple colour-coded scale), and 1:100,000 and 1:200,000 maps of the Congo.

This first journey to Central Africa also provided Gourou with material for a monograph on the population density of the two colonies, published by the Royal Belgian Colonial Institute. He worked in different parts of the Congo in 1952, 1955, 1958 and 1959, and returned to Ruanda-Urundi in 1957. Between 1949 and 1960 he also compiled yearly reports on publications relating to the Belgian Congo. From 1952, CEMUBAC funded graduate students, a number of whom worked under Gourou, and there were a few small Belgian specialist research institutes in the region, although not universities until the mid-1950s. In 1957–1958, and against the backdrop of rising Hutu and Congolese nationalism, Gourou sought to put together a programme of multidisciplinary research in eastern Congo. This was the apex of his research career in tropical Africa. Following independence, the remit of CEMUBAC was pulled back largely to medical research and the provision of emergency aid.

Gourou always travelled to the region by plane, his student Henri Nicolaï told us. A regular air service between Brussels and Léopoldville began in 1946. He did not have a permanent base whilst there. He was on the move a lot, and used vehicles provided by the colonial authorities, Belgian companies, and CEMUBAC doctors. The information he obtained was often generated with the help of local civil servants, and through their liaison with 'native' informants. However, Gourou liked to work alone and with his students. He presented himself in the mould of the detached expert, and, as before, it was in this manner that his work both contributed to and questioned the colonial project.

The *Travel Guide to Belgian Congo and Ruanda-Urundi*, published from 1949 to 1958, did much to shape Belgian perceptions of their colonial domain. The *Guide* envisaged tourism to the Congo in an exotic sport-hunting and adventure mode, Nicolaï relates, and was fuelled by myths about the Mountains of the Moon and the Source of the Nile, and with Ruanda dubbed *la terre des mille collines* (the land of a thousand hills). Produced by the colony's official information services, the *Guide* also sought to extol the results of Belgium's civilising mission.[28] These two imaginative geographies were combined through the claim that the Congo had no history or artistic traditions, and that its natural environment overwhelmed the human scene. The emphasis on natural beauty was expressed in vivid and lyrical descriptions of rivers, waterfalls, and jungle and gorilla landscapes. The region became the preserve of safari (hunting and fishing) tourists. In the third edition, for example, 138 pages are devoted to hunting, with a host of recommendations for the traveller, including advice on attire and the conservation of ammunition in a tropical climate.

On the other hand, the decennial colonial plan for Belgium's economic and social development of Congo was published in 1949, and one for Ruanda-Urundi in 1951. Major public and private investments were made in the areas of energy, communications, health and education. The Belgians also devised an original

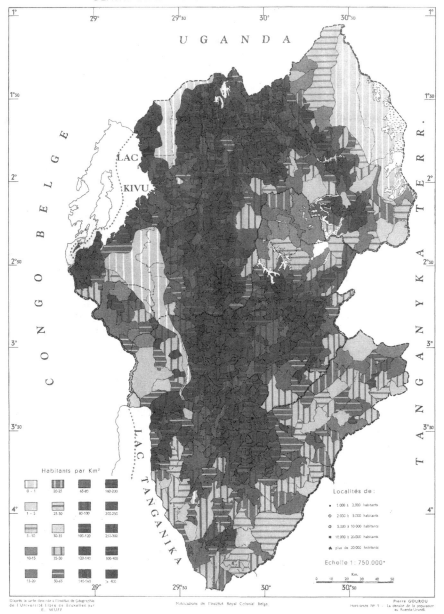

Figure 7.1 The density of population in Ruanda-Urundi, c. 1950. From *La densité de population au Rwanda-Burundi. Esquisse d'une étude géographique.*© RAOS

paysannat – agricultural re-organisation – scheme in which the colonial authorities endeavoured to look past received imperial wisdom regarding the inefficiency and environmental drawbacks of traditional shifting cultivation, and 'rationalise' and 'discipline' (terms which abound in colonial literature on the question) it as an agricultural system, with 20-metre plots of forest and field and a regimented cycle of cultivation and fallow. Contrary to colonial opinion elsewhere in Africa and Asia, shifting cultivation was deemed a means of preserving soil fertility, and more immediately of giving agricultural change an indigenous face and averting the disabling (and usually counter-productive) effects of sizeable peasant relocation. There was a gulf between plan and reality, but as Sautter attests the *lotissements agricoles* of the Belgian Congo and parts of Ruanda-Urundi were a far cry from the vast and wasteful vanity development projects that were being fashioned in other parts of tropical Africa.[29]

Gourou's fact-finding mission was at odds with the imagery to be found in the *Guide*, for a start in the way he disputed the exoticism and environmental determinism it emitted, with dramatic landscapes and wild animals towering over human life. "The colonizing and tutelary powers are changing the nature of their concerns," Gourou observed, and in addition to their "successes in the areas of communications, hygiene and the maintenance order . . . [they] now need to take into account national political demands."[30] He noted Hutu disdain of the dominant and traditionalist Tutsi monarchy, through whom the Belgians ruled, but also colonial 'successes' in education and agricultural reorganisation.[31] However, his prime concern was with how 'national political demands' turned on basic geographical problems of population density, land use, and rural living standards, and he hoped that the study of such matters would yield new forms of "interdependence" between metropole and colony, however difficult this might prove.

Gourou's Rwanda: an incomplete geography?

Various scientific organisations were established in Brussels and Leuven to probe these problems: the National Institute for the Agricultural Study of Belgian Congo (1933); the Institute of National Parks of the Belgian Congo (1934); and the Institute for Scientific Research in Central Africa (1947). A three-volume *Encyclopedia of the Belgian Congo and Ruanda-Urundi (1951–1952)* stemmed from these initiatives, and in 1949 the Royal Academy of Colonial Sciences began publishing maps and charts that would be collated in the *General Atlas of the Belgian Congo and Ruanda-Urundi*. Gourou helped to fashion this scientific-colonial nexus. During the 1950s he wrote regular bulletins on 'progress in the geographical knowledge' pertaining to the highland areas of central and east Africa, and with a focus on Belgian Ruanda-Urundi. They were disseminated through the Société belge d'études géographiques, and saw his efforts as an "attempt at reasoned and methodical description and, then, an attempt at explanation that is as comprehensive as possible."[32] Among other things, he sought to overhaul patchy American population data on the region.

While his research was funded by the Institut Royal Colonial Belge, Gourou was not a peon of the colonial establishment. He rejected "categorical statements," included those about ethnicity that imbued the way the Belgians governed the region, and his labours came to fruition in his 1953 report, *La densité de la population au Ruanda-Urundi*, which the American geographer Wilbur Zelinsky hailed as the first-ever book-length study of a nation's population geography, and the French tropical geographer Jean Gallais praised for opening up the question of how the highland areas of Central Africa, stretching from Zambia to Ethiopia, functioned as "ethno-demographic bastions," and ones whose tropicality was tempered by altitude.[33]

The Belgian anthropologist Jacques Maquet noted that while Gourou's ostensible aim was to explain the high population densities in this colonial region, his book was also a valuable – albeit "non-theoretical" – critical study of the means by which the Belgians had sought to raise the living standards of a growing population with a meagre subsistence, and in a land with thin soils that peasants used resourcefully.[34] Gourou declared that the tropical uniqueness of Ruanda-Urundi in an African context rested not simply on the fact of its "high density of its population" but more particularly on the recognition that these high densities could be found at relatively high altitudes, over 1,500 metres above sea level, which fostered "salubrity" (the absence of malaria and tropical diseases afflicting cattle), and on the "very rugged and fragmented" relief surface on a pre-Cambrian shelf which had generated "a soil poor in quality because of its fineness and fragility."[35] The surface area of the country is almost entirely above 1,000 metres. Altitude did not have a determining influence, Gourou argued. Rather, its influence was mediated by three other factors. First, tropical diseases afflicting humans and cattle (particularly trypanosomiasis, affecting both) do not abound above 1,000 metres (and with cattle breeding virtually impossible below 1,000 metres). Second, by establishing villages on hill slopes and by avoiding marshes, the peoples of Ruanda-Urundu kept malaria at bay. And third, the more abundant and regular rainfall in high areas was favourable to agriculture and the harvesting of two annual crops.

"The abundance of people confers on most of Ruanda-Urundi a very peculiar landscape," Gourou observed, "and one unexpected in Black Africa": a "human society" which is also a "bovine society," and a landscape where "the techniques used to exploit nature have never been geared to mitigating the pain of humans. Big cattle were never used for plowing or traction. Besides, the wheel was not known."[36] There were only 15 settlements with more than 2,000 people, and the highest rural population densities were to be found in the central and western (plateau and valley) regions of the colony, and around the capital Kigali. Cultivators made good use of the relief and thin soils, Gourou continued, growing crops in intricately gradated banks and furrows along steep valley slopes that were irrigated by rain runoff and fertilised with animal and human waste. The entire land system was based on customary traditions that ignored the colonial land register, and different crops (cereals, tubers and legumes) thrived at different altitudes (sweet potato, manioc and bananas between 1,500 and 2,000 metres, and peas, maize and sorghum at higher altitudes). Cultivators also kept chickens, goats, cattle, sheep and pigs.

Gourou's core findings and data have had a long shelf life in French and African demography and geography, and like other geographers working in similar colonial situations, he occupied a lateral position, working neither wholly inside nor completely outside colonial discourses and practices.[37]

With a total population of around 2.5 million around 1950, in a territory of 24,000 km², with an average population density of around 100 inhabitants per km², and around 65 per cent of the population in densities of between 50 and 150 per km², and those densities reaching 200–400 per km² in the colony's most populous localities above 1,500 metres, Ruanda-Urundi was one of the most densely populated regions of Africa. Uganda and the Congo were far behind, with average population densities of 25 and 19 per km², respectively. Ruanda-Urundi was 90 per cent agricultural, and to use categories that we shall shortly deconstruct, the dominant pastoralist Tutsis comprised around 10 per cent of the population, the subordinate and predominantly agriculturalist Hutus comprised around 85 per cent, and a third group, the Twa, around 5 per cent. Tutsis and Hutus lived side by side, and the Belgian authorities ruled through the aristocratic Tutsis, whose material and symbolic power centred on cattle.[38]

Neither ethnicity nor colonialism are much discussed in Gourou's discussion of his population data and maps, and the patchy Belgian censuses from which he drew much of this data did not gather information on ethnicity. Yet Gourou used *sous-chefferies* (the traditional-tribal administrative sub-districts used by the Belgians for policing the colony) as the cartographic platform for his population mapping, and for this reason alone his work does not sit in a power vacuum. His research gathered meaning in a colonial situation that he sensed would not last long, and he raised questions about both the direction of Belgian colonial policy and its impact on tribal-customary traditions. In the latter regard, he was critical of the 'ubuhake contract' – the traditional social (monarchical-male) order, reinforced by the Belgians, which revolved around the distribution of cattle and determined that Hutus were entitled to use Tutsi cattle in return for tribute (chiefly in the form of personal or military service).[39] Nicolaï told us that Gourou was "naturally worried about the way in which decolonisation would proceed" on account of the way the Belgians had governed through the Tutsis and precipitated Hutu resentment.[40] Gourou suggested that Belgium's promotion of Tutsi dominance through its system of district administration detracted from the historic ties between Tutsi, Hutu and Twa groups through language, custom, intermarriage, religion (Christianity), pastoral-agricultural exchange, and a clan system that regulated these subsistence and social relations.

Gourou avoided the terms Tutsi and Hutu. He wrote in a more generic fashion about pastoralists and cultivators, and suggested that poverty (rather than ethnicity) was the chief source of tension and conflict between different ethnic and tribal groups, and between them and the Belgian establishment, partly to the chagrin of Maquet, who observed:

> Gourou's explanation seems valid but incomplete. Because if it indicates the geographical conditions which made possible the creation and conservation of such a large population, it does not take full account of its formation. For

if the geographical factor (altitude) examined does not determine things, but is a way of thinking about certain cultural possibilities and the exclusion of others, would it not be preferable to render an explanation that considers the determining role of culture? The population of Ruanda-Urundi is dense because the African societies that developed in this region had a social system that prised fecundity as a cultural value. Altitude simply allowed the realization of this cultural value.[41]

We shall see that Maquet was edging towards a rather different, racial, explanation, and it is important to stress that, wittingly or unwittingly, Gourou kept his explanation within the bounds of geography.

At the same time, he was aware that previously fluid ethnic distinctions were hardening, and in his book he noted that Belgium's 1951 "Ten-year Plan" for the economic and social development of the colony simplified the social and political complexity of cattle in rural relations, and bypassed how both population pressure and cattle-herding detracted from agricultural production and fuelled rural poverty.[42] And while generally supportive of Belgium's *paysannat* projects, he was quick to suggest that they focused too narrowly on shifting cultivation. While emigration was perhaps a prudent peasant strategy as well as colonial means of reducing population pressure on the land (by 1962 around 350,000 Rwandans resided in Uganda, and around 250,000 in Congo), Gourou noted, it risked depleting the country of valuable labour.[43]

'Tropical Nazism'

Nicolaï surmises that Gourou "never had any illusions" about the political overtones of his research, even if he addressed them only indirectly. But whatever chimeras we might locate therein, his work on Ruanda-Urundi (and which from this point we shall refer to as Rwanda and Burundi) is poignant for a wider reason. It prompts us to reflect on what Homi Bhabha has written about the way violence enters into the constitution of the humanities. Bhabha uses the Rwandan genocide of 1994, and Rwandan Truth and Reconciliation Commission of 1999–2002, to explore the epistemic difficulties and ethical anxieties involved in fashioning what he calls a "proleptic humanism": a humanism that is able to see where trouble – ethnic trouble, gender trouble, nationalist trouble, genocidal trouble – comes from, and how to respond to it with analytical and moral resolve.[44] Bhabha suggests that the Rwandan genocide not only presses the capacity of the humanities to explain violence, but also, and more fully, reveals that there would be no humanities without the corporeal and epistemic violence – cuts in the world and flesh – from which its categories of research, criticism and reflection are drawn.

The Rwandan genocide was triggered by the deaths of President Juvenal Habyarimana of Rwanda and President Cyprien Ntarymira of Burundi in a plane crash, in April 1994, while they were travelling to Tanzania for a peace summit that aimed to tackle a host of problems created by the 1990 invasion of Rwanda by the exiled and predominantly Tutsi Rwanda Patriotic Front (RPF).

Hutu radicals declared the crash the work of the RPF, and a frenzy of violence ensued. By that time the average population density of Rwanda had more than doubled, to 212 per km^2.[45] Almost 1 million Rwandans (over 10 per cent of the country's population), and up to 800,000 of them Tutsis (between 75 and 80 per cent of the Tutsi population, which then comprised between 8 and 12 per cent of the total population), were slain and maimed in a three-month spree of killing, rape and torture that stemmed from a longer civil war and process of Hutu demonisation of Tutsis.[46] If there was ever a post-war case of emotional life overwhelming intellectual life with such speed and ferocity, this, from the vantage point of the humanities, was it.

In his essay "My three identities" – an identity "first called negritude, then national sentiment, then authenticity" – the Congolese writer and politician Henri Lopes suggests that the search for African origins can easily precipitate a genocidal "tropical Nazism."[47] He deploys this troubling expression not simply to capture the horror of killing and succumbing to hatred, but also in order to defamiliarise African understanding of the Rwandan genocide and see it as part of a wider and deeper failing in the human condition. And for Lopes this wretched tropicality is not just a matter of blood, soil and roots; it is also about categories and the disdain they can wreak.

For Bhabha, the Rwandan genocide brings to the fore the problem of what he calls "neighbourliness": how the humanities might promote more equitable terms of representation and forms of identity that embrace an ethics of relationality rather than division and come to terms with how identity is formed through the construction and affirmative of difference, albeit usually on an unequal basis. The fact that neighbours turned on and killed neighbours in such brutal and inexplicable ways – and in densely populated areas of the country, around the capital Kigali and in the south of the country (Butare and Gikongoro), where 'Hutu' and 'Tutsi' had long coexisted – obviates the notion that there is some ethically robust or stable outside position from which to account for the atrocity or prevent others genocides from happening. But search for explanations we must, Bhabha insists, and as scholars such as René Lamarchand, Jean-Pierre Chrétien, Mahmood Mamdani, David Newbury, James Tyner and Peter Uvin, as well as Bhabha, have argued, we need to think about the legacies of colonialism and complicity of academic researchers and knowledges in the story of how Hutu/ Tutsi difference was observed and fomented mass hatred and psycho-political processes of demonisation.[48]

The Rwandan genocide "is neither reducible to a tribal meltdown rooted in atavistic hatreds nor to a spontaneous outburst of blind fury," Lemarchand argues.[49] More recently, Tyner has affirmed that the genocide "was in fact a planned – and thoroughly modern – corporeal annihilation" with roots in "colonial constructions" of Tutsi/Hutu difference, and in post-independence Hutu historiography in which Tutsis are represented as aggressive invaders and Hutus as a peaceable people who are indigenous to the region.[50]

There is a large and diverse literature on the different roles that colonial and academic knowledges and categories, and Rwandan nationalism and governance, played in precipitating the genocide. However, Ben Kiernan seeks to show that there is common denominator to different perspectives on the Rwandan and other genocides. They are each characterised by what he calls "ideological levelling" – the placing of entire populations into categories of self and other, friend and enemy, and perpetrator and victim.[51] Tropicality does the same, placing the world in a temperate/tropical binary. However, the source and expression of (in this case geographical) violence is not the same as that of genocide. Is there, then, a connection between tropicality and genocide? How far can the link between them be pressed?

Gourou recognised that population research in this and other tropical settings had political connotations, particularly by 1960, when the Belgian Congo gained independence, and Belgian politicians started to think about a 'planned independence' for Rwanda and Burundi. Hutus and Tutsis were deemed discrete populations vying for power, and the Belgians exploited this purported division for their own ends. As Frederick Golooba-Mutebi observes:

> Rwanda entered independence following a transition marked by violent internecine conflict. The conflict was stoked by the departing colonial rulers as they sought to place control of the levers of state in the hands of an ethnic majority, which they had hitherto marginalised in favour of a minority they now sought to exclude. It carried on into the country's post-colonial politics. For nearly three decades Rwanda's postcolonial rulers presided over an ethnocracy that perpetuated the negative colonial legacy of ethnic division.[52]

Here and elsewhere in tropical Africa (the Congo, Kenya, Uganda) colonial rule "prepared the ground for ethnicity-inspired resentments and, during its dying years and after independence, ethnicity-focused politics."[53]

It is important to register that Gourou eschewed ethnic categorisation and record that some have sought to exonerate him from any sort of complicity in colonial and post-colonial turmoil on this basis. As François Bart reflected in 2000:

> We should take from his 1953 book his constant refusal of simplistic determinisms. . . . Gourou makes a vibrant plea for the suppression of the ubuhake contract, against the draining of the marshes, and against [Belgian] resettlement measures. Since then, the ubuhake has disappeared, but the majority of marshes have been drained; as for regrouping of the habitat, is has remained a recurrent theme in the discourse of 'developers' and politicians. The failure of villagisation in Tanzania does not seem to have taught any lessons to the new Rwandan authorities, since the creation of villages seems to be on the agenda again in the wake of the demographic upheavals caused by the 1994 genocide. Pierre Gourou's arguments therefore remain pertinent today. . . . He reminds us that the key to the contemporary problems facing Rwanda and Burundi is to be found in poverty.[54]

Only after the genocide did Gourou reflect on Belgian colonialism and its legacies. He wrote of how the Belgian administration had "intervened directly in the management of Rwanda, opening up roads, creating infirmaries and schools, and through its health and education efforts, which were assisted by the Catholic Church." All of this was "warmly welcomed by the Hutus," he continued, and in the process

> Hutus acquired a new consciousness of their number and took great advantage of the [Belgian] administration. . . . With the increasing number of better educated and treated young Hutus, resentment grew among the Tutsis, which expressed itself in riots and massacres. The democratic constitution of Rwanda gave power to the majority Hutu. Rancour against the Tutsis reached its paroxysm in April 1994 when, it was believed, the Hutu president was assassinated.[55]

A few years before the genocide he observed that "Rwanda is under serious threat of hunger if not famine" because efforts to intensify agricultural production had failed to keep pace with population growth; and shortly after it he lamented: "How can you not feel a certain disillusionment and pessimism? The genocide has, in fact, revealed that the very big and fine attempt by the Hutu youth to liberate itself democratically from what it considered to be an unbearable constraint imposed by the Tutsis finally led to an immense catastrophe which plunged the Hutus into an abyss of misery."[56] The catastrophe defied logic, and he lamented "a deserted country, abandoned fields. A countryside without peasants, fields without crops, schools and churches closed."[57]

A categorical or untimely geography?

A 'very big and fine' attempt at liberation not only back-fired; it also defied the ability of humanist categories to comprehend what had transpired. But Gourou had a stab. The genocide "teaches us at least one lesson," he thought: "we must avoid rapid and brutal reforms of African 'encadrements.' The Rwandan affair would perhaps not have taken place if European influence had not made the Hutus understand clearly the injustices of which they were victim."[58] He hinted that both the Belgian colonial regime, and the Hutu regime that took power in 1962 after a three-year revolt against colonial rule and Belgian-backed Tutsi dominance, were partly to blame for the violence and slaughter. Whereas German colonisers had not intervened in the countryside and "left the Tutsis to their classic intrigues," Belgian policies had been divisive.[59] Following independence around 450,000 Hutus were resettled to Tutsi-dominated localities, especially along Rwanda's borders, in order to guard against Tutsi militia drawn from the large ranks of Tutsis who had fled the country (mainly to neighbouring Burundi, Congo, Tanzania and Uganda). The drive of the RPF to repatriate exiled Tutsi to these areas fuelled resentment and anxiety among their Hutu occupants, and helps to explain the 1994 slaying of Tutsi settlers in those areas.

However, these remarks come from hindsight. Colonialism did not seem entirely defunct to Gourou in the 1950s, and he did not return to the region for fieldwork after 1957. His 1960 and 1961 'progress' reports on geographical knowledge do not pick up on the rising tide of Hutu nationalism or Belgian resettlement projects.[60] By the mid-1960s his student Nicolaï had taken over the task of compiling this annual report. In his 2009 report, surveying work on the region between the years 1993 and 2008, Nicolaï laments the paucity of geographical scholarship on the genocide and preponderance of writing on the subject by anthropologists and political scientists.[61] Much of this writing probes the role of the post-independence state, and social and political categories, in fomenting and organising political and ethnic violence. Much of it also situates the genocide in wider histories and conceptions of mass hatred and killing.

The Belgians accepted the ancient Hamitic hypothesis that the Tutsis were superior to the Hutus because they descended from ancient Egypt, and much Belgian colonial investigation and writing on Tutsis and Hutus cultivates this idea of difference, which boosted Tutsi self-worth and fuelled Hutu resentment. Jinmi Adisa adds that "the Tutsi tradition of hierarchical administration readily lent itself to the organization of colonial districts" (*chefferies* and *sous-chefferies*)[62] While Gourou's study was not framed or fuelled by ethnicity or race, he bought the colonial idea that, as Mamdani expresses it, "the Tutsi were superior because they came from elsewhere."[63] In so doing, Gourou lost sight of the colonial construction and politics of difference. "The social and political organisation of Ruanda-Urundi," he observed, "is based on a remarkable pastoral organisation and monarchical and feudal regime" that stemmed from a "conquering aristocracy. . . . Definitely, the responsibility for the high density of the population in Ruanda-Urundi lies with the institutions of the Tutsi, which have been imposed on the Hutu and accepted by the latter."[64]

Adisi distinguishes between "immediate and remote causes" of the 1994 genocide, and underscores the importance of placing them within "the context of the regional peace effort to deal with . . . [a] 35-year old refugee problem."[65] The genocide can be traced back not just to the Hutu uprising of 1959–1962, which culminated in the overthrow of the Tutsi monarchy and creation of an independent Rwanda under Hutu control, but also to earlier colonial processes of categorisation and a political culture of ethnic bipolarity – of alien/native and purity/danger. Such processes spawned what Arjun Appadurai terms "ethnocidal violence" by making minorities "objects of fear and of rage."[66] He sees Nazi Germany and the Holocaust as the grimmest instances of ethnocidal reaction to endangered purity (genocide in the name of an Aryan race), but shows that it was writ large in the complex links between modern European colonialism and the operations of the modern nation-state.

While Belgian census-taker did not ask about ethnicity, there is no question that Tutsi, Hutu and Twa were classified as 'three races,' and with the Tutsi as a dominant-alien-minority. Racial difference was institutionalised from 1929 onwards, in colonial law, in political representation, and through taxation and labour regimes that impacted adversely on Hutus. Mamdani argues that Tutsi and Hutu identities became "frozen" through these projects of colonial governmentality.[67] Placing this in a wider context, Appadurai argues that "minorities and majorities are recent historical inventions, essentially tied up with ideas about nations, populations, representations, and enumeration," and that colonialism played a huge role in developing associated "techniques of counting, classification, and political participation."[68]

Maquet left a decisive mark on this scene. Writing in 1954 – and releasing an accompanying ethnographic film – of "the social structure" of the ancient Kingdom of Ruanda, he noted: "The Batutsi are intelligent (in the sense of astute in political intrigues), apt to command, refined, courageous, and cruel. Bahutu are hard-working, not very clever, extrovert, irascible, unmannerly, obedient, physically strong."[69] Mamdani and Uvin both talk of how Maquet and other scholars helped to fashion a bio-political order in Rwanda that was imposed on more fluid understandings of identity. "After World War II," Uvin notes, "the scientific enterprise of categorising Hutu and Tutsi moved into full gear," and he names Gourou's population and agronomic mapping, Jean Hiernaux's anthropometric study, along with Maquet's work on the "feudal" essence of Hutu and Tutsi identities, as withering examples.[70] Yet Uvin admits that only the latter two undertook their fieldwork with "the question of the distinctions between, and origins of, the Hutu and the Tutsi" foremost in mind.[71] In other words, Gourou perhaps occupies a more ambiguous position.

Gourou had toyed with racial categories. "Human geography cannot ignore the psychology of the Blacks," he wrote in 1952 of the Congo: "If the latter truly have a special psychology which puts them in an inferior position in terms of productivity and intellectual and material progress, human geography must take account of this in its explanations."[72] However, his 1953 book eschews racial reasoning. It focuses on questions of population and has routinely been used since the 1950s and to this day as a source of putatively non-dogmatic information about relations between people and land that feeds into a critical appraisal of ethnocidal politics and violence.[73] Yet Gourou never went as far as Pélissier in insisting that "Tutsi and Hutu are equally native [to Rwanda] and the opposition between cattle raisers and farmers is an erroneous one, created in the second part of the 19th century."[74] That for all his critical acuity, Gourou did not interrogate this opposition, even if he did not organise his work, perhaps underscores its potency in the Belgian colonial mind. Facil Tesfaye suggests that while Gourou's mapping was "essentially agronomic," it is not difficult to infer from it a division of land and society along ethnic lines, with the Tutsi more prominent in the pastoral belt in the middle of the colony and around Kigali, and with Hutu cultivators living side by side with Tutsis along surrounding hill slopes.[75] Gourou's study was published shortly before Maquet's and does not discuss the *ubuhake* (clientage) system in Tutsi-Hutu relations, Maquet's focal point, in any detail. Bart alights on some fleeting remarks about the matter in Gourou 's study. Concomitantly, Gourou's work was not subjected to the same scrutiny as Maquet's – scrutiny which revolved around whether "the proposition that the Tutsi and Hutu are entities 'invented' by the Belgians" should be accepted or refused, and whether, in fixing understanding of a Tutsi order based on courage (*ubutware*), manliness (*mugabo*) and self-mastery (*itonde*), Macquet acceded to what Claudine Vidal described as a "fetishism of the cow."[76] While such orientations and debates moved in essentialist directions, and pivoted on questions of enmity, Gourou intimated that Hutu/Tutsi relations were not based on mutual fear and hatred.

There is much to Mamdani's observation that "the form of [colonial] rule shaped the form of revolt against it."[77] Belgian reliance on the Tutsi aristocracy, and then acceptance by 1959 that power in an independent Rwanda would need to rest with the Hutu majority, followed a wider colonial pattern: "anti-colonial struggle was first and foremost a struggle against the hierarchy of the local state, the ethnically organized Native Authority that claimed ethnic legitimacy. Indirect rule at once reinforced ethnically bound institutions of control [far beyond their customary reach] and led to their explosion from within."[78] Uvin adds that "After independence, data on the ethnic/racial categories became the subject of intense passion and political importance," and with the nature and consequences of such debates varying "dramatically" between Burundi and Rwanda.[79] And Luc de Heusch (who had helped Maquet to make his 1954 film) averred that from 1959, "Hutu leaders, brought up in the hothouse of a Church concerned about its privileges, persuaded a growing number of illiterate Hutu that they had been 'colonized' by the Tutsi, who were described as foreign invaders that [sic] had imposed a 'feudal' rule on them."[80]

However, Paul Magnella suggests that when it comes to the machinations at the end of empire and their correspondence with the 1994 genocide, Mamdani and other scholars "confuse proximate causes with ultimate ones."[81] Magnarella returns to the demographic and agronomic scene Gourou bequeathed us, arguing that the genocide would not have happened without the exacerbation of inequalities in the distribution of land and resources, and histories of displacement and resettlement, over the previous 50 years. Competition over land became particularly acute. Gourou had helped to steer discussion of the region's woes away from the psycho-political orientation of anthropologists such as Macquet, and Hutu ideologues, and towards the delicate and explosive balance between people and environment in a colonial situation of high population density. More recently, Ronald Pourtier has argued:

> The analysis of the [genocidal] situation in central Africa and its perspectives would not be complete without consideration of what constitutes certainly the fundamental factor of destabilisation of the region: demography.... The balance between population and resources is threatened by the dizzying growth in the number of people in tiny spaces.... One cannot help thinking that the genocide of 1994 is not unrelated to a brutal form of demographic regulation, as were once famines. You killed your neighbour with an eye on their land.[82]

Yet we wonder about the critical utility of trying to distinguish between 'proximate' and 'ultimate' causes – and by implication 'good' and 'bad' explanations. We concur with Bhabha that genocides exceed the capacity of economic, political or spatial rationalities to fully explain them or know how to respond to them. He insists that in the Rwandan situation there is a "third party" – a "phantasmatic object of thirdness" – bestriding Tutsi/Hutu difference and making "existential fear" at once everywhere and nowhere. There is an "indeterminacy and contingency" in the hierarchical distribution of cause and effect that makes it difficult to know how the future might have been anticipated and why it could not have been seen. This is not to give up on explanation, he suggests, but to continue to wonder, in more unclassified and conditional ways.

In 1991 Gourou observed that "independence did not modify the political map created by the colonial episode. Consequently, post-independence governments face difficulties spawned by [an older] and very complicated ethnic map."[83] However, he paid little attention to how Hutu consciousness developed not just on Rwandan soil but also among hundreds of thousands who left Rwanda from the late 1950s, some looking for work, many as refugees. Nor did he fully see how ethnic bipolarity had been fuelled by colonialism and post-independence regimes. The complicated ethnic map that came into view after independence was partly a product of the colonial episode and not simply about the re-awakening of a pre-colonial cartography.

In the Rwandan situation Gourou's impure and worldly geography perhaps did not rest directly on a failure of intellect or absence of foresight. Rather, we find in it traces of how Bhabha sees the genocide: as "a past that was refusing to die confronting a future that was waiting to be born."[84] In Gourou's case, that past was rooted in traditional *encadrements* that were imperilled by colonial and indigenous dynamics. Ultimately, it was a past in which, as Tyner puts it, "mass violence results from the imposition of state sanctioned normative geographical imaginations that justify and legitimate unequal access to life and death."[85] This geographical imagination whittled Rwanda down to a timeless and incendiary ethnic geography. However, Gourou's 1950s fieldwork and subsequent reflections on the genocide point to more convivial and neighbourly geographies. Bhabha argues: "Our ethical stand is always untimely. There is no proper time to make the right ethical judgement with all the facts."[86] We cannot tell exactly how aware Gourou was in the 1950s of the hostile environment that is implicit in Maquet's reading of his book, but we might return to his work to re-imagine a realm of co-existence that was waiting to be born.

Geographical warfare in the tropics

As various sections of this chapter intimate, Gourou's post-war tropicality was neither just pessimistic nor a throwback to dated pre-war ideas, as David Arnold suggested in his formative essay (see Chapter 1). Rather, it was polyvalent, and grounded in a range of post-war places and problems. We have seen that his tropicality was in turns strident, haughty, aloof, sceptical, insightful, ambivalent, and partially blind, and that Gourou made his own way through the dominant – Western, Cold War and anti-colonial – ideologies of his day. In Chapter 2 we looked at how elements of Gourou work got translated into the language of Vietnamese nationalism and mined for anti-colonial purposes. We shall end this chapter by returning to Vietnam in the early 1970s and showing how Gourou's 1936 study of the peasants of the Tonkin Delta was used as a tool of anti-imperial resistance.

In 1972, Yves Lacoste, a Moroccan-born radical geographer trained at the Sorbonne under the Communist Party figurehead of French geography, Jean Dresch, was in the Tonkin Delta with an International War Crimes Tribunal investigating claims that the U.S. Air Force had been bombing the dikes, thus contravening international laws of war by targeting civilians and threatening catastrophic flooding. Lacoste was escorted to the region by the North Vietnamese military in armoured vehicles supplied by China and the Soviet Union, with Gourou's *Les paysans du delta tonkinois* (hereafter *Les paysans*) in hand, and a year after major flooding in the region. While there had been some development of the hydraulic system of irrigation and flood protection since 1945 Lacoste surmised that its basic contours had not changed much since Gourou's days and that his book remained the most exhaustive treatment of the region. Lacoste also knew that an unauthorised 1955 English translation of *Les paysans* had been commissioned by the Human Relations Area Files project (an American inter-university consortium based at Yale that facilitated cross-cultural research), and conjectured that copies of Gourou's fastidious maps and figures of villages and dikes in *Les paysans* had wound up at the Pentagon.[87] Gourou wrote to Gottmann in December 1956 that it was only by chance that he had discovered this bootleg translation in a catalogue, on sale for six dollars.[88] A U.S. State Department official later divulged to Lacoste that he had guessed correctly about the fate of Gourou's maps.[89]

Lacoste's "on-the-site" analysis of American bombing of the dikes during the summer of 1972 spawned a blistering exposé which was initially published in the French daily *Le Monde* and newspapers around the world in August 1972, and a longer version of which (with a detailed map and photographic evidence of where the American bombs had fallen) was presented at various conferences and in the American Marxist weekly *The Nation*. His findings reached the floor of the U.S. Congress and prompted the Pope to protest. While the White House and Pentagon claimed they were targeting only military and industrial installations, and that any damage to the dikes was incidental ('collateral damage,' to use military parlance), Lacoste used Gourou's work to assert that there was a concerted attempt to hit the dikes in places that would cause maximum damage to surrounding areas. There was a precise geographic logic to the bombing and Lacoste coined the expression "geographical warfare" to describe it. He also drew on Gourou's aesthetic imagery of the dikes holding together a beautiful yet vulnerable rice landscape where people lived in complex and delicate harmony with nature. The American bombing was a war crime, Lacoste proclaimed, because it was designed to destroy a whole way of life.

In short, Lacoste politicised *Les paysans* in a different way than it had been politicised during the 1930s – an anti-imperial way – and loaded his exposé into an explosive 1976 book, entitled *Géographie, ça sert d'abord a faire la guerre* (Geography serves firstly to wage war), where he traces this adage from Strabo and Herodotus through to the Vietnam War. We thus return to the idea of a resistant – or militant – tropicality, and with Vietnam one of the key arenas of anti-imperial struggle in which 'the tropics' were shaped, harnessed and re-invented as a zone of resistance, and tropicality was refashioned as a mode of subversion rather than othering.

But once more, this picture is punctuated by ambivalence. If Gourou's work got inadvertently caught up in the Vietnam War, and facilitated Lacoste's intervention, he was criticised for his public silence over the war itself, and during the 1970s his tropical geography was challenged by a young generation of *tiers-mondistes* who believed that, in addition to condemning the tropical zone to perpetual backwardness, Gourou did not question the roles that the state and global capitalism played in its exploitation. Lacoste started the radical geography journal *Hérodote* in 1976 and began the inaugural issue (best known for its 'geography' interview with Michel Foucault) with a further laudatory retrospective on Gourou's thesis about the Tonkin Delta. But the issue also contained a critical commentary by two young cinematographers in which – and akin to the dramaturgical image that Said used just two years later, describing the Orient as a stage affixed to Europe – they question how Gourou's "inaugural landscape literally functions [in *Les paysans*] as the opening of an opera" – an opera about a resplendent and supine region at the aesthetic behest of a temperate West high on exoticism.[90] This re-engagement with Gourou marks the beginnings of a longer and more wide-ranging critical appraisal of his work and legacy. This critique was at its most intense in the mid-1980s, when radical geographers Bruneau and Courade questioned the ethos and remit of tropical geography, although neither Césaire nor Said's *Orientalism* feature.

Sylvie Fanchette recalled that "the 1980s revolt against the old descriptive tropicalist geography" became particularly intense in ORSTOM, and Georges Courade remembered well the closing speech that Gourou gave at a 1978 ORSTOM conference in Ouagadougou (Burkina Faso) on "agrarian space and development in tropical Africa."[91] This was one of Gourou's most trenchant evaluations of the power and pitfalls of both liberal and Marxist development discourses, and he offered a more upbeat assessment of tropical Africa's future than the one he had presented 30 years before. But he warned of how easily faith got the better of facts, and prejudice got in the way of noble intentions. "Today the means exist to master tropical insalubrity and increase productivity, but there are still many obstacles. If a part of the tropical world is kept in a state of backwardness, this can provide a basis for climatic racism. The answer to this is economic renewal. Here the honour of the human race and scientific honesty meet. . . . Ultimately, it is for the African peasants to take up the challenges of development."[92] Courade recalled that a large crowd of *confrères* huddled in a small lecture hall to hear the words of their " master ," but that the speech was also something of a swan song for Gourou.[93]

In 1984 Gourou himself was pressed about the politics of his work in a long interview, organised by Lacoste, which was published in *Hérodote*.[94] It was here that Gourou quipped "I would rather see a peasant than a politician" and asserted "the term judge of the world expresses my attitude to what I see . . . I try to understand without approving or condemning."[95] Something of Dalat may still have lain behind the remark, but the radicalism of the 1960s and its impact on geographers was, by 1984, a much closer *agent provocateur*. Gourou did not think it was his job to protest. Observe – yes. Question and reflect – perhaps. But intervene in politics (like Lacoste) – no. Debate about the utility of tropical geography pointed to wider ideological schisms within French geography, and a generational shift in attitudes towards the relationship between activism and the academy.[96] These schisms and shifts constituted their own 'colonial situation' at the tail end of Gourou's career: an ensemble of relations between advocates and detractors from different political and intellectual positions. Gourou, then in retirement, largely looked on with diffidence, as if to suggest, through a false dichotomy, that geography was momentarily being colonised by theoretical and political fads (such as 'under-development') but would come to its senses (or 'decolonise' itself). His relationship with Lacoste is an interesting case in point.

Lacoste told us that he came to regret some of the criticism of Gourou's project. His beef was with *Les pays tropicaux*, which "made no mention of colonisation and independence struggles," rather than *Les paysans* or even Gourou's semi-autobiographical *Terres de bonne espérance*.[97] Gourou, however, seemingly did not quite know how to respond to the *rapprochement* solicited by Lacoste during the 1980s, not least in the *Hérodote* interview, and nor, seemingly, did others. Georges Condominas told us that when he sat on the jury that examined Lacoste's thesis in 1980, with Dresch and Pierre George, he provoked protests from the other two by praising Gourou's work.[98] This anecdote points to a wider ideological rift between the Collège de France and Sorbonne (on the other side of the Rue St Jacques), and with the latter viewing the former as elitist and stuffy. Lacoste told us that he initially thought Gourou did not like him but later thought this was a misunderstanding.[99] Their relationship raises wider questions about what counted as a radical, progressive, *passé* or reactionary geography, and as we have sought to show in this chapter when it comes to individuals and their projects the answers to such questions are often more complex and ambivalent than first meets the eye.

In 1968, for instance, Gourou noted with regard to the search to find a replacement for the retiring Roger Dion at the Collège de France that "to belong to the old guard of geography does not make me insensitive to the feelings and aspirations of the new guard."[100] He later acknowledged the good use that Lacoste's 1972 exposé of the American bombing of the dikes had made of his work and lamented how his maps and study had been used by the U.S. Air Force.[101] When, in 1980, Gourou discovered that Lacoste would be in Brussels to give a lecture, he invited him to lunch. He greeted his guest, Lacoste told us, at the top of the staircase to his upper floor apartment with a piece of paper clenched in front of him. It was a letter from Võ Nguyên Giáp, the revered military commander of North Vietnam and former exponent of militant tropicality, whom Gourou had taught in Hanoi, and who visited Gourou in his apartment few years later (see Chapter 2). Lacoste implied that Gourou brandished it in order to flag his political credentials.[102]

By that time, tropical geography was on the wane, and the economic dynamism of the Pacific Rim countries, and Brazil, showed that the tropical zone could boom, something which Gourou's had also come to acknowledge. As for Communist Vietnam, collectivisation had not led to abundance: instead, the country imported millions of tonnes of grain. The communist regime, like its East European counterparts, was beset by difficulties, and Gourou pointed to some of the country's ongoing agrarian problems in his last two books, *Terres de bonne espérance* (1982) and *Riz et civilisation* (1984). He suggested that collectivisation, state irrigation and drainage projects had failed to deliver higher yields. New *encadrements* (especially new land-holding relations) were necessary to fulfil the great agricultural potential of the Tonkin, and especially Mekong, Deltas.

Lastly, Lacoste wrote to Gourou in December 1987, now perhaps in his own spirit of *perestroika*, asking for comments on an article he had written on "your friend Braudel," and noting: "What you say about *encadrements* is of capital importance. What do you think about Braudel's misplaced criticisms of your 'determinism'? I would like to visit you in Brussels to speak about various things, including an issue of *Hérodote* which we are planning on Monsoon Asia."[103] Lacoste got a tardy reply three years later: "As you know, Braudel was a fraternal friend of mine: we lived in a community of open-mindedness. . . . He accepted my conception of geography." Gourou signed off by noting that he was finished a book that reflected on the fate on tropical Africa, and with his own concern: "I don't understand why geography is currently so under attack [at school and university level in France]. How many now leave *lycée* without knowing what the word *tropique* means."[104] It had been supplanted by the term 'Third World,' and its meaning and specificity had become buried in languages of development, under-development, and globalisation (*mondialisation*). In 1982, he asked, ostensibly with Lacoste's exposé in mind, "What ever happened to the Annamite houses I studied and wrote about after this senseless war? Or the charming landscape that had the misfortune of being on the frontier between North and South Vietnam?"[105] He had not been to Indochina for nearly 40 years, and communist North Vietnam had long been closed off to Western researchers. But this query might also be taken as a further hint – refrain in this book – that the 'tropics' is a discursive construct and one which Gourou had helped to shape. What had happened to the Tonkin Delta, and Annamite houses, of his notes? He had affected the tropics and was steeped in its 'colonial situations.'

Notes

1 Georges Balandier, "The colonial situation: A theoretical approach," [1951] reprinted in Immanuel Wallerstein ed., *Social Change: The Colonial Situation* (New York: John Wiley & Sons, 1966), 34–61, at 60–61.
2 On this important point, see Frederick Cooper, *Colonialism in Question: Theory, Knowledge, History* (Berkeley: University of California Press, 2005).
3 Balandier, "The colonial situation," 34–37.
4 Georges Balandier, "General report on the roundtable organised by the International Research Office of the Social Implications of Technological Change (Paris March 1954)," *UNESCO International Social Science Bulletin* 6, no. 3 (1954): 372–384, at 373; Balandier, "The colonial situation," 41.
5 Pierre Gourou, *Les pays tropicaux: Principes d'une géographie humaine et économique* (Paris: Presses Universitaires de France, 1947), 148.
6 Balandier, "General report," 373.
7 For a thought-provoking way into this literature and debate, see Matthew Lange, *Lineages of Despotism and Development: British Colonialism and State Power* (Chicago: University of Chicago Press, 2009).
8 Edward W. Said, "Intellectuals in the post-colonial world," *Salmagundi* no. 70/71 (1986): 44–64, at 53.

9 Michel Bruneau and Georges Courade, "A l'ombre de la 'pensée Gourou'," *Espaces Temps* 26–28 (1984): 67–78, at 67.
10 Gourou, *Les pays tropicaux*, 181, 173–174 [slightly different wording in the 1953 English translation: 147–148, 141].
11 Gourou, *Les pays tropicaux*, 137 [113].
12 ACDF: Pierre Gourou, "Leçon inaugurale du cours de Pierre Gourou, 4 décembre 1947."
13 Gourou, *L'Afrique tropicale: Nain ou géant agricole?* (Paris: Flamarion, 1991), 224.
14 Gourou, *L'Asie* [1953] (Paris: Hachette, 1957 edition), 334.
15 Ranajit Guha, "Not at home in empire," *Critical Inquiry* 23, no. 3 (1997): 482–493, at 482.
16 Guha, "Not at home in empire," 482.
17 Pierre Gourou, *La Densité de population au Rwanda-Burundi. Esquisse d'une étude géographique*, Institut royal colonial belge, Section des Sciences morale et politique, Mémoire XXI, 6, Bruxelles, 1953, 150–152.
18 On this imagery, which continues to this day, see René Lemarchand, *Rwanda and Burundi* (London: Pall Mall Press, 1970), 15–20.
19 Gourou, *La densité de la population au Ruanda-Urundi*, 188.
20 Gourou, *La densité de la population au Ruanda-Urundi*, 189.
21 Pierre Gourou, "Les Kikuyu et la crise Mau-Mau: une paysannerie au milieu du XXe siècle," *Les Cahiers d'Outre-Mer* 7, no. 3 (1954): 317–341, at 328.
22 Pierre Gourou, "Le 'plan des arachides': Une expérience d'agriculture mécanisée en Afrique orientale," *Les Cahiers d'Outre-Mer* 30, no. 1 (1955): 105–118, at 106; Corey Ross, *Ecology and Power in the Age of Empire: Europe and the Transformation of the Tropical World* (Oxford: Oxford University Press, 2017), 355.
23 Ross, *Ecology and Power in the Age of Empire*, 358–358.
24 Auguste Chevalier, *Révolution en agriculture*, cited in Ross, *Ecology and Power in the Age of Empire*, 359.
25 Pierre Gourou, "Angola et Moçambique: etudes la géographie regionale," *Revue Géographique de Montréal* 22, no. 1 (1968): 5–20, at 5, 19.
26 Gourou, *L'Afrique tropicale*, 246.
27 Robert Delavignette, "L'Afrique de Pierre Gourou et le point de vue de l'administrateur," in *Etudes de géographie tropicale offertes à Pierre Gourou* (Paris: EHESS, 1972), 279–286, at 286.
28 The discussion in this paragraph is drawn from: Henri Nicolaï, "Un guide colonial: Le Guide du Voyageur au Congo belge et au Ruanda-Urundi," *Belgeo* [En ligne], 3 2012, http://journals.openedition.org/belgeo/7161.
29 Gilles Sautter, "'Dirigisme opérationnel' et stratégie paysanne, ou l'aménageur aménagé," *L'Éspace Géographique* 7, no. 4 (1978): 233–243, at 238–239; Ross, *Ecology and Power in the Age of Empire*, 359–361.
30 Gourou, *La densité de la population au Ruanda-Urundi*, 150.
31 Gourou, *La densité de la population au Ruanda-Urundi*, 150–152.
32 Pierre Gourou, "Progrès de la connaissance géographique du Congo belge," *Bulletin de la Société belge d'études géographiques* 19, no. 2 (1950): 111–133, at 111–112. For an overview of geographical interest in highland Africa, see François Bart, "La montagne au cœur de l'Afrique orientale," *Les Cahiers d'Outre-Mer* no. 235 (2006): 307–322.
33 Wilbur Zelinsky, "Review of *La densité de population au Congo belge*," *Geographical Review* 44, no. 4 (1954): 612–613; Jean Gallais, "Pôles d'États et frontières en Afrique contemporaine," *Les Cahiers d'Outre-Mer* no. 138 (1982): 103–122, at 118.
34 Jacques J. Maquet, "Review of *La densité de la population au Ruanda-Urundi*," *Africa* 24, no. 3 (1954): 276–277, at 276.
35 Gourou, *La densité de la population au Ruanda-Urundi*, 3, 33, 82.
36 Gourou, *La densité de la population au Ruanda-Urundi*, 102, 114.

37 See, e.g., Robert E. Ford, "The dynamics of human-environment interactions in the tropical montane agrosystems of Rwanda," *Mountain Research and Development* 10, no. 1 (1990): 43–63.
38 Gourou, *La densité de la population au Ruanda-Urundi*, 4, 25.
39 See Jinmi Adisa, "Antecedents of the 1994 refugee movement in Rwanda," in *The Comfort of Strangers: The Impact of Rwandan Refugees in Neighbouring Countries* (Ibadan: Institut français de recherche en Afrique, 1995), 11–24; Jean-Marie Kamatali, "Rwanda: Balancing gender quotas and an independent judiciary," in Gretchen Bauer and Josephine Dawuni eds., *Gender and the Judiciary in Africa: From Obscurity to Parity?* (Abingdon: Routledge, 2016), 137–153.
40 Henri Nicolaï, Interview with the authors.
41 Maquet, "Review of *La densité de la population au Ruanda-Urundi*," 277.
42 Gourou, *La densité de la population au Ruanda-Urundi*, 178–194.
43 Gourou, *La densité de la population au Ruanda-Urundi*, 185.
44 Homi Bhabha, "The humanities and the anxiety of violence," Keynote address to the 'Changing the Humanities / The Humanities Changing' conference (2009), University of Cambridge, http://sms.cam.ac.uk/media/1184691.
45 See John May, "Policies on population, land use, and environment in Rwanda," *Population and Environment* 6, no. 4 (1995): 321–334.
46 On these scenarios, and the difficulties in generating precise figures, see Marijke Veroorten, "The death toll of the Rwandan Genocide: A detailed analysis of Gikongoro Province," *Population* 60, no. 4 (2005): 331–367.
47 Cited in Roxanna Curto, "A particular universalism: The 'francophonie' of Henri Lopes," in H. Adlai Murdoch and Zsuzsanna Fagyal eds., *Francophone Cultures and Geographies of Identity* (Newcastle: Cambridge Scholars Publishing, 2013), 32–50, at 41.
48 Jean-Pierre Chrétien, *Le Défi de l'ethnisme. Rwanda et Burundi: 1990–1996* (Paris: Karthala, 1997); Luc de Heusch, "Rwanda: Responsibilities for a genocide," *Anthropology Today* 11, no. 4 (1995): 3–7; Mahmood Mamdani, *When Victims Become Killers: Colonialism, Nativism, and the Genocide in Rwanda* (Princeton: Princeton University Press, 2001); René Lamarchand, "Rwanda: The rationality of genocide," in Roger W. Smith ed., *Genocide: Essays Toward Understanding, Early-Warning, and Prevention* (Williamsburg, VA: Association of Genocide Scholars, 1999), 17–25; David Newbury, "Understanding genocide," *African Studies Review* 41, no. 1 (1998): 73–97; Peter Uvin, "Reading the Rwandan genocide," *International Studies Review* 3, no. 3 (2001): 75–99.
49 Lamarchand, "Rwanda," 17.
50 James Tyner, *War, Violence and Population: Making the Body Count* (New York: Guilford Press, 2009), 152.
51 Ben Kiernan, *Blood and Soil: A World History of Genocide and Extermination from Sparta to Darfur* (New Haven, CT: Yale University Press, 2007), 564.
52 Frederick Golooba-Mutebi, "Collapse, war and reconstruction in Rwanda: An analytical narrative on state-making," Working Paper No. 28 (series 2), 2008 (London: Crisis States Research Centre, UK Government Department of International Development), 1.
53 Frederick Golooba-Mutebi, "Collapse, war and reconstruction in Uganda: An analytical narrative on state-making," Working Paper No. 27 (series 2), 2008 (London: Crisis States Research Centre, UK Government Department of International Development), 4–5.
54 Bart, "A propos de Pierre Gourou," 127.
55 Pierre Gourou, "Géographie et civilisation en Afrique noire tropicale pluvieuse," *Revue Belge de Géographie* 64 (nouvelle série), no. 2 (1998): 182–192, at 189.
56 Pierre Gourou, *L'Afrique tropicale*, 79; Gourou, "Géographie et civilisation," 188.
57 Gourou, "Géographie et civilisation," 190.
58 Gourou, "Géographie et civilisation," 190.
59 Gourou, "Géographie et civilisation," 189.

60 See Jennifer M. Olsen, "Behind the recent tragedy in Rwanda," *GeoJournal* 35, no. 2 (1995): 217–222.
61 Although he does reference important exceptions, such as the work of Jennifer Olson. Henri Nicolaï, "Geografia, popolazione e violenza nelle regione dei Grandi Laghi," *Terra d'Africa* VII, no. 1 (1998): 17–56; Henri Nicolaï, "Progrès de la connaissance du Congo, du Rwanda et du Burundi de 1993 à 2008," *Belgeo*, 3–4 (2009): 247–404.
62 Adisa, "Antecedents of the 1994 refugee movement in Rwanda," 14.
63 Mamdani, *When Victims Become Killers*, 80.
64 Gourou, *La densité de la population au Ruanda-Urundi*, 33–34, 103.
65 Adisa, "Antecedents," 11.
66 Arjun Appadurai, *Fear of Small Numbers: An Essay on the Geography of Anger* (Durham, NC: Duke University Press, 2006), x, 49.
67 Mamdani, *When Victims Become Killers*, 101.
68 Appadurai, *Fear of Small Numbers*, 50.
69 J-J. Maquet, *The Premise of Inequality in Ruanda* (Oxford: Oxford University Press, 1961), 185; and see J-J. Maquet, "The kingdom of Ruanda," in Daryl Forde ed., *African Worlds* (Oxford: Oxford University Press, 1954), 164–189.
70 Peter Uvin, "On counting, categorizing and violence in Burundi and Rwanda," in David Kertzer and Dominique L. Arel eds., *Census and Identity: The Politics of Race, Ethnicity, and Language in National Censuses* (Cambridge: Cambridge University Press, 2002), 148–175, at 156. Jean Hienaux, "Analyse de la variation des caractères physiques humains dans une région de l'Afrique centrale: Ruanda-Urundi et Kivu," *Annales du Musée Royal du Congo Belge* no. 3. (1954): 131.
71 Uvin, "On counting, categorizing and violence," 156.
72 Pierre Gourou, "Progrès de la connaissance géographique du Congo belge," *Bulletin de la Société belge d'études géographiques* 21, no. 2 (1952): 373–404, at 376.
73 For example, Stanford political scientist Lachlan McNamee relies on Gourou's data from the period 1949–1952 on Tutsi pastoral dominance, as measured by cattle per 1,000 inhabitants, and its areal expression in chieftaincy boundaries, as core geographic data with which to test part of his hypothesis that there is a direct relationship between resettlement and political violence: "Mass resettlement and political violence: Evidence from Rwanda," *World Politics* 2018 in press, https://people.stanford.edu/lmcnamee/mass-resettlement-and-political-violence-evidence-rwanda.
74 Pélissier, cited in Jean-Pierre Chrétien, J-P. and Kabanda, M. 2013, *Rwanda. Racisme et génocide: L'idéologie hamitique* (Paris: Belin, 2013), 23.
75 Facil Tesfaye, "Statistique(s) et génocide au Rwanda: Sur la genèse d'un système de catétorisation 'génocidaire'," Unpublished Masters in Political Science, Université du Québec à Montréal, 2007, 71, https://archipel.uqam.ca/4759/1/M9995.pdf.
76 The quotation is from Heusch, "Rwanda," 3. For any earlier and important review of Maquet's influence, see René Lemarchand, "Power and stratification in Rwanda: A reconsidertation," *Chaiers d'Études africaines* 24, no. 4 (1966): 592–610.
77 Mamdani, *When Victims Become Killers*, 9.
78 Mamdani, *When Victims Become Killers*, 9.
79 Uvin, "On counting, categorizing and violence," 156.
80 Heusch, "Rwanda," 4.
81 Paul Magnarella, "Explaining the Rwandan genocide," *Human Rights and Human Welfare* 2, no. 1 (2002): 25–34, at 31. He is also critical of: John A. Berry and Carol Pott Berry eds., *Genocide in Rwanda: A Collective Memory* (Washington, DC: Howard University, 1999). Lee Ann Fujii argues something similar to Magnarella in *Killing Neighbors: Webs of Violence in Rwanda* (Ithaca, NY: Cornell University Press, 2009).
82 Roland Pourtier, "L'Afrique centrale dans la tourmente: Les enjeux de la guerre et de la paix au Congo et alentour," *Hérodote* 111, no. 4 (2003): 11–39, at 35.
83 Gourou, *L'Afrique tropicale*, 131.
84 Bhabha, "The humanities and the anxiety of violence."

85 James Tyner, *Genocide and the Geographical Imagination: Life and Death in Germany, China and Cambodia* (New York: Rowman and Littlefield, 2012), 25.
86 Bhabha, "The humanities and the anxiety of violence."
87 Melvin Ember, "Evolution of the human relations area files," *Cross-Cultural Research* 31, no. 1 (1997): 3–15. We deal with this story in more detail in: Gavin Bowd and Daniel Clayton, "Geographical warfare in the tropics: Yves Lacoste and the Vietnam War," *Annals of the Association of American Geographers* 103, no. 3 (2013): 627–646.
88 BNF: Fonds Gottmann: Gourou to Gottmann, 26 December 1956.
89 Lacoste, Interview with the authors.
90 Dominique Chapuis and Maurice Ronai, "Morceaux choisis," *Hérodote* 1, no. 1 (1976): 118–124, at 118–119.
91 Sylvie Fanchette, Interview with the authors; Georges Courade, Interview with the authors.
92 Pierre Gourou, "Pour une necessaire amelioration de la condition paysanne en Afrique noire," in *Maîtrise de l'espace agraire et développement Afrique tropicale: logique paysanne et rationalité technique* Colloque de Ouagadougou. (HAUTE VOLTA), 1978/12/04–08 (Paris: ORSTOM, 1979), 567–573.
93 Georges Courade, Interview with the authors.
94 See, especially, Olivier Dollfus, "Le regard attentive et le regard sélectif: Jean Dresch et Pierre Gourou," *Hérodote* 33–34–35, nos. 2–3 (1984): 73–88; Michel Bruneau et Georges Courade, "Existe-t-il une géographie tropicale A la recherché du paradigm de Pierre Gourou," *L'Espace Géographique* 4, no. 4 (1984): 306–316.
95 Pierre Gourou, "La géographie comme 'divertissement'? Entretiens de Pierre Gourou avec Jean Malaurie, Paul Pélissier, Gilles Sautter, Yves Lacoste," *Hérodote* 33, no. 1 (1984): 50–72, at 53, 68.
96 That this issue was as much generational as ideological, and one that imbues the way younger and older Marxist academics responded to May 1968, in more activist and contemplative ways, respectively, is taken up in Richard Vinen, *The Long '68: Radical Protest and its Enemies* (London: Allen Lane, 2018).
97 Yves Lacoste, "Géographie coloniale et géographie académique: approche épistémologique," in Michel Bruneau and Daniel Dory, eds., *Géographies des colonisations XV-XX siècles* (Paris: L'Harmattan, 1994), 343–348. Yves Lacoste, Interview with the authors. For an interesting reflection, see Béatrice Giblin, "*Hérodote*, une géographie géopolitique," *Cahiers de géographie du Québec* 29, no. 77 (1985): 283–294.
98 Georges Condominas, Interview with the authors. Lacoste surmises that rivalry with Dresch stemmed from the fact that, as a socialist rather than a member of the Communist Party, Gourou had been linked to national and international policy initiatives surrounding decolonisation between 1944 and 1947 and was in "the brain trust of General Leclerc." Lacoste, "Géographie coloniale et géographie académique," 345.
99 Yves Lacoste, Interview with the authors.
100 Gourou, cited in Olivier Orain and Marie-Claire Robic, "La géographie au Collège de France (milieu xixe-milieu xxe siècle): Les aléas d'une inscription disciplinaire," in W. Feuerhahn ed., *La politique des chaires au Collège de France*, Collège de France, 2017, open edition, https://books.openedition.org/lesbelleslettres/176#ftn46.
101 Kleinen, Interview with Gourou, n.p.; Gliberte Bray, Interview with the authors.
102 Lacoste, Interview with the authors. Gourou's daughter told us that the letter contained information on Annamite villages that Giáp had collected for his teacher's second thesis, and which proved vital to his revolutionary evaluation of the peasant condition. Gilberte Bray, Interview with the authors.
103 AULB: Fonds Pierre Gourou PP153 Yves Lacoste to Pierre Gourou, 24 December 1987.
104 PGB: Pierre Gourou to Yves Lacoste, Bruxelles, 16 February 1990.
105 Gourou, *Terres de bonne espérance, le monde tropical* (Paris: Plon, 1982), 16–17.

8 *Fin de la tropicalité* (as we knew it)?

Gourou's tropics

While Pierre Gourou's worry in his melancholic 1990 letter to Yves Lacoste (see Chapter 6) that the word *tropique* was disappearing from French schools may have been well placed, the term never left academic study and debate, and sprang back to life at the end of the century in the guise of tropicality. In Chapter 1 we explained how Gourou enters this frame via two critiques of his work that appeared fifty years apart, and in different anti-colonial and postcolonial climates. In these last few pages we reflect on what we have made of Gourou's "impure and worldly geography," and whether tropicality in a wider sense is dead or markedly different than yesteryear, and partly in light of the fact that Orientalism has become emboldened in recent decades.[1]

Gourou had a significant influence on twentieth-century tropicality and cut a particular path through it. His tropicality pivoted on a prejudice of low expectation regarding the prospects of development in the tropical world, chiefly on account of its insalubrity and thin soils, but also, in some parts of the tropics, due to great number of people in tiny spaces. "Constant heat and high humidity create a nature different from the temperate zone," he told his Collège de France class in 1947; and "insalubrity has been an obstacle to inhabiting and cultivating the tropical world, as has the poorness of the soil. The Tropics have been inhabited on the whole by ineffective civilisations."[2] To hark back to Said, the 'trouble' usually set in when Gourou was writing about the tropics with broad brush strokes, as zone set apart from the northern temperate zone (an image encapsulated, we noted in Chapter 1, on the cover of his 1947 *Les pays tropicaux*). At smaller scales, however, his tropicality reached across of the positive and negative spectrum of tropicality, and particular places and regions – the Tonkin Delta, Japan, and Belgian Ruanda-Urundi – became privileged as models of particular kinds of tropical problems and vistas. Gourou's work also crossed the ends of empire and birth of a plethora of independent nation-states in tropical Asia and Africa (post-war decolonisation), and the trauma of World War II had a decisive impact on how he responded to this major shift in political arrangements and perception regarding what the term 'tropics' stood for.

He both marvelled at and pitied the tropics, and strove for a critical but humane voice, and it was a voice that was at times contradictory. He was both unsparing in his assessment of the limits to tropical development and lyrical in his admiration for traditional *encadrements*. He romanticised the Tonkin Delta and what he saw as the rich 'vegetal' civilisations of the Far East, which had eked out a meagre existence in ornate rice landscapes. He deemed tropical Africa and the Amazon as more primitive, more forbidding to European colonisers, with populations that were unable to grow and progress without outside help, and with agrarian civilisations that were ill-placed and in many ways not willing to take on board Western-style development.

Either way, the tropics would never fully emulate the temperate West, and through the book we have sought to consider when, where, and for what reasons an earnest knowledge either lapsed or could get sucked into something more disingenuous, and thus to pose the question of whether there are 'impure and worldly geographies' in all us. On the one hand, today's geographers and tropical researchers work in an age that is hyper-sensitive to the locational basis of knowledge and research, to how particular peoples and entities become arbiters of truth, and to how and why knowledge is manipulated and thus arbitrary. On the other hand, during his lifetime Gourou clung to the idea that while there is bias in all knowledge, there are still geographical facts and realities, and patterns and process, about which it is possible to generate a credible and durable knowledge. The clung tightly to this idea – not necessarily of an objective and disinterested knowledge, but certainly of a credible and sure-footed knowledge. The 'impure and worldly geography' we have unearthed does not have Gourou marching with the army or upon direct orders of the state. Rather, it operates lateral to the state, and there is much in Aimé Césaire's argument that Gourou wielded a "watchdog" knowledge that betrayed a set of cultural assumptions about the superiority of the temperate world over its tropical counterpart.[3]

Towards the end of his 1982 semi-autobiographical book *Terres de bonne espérance*, Gourou declares that "my tropics are attractive; when they exude sadness it is not for 'tropical' reasons but because of the unfortunate condition of the inhabitants, due to insufficient techniques and an unhappy history."[4] Gourou here seemingly eschews the essentialist and condescending tone of a swath of Western thinking about the tropics and goes on to argue that a humanistic appreciation of this zone should start with civilisations and not with nature. Yet it is not hard to sense from this same passage and much else Gourou wrote that his humanism was encompassed by a more entrenched body of ideas about the inherent backwardness of the tropical world. If the tropics were 'sad,' then for Gourou it was partly because only a few tropical civilisations had attained one of the West's principal achievements: a productive relationship with nature. And if the tropics were attractive, he suggested in this book – partly reversing what he had argued over the previous four decades – then it was because tropical peoples and regions might spur economic growth by adopting Western science and technology. Either way, the tropical world is constructed as the preserve of the Western scientist and scholar.

Gourou's tropicality was part and parcel of a twentieth-century culture of Western dominance that morphed out of empire and into a new era of development. But this was a troubled culture, and dominance was expressed and became fractured in various ways. Western 'civilisation' was beset by totalitarianism and war, and for scholars of Gourou's generation and humanist mind-set, exaltation of the West came with intense scrutiny and criticism of the violence perpetrated in the name of progress and development. Gourou commiserated with peasant subjugation and suffering at the hands of alien – colonial, wartime, revolutionary and development – dynamics and projects that were not well adapted to the tropics. Of course, his assessment of the ghastly flavour and ill effects of many "Western recipes" hinged on his own evaluation of "tropical conditions" (see Chapter 6). He dealt more directly and extensively with questions of colonisation and decolonisation than many commentators think, but he was more interested in the idea and workings of 'traditional' ways and saw modernity as a spoiler. Yves Lacoste perhaps got the measure of Gourou and other European geographers of his generation who pursued a career in the colonies and tropics when he characterised them as "anti-colonialist colonial geographers" – as democratic, and certainly critical of empire, but as never fully relinquishing "the values and principles of the metropolis."[5] A commitment to reason and order, and the quest to raise living standards, were paramount values and principles.

Empire – or, more precisely, the West's attempt to harness tropical resources and commodities, and concomitant efforts at territorial and labour control – has been central to the way the story of tropicality has been told, and forms a significant part of Gourou's story. Gourou's friend from his Indochina years, Joseph Inguimberty, confided to his children: "I was right-wing 'like Gourou': colonisation was a positive thing to us."[6] Yet as we have shown, Gourou was never a handmaiden of empire in any straightforward sense. Between 1936 and 1947 he was drawn into the armatures of politics and international relations, but as Pélissier confided, became disillusioned about how scholars and their expertise had been pressed into the service of the state and political agendas. His interests and concerns brushed against the grain of how tropical colonies were valued and tropical ecologies were exploited, and from 1945 he became increasingly preoccupied, and again often without towing a dominant Western (imperial or disciplinary) line, with the readiness of the tropical world for decolonisation and development. He quashed excitement and optimism surrounding the idea that the true potential of tropical nature might be unleashed if only Western capital and know-how came flooding in, and placed the challenges involved in raising living standards into sharp relief.

While the ruses and demise of empire have a central place in Gourou's impure and worldly geography, we have also drawn attention to the significance of war – two world wars, and colonial warfare in Indochina and elsewhere in the tropics – and think that it should be accorded a more important place in the genealogy of twentieth-century tropicality than it has hitherto. If a prejudice of low expectation constituted an elementary cognitive aspect of Gourou's tropicality, then war and empire were vital contextual and experiential factors that complicated that tropicality in equally important ways. War cast a long emotional shadow over how and why many geographers of Gourou's generation venerated order, reason and stability, and in our view Gourou's pessimism cannot be extricated from this shadow and what Febvre termed "the dark side of modernity" (see Chapter 6). Gourou insisted that the tropical world was essentially stuck, or at least could not be pushed beyond a certain point, because the idea that tropical nature could be conquered by Western science and technology could quickly reap misery, chaos and despair. This was the affective shape of tropicality as Gourou purveyed it in the post-war era, and it should not necessarily be judged as a paternalist or backward-looking stance. It sprang from what Gourou described as "extraordinary agitation[s] on the human soul" (see Chapter 6).

To labour the notion that some of that pessimism now seems entirely misplaced would be to wrench his work out this context, and away from its war footing. However, it is also important to note that some of Gourou's views and prognoses were erroneous, even during his lifetime. The population of Africa was around 220 million in 1950 and had reached 650 million by the time he published *L'Afrique tropicale* in 1991 (and Rwanda's population was around 2 million and 7 million, respectively).[7] The population of the Tonkin Delta was around 7 million when Gourou completed his study in 1936, and stood at over 17 million by the time he died. In spite of his warnings about the dangers of overpopulation, there had been rapid population growth in Vietnam, particularly during the 1950s, and there was further demographic expansion from the 1970s, which was spurred by the introduction of higher-yield rice varieties cultivated in more stable conditions and under shorter cycles, and latterly by processes of land de-collectivisation that produced "what appears to be . . . a relatively equitable land-holding peasantry," as Michael Watts saw matters in 1998.[8] Gourou also changed his mind on some basic things, arguing in the first (1947) edition of *Les pays tropicaux* for the need to stimulate tropical exports in order to boost living standards, and in the last (1966) edition for local food production for protected African markets.[9]

A consideration of Gourou's post-war research and the 'colonial situations' in which he worked brings relations between the material and epistemological violence of tropicality to the fore. Gourou's study of late colonial Ruanda-Urundi, and later reflections on the Rwandan genocide, deflect attention from the ethnic hatred and horror that unfolded, and from the complicity of Western knowledge and categories in its fomentation. Yet his scepticism, even fatalism, regarding the problems facing traditional agricultural regimes in Rwanda and elsewhere in tropical Africa might be seen as an antidote to ethnological schemes of colonial and post-independence thought that often papered over the way conflict and violence were rooted in basic geographical questions of land, population and resources, and problems of what Georges Balandier described as "adjustment to place."[10]

Again, Gourou's sensitivity to the distributional aspects of such questions did not just stem from disciplinary protocol. In wartime Bordeaux he had grappled with how food shortages and rationing had been exacerbated by 'administrative disorder,' and this concern with organisation never left him (see Chapter 5). So were violence, hatred and horror out of Gourou's disciplinary line of sight?[11] He and other geographers of his generation who had witnessed or served in World War II, and had encountered cruelty and privation first-hand either there or in the colonial world, rarely addressed violence and hostility openly in their work, and many of them came to begrudge latter-day radical orientations that located violence in writing, discourse, systems of knowledge and representation. This experience did not create a traumatic 'past is a foreign country' loop (or 'groundhog day') from which these geographers were unable to escape. However, we think that their veneration of peace, stability, harmony and beauty in the relations between people and environment, and diagnoses of danger and trouble, were profoundly affected by war and conflict. Their "proleptic humanism," as Homi Bhabha conjures with the tasks of a "critical humanities," lay in a grounded concern with the idea of dwelling (land, place, settlement) and existential security in that dwelling (which is part of what Gourou meant by *encadrement*).[12]

The last book Gourou wrote was a short story book for his grandchildren entitled *Les aventures de Kataboum*. The story is about a young and impetuous but kind-hearted rhinoceros whom strays from his family and village into the tropical bush, there to encounter annoying ants, cheeky monkeys, threatening crocodiles, and rhinoceros poachers with rifles. Kataboum unwittingly protects his family and his tropical milieu from the poachers and Gourou writes that "it was thanks to Kataboum's disobedience that his family was saved."[13] This same adventurous spirit of non-compliance also applies to Gourou. He eschewed dogma and orthodoxy, but at the same time was stubborn in some of his convictions, and his impishness was sometimes overridden by convention. Moreover, while he was deemed a tropical expert by national governments, international organisations and an international academic community, by the 1950s he could not assume that his advice would be heeded, or, as Césaire pronounced, that his views would be seen as unimpeachable. Gourou knew that a tropical world that had long been ordered and exploited by empire was changing fast, and that categories of thought were changing with them. He became an object of anti-colonial ire, and caught in a thicket of shifting values and priorities with respect to how and for whom one observed, experienced, represented and explained 'the tropics.'

Following rationalist and humanist traditions of European thought stretching from Descartes, Montaigne and Valéry (from whom he quoted frequently in later years) to Bloch, Febvre and Braudel, Gourou thought that his job was to demonstrate and explain, and only judge on the basis of facts and evidence that had been carefully assembled and sifted. In 1977 the American geographer Baruch Boxer reflected that "The beauty of Pierre Gourou's work is that it makes the hackneyed phrase 'man-land relations' fairly sing with meaning. Anticipating by several decades current interest in such research areas as cultural ecology, energy flow analysis, and natural resources management, his studies of the human geography of the tropics brilliantly combine the sensitive insights of a humanist with the skills of a trained ethnographer and the cautious empiricism of a scientific observer."[14] Césaire's beef was that this Western tradition of science and humanism failed to live up to its own fair-minded pretensions: it also warranted wanton generalisation and moralising judgement. While under no illusions about the ugliness of empire, Gourou's politics and aesthetics were still imperialist – presumptive, exoticist and domineering – Césaire opined.

Exoticism permeated Gourou's suspicion of the development mind-set that gradually supplanted empire from the 1940s and worked its way in the thinking and ambitions of newly independence countries as well as Western powers in imperial retreat and international organisations. He made his views plain in a report for UNESCO that was published in the summer of 1968, when the international air was thick with talk of 'under-development,' and the streets of Paris were full of student protesters and talk of revolution.[15] In that report he questioned the saliency of the idea of under-development (although with a skewed understanding of what the term signified) and warned about the perils of revolution. He stuck to his guns about the need for gradual reform, even if it meant being accused of standing in the way of modernisation, or of being seen as curmudgeonly, or as an exoticist bent on harbouring traditional tropical civilisations from the modern world for his own aesthetic pleasure and *divertissement* (distraction). There was a deeply baroque side – indeed reverie – to Gourou's tropicality: a contemplation of harmony and search for symbolism in tropical landscapes and peasant civilisations.

Césaire thought that Gourou's *Les pays tropicaux* often left the bounds of fact and veered into more insidious and self-indulgent opinion and symbolism. Gourou was wedded to a humanist (compassionate and studious) view of the world, and his curiosity sprang from a deep respect for the variety of life in the tropics. But his humanism was marred by colonialist conceits and contradictions, and imbued with romanticist and primitivist fixations that worked to keep the tropical world in its place as the exotic object of a Western and colonising gaze, rather than as a zone aspiring to freedom, as Césaire himself sought to re-imagine the tropics in the wartime revue *Tropiques* (see Chapter 5).

While Gourou's work was undoubtedly affected by 'worldly' forces of empire, war, development, and latterly the radicalism of the 1960s, he was also, as Febvre related, 'a man': a principled individual who was never a stooge to any one ideology, worldview, or system. While growing international and multidisciplinary networks of knowledge exchange and scholarly collaboration influenced how the zonal and comparative imaginary at the heart of *Les pays tropicaux* became articulated, Gourou was a geographer at heart, and the politics of internationalism and globality were ancillary. Moreover, geography itself had a long and strong comparative bent, and there were much longer-standing traditions of thinking about the 'torrid zone.' Gourou was steeped in, and added significantly to, the French geographical project pioneered by Vidal de la Blache. Accordingly, he was as interested in the evolution of landscapes as he was in their organisation. The one side of his geography was unthinkable without the other one, and Febvre, who orchestrated his election to a chair at the Collège de France, and his friend Braudel, had as much, if not more, influence on his thinking than Vidal or his supervisor, Albert Demangeon, did.

From this geography sprang a core belief that the tropical world could not be shoved into modernity quickly or *in toto*. There were different modes of tropical cultivation, and important differences in how tropical regions had been colonised and the extent to which their traditional landscapes and ecologies had been transformed. In general, Gourou was wont to suggest, the lighter the colonial touch the better states fared in independence. Yet he and other geographers also cultivated what we have described as a 'phantom tropics,' with stereotypes and assumptions about this part of the world as deficient and dependent, inferior and insalubrious, as essentially agrarian and traditional rather than urban and modern, and with important implications for how 'the tropics' continue to be seen.

Yet Gourou's 'impure and worldly geography' was neither simply set in stone, nor the expression of older (pre-war and imperial) ways of thinking. It was the product of specific historical forces, political expediencies, institutional and geographical circumstances, and webs of knowledge and power. Much of what he wrote hinged on the time and places in which he worked (French Indochina, the Belgian Congo and Ruanda-Urundi), and particular events, moments, trips, relationships and memories – the water buffaloes of the Tonkin Delta; his experience of German occupation and encounter with Nazi geographers in 1938; his attendance at Institute of Pacific Relations conferences; his contribution to the 1949 *France Illustration* special issue and friendship with Joseph Inguimberty; his sojourn in Dakar, and trips to Sao Paulo and Lucknow. Yet much in the life and trajectory of twentieth-century tropicality – and by extension the politics of empire, development, independence and internationalism – threaded their way through Gourou's work and projects.

There was a complex politics to Gourou's tropical geography and tropicality, and part of our aim in this book has been to broaden understanding of what, in Gourou's case, counts as a 'politics of knowledge.' On the basis of the above analysis, we think it is misplaced to suggest, as critics of Gourou's work have in the past, that he was oblivious to questions of power. We think that some of the more obvious ways in which he evaded 'the political' were a purposive – in turns humanist, scientific, aesthetic, emotional, practical and judgemental – response to the issues of his time and circumstances in which he found himself. Gourou's career is fascinating for what it says about how academics working in the middle decades of the twentieth century were drawn into what Febvre pinpointed as stresses and strains of 'intellectual life' and 'emotional life' (see Chapter 6). Febvre knew that one would be judged and remembered not only by one's ideals, but also by how one's life was lived and one faced up to the challenges of history. In this sense, Gourou was not just a scholar but also 'a man' to Febvre and Bloch.

In Chapter 1 we framed our discussion of Gourou and tropicality as a conversation between the past and present, and between different ages and disciplinary positions. Some of our chapter titles and themes – especially of war and conflict, networks and internationalism, and agitation and affect, but also, in more general terms, of what it means to exoticise or tropicalise – are hewn from contemporary climates of feeling and concern, and we have sought to show that such sensibilities have rich and important histories that might usefully feed into present-day theory and debates, and not just about tropicality. We will end the book with two final thoughts about the legacies of Gourou's entanglement with tropicality: the first concerning the ongoing relevance of his critique of environmental determinism;

and the second concerning the rediscovery of his work in twenty-first century Vietnam.

Tropicality – but not as we knew it

The year after David Arnold published his important survey of tropicality and Gourou's complicity in it, in 2000, the prominent American economist and UN adviser Jeffrey Sachs published an essay entitled "Tropical underdevelopment" in which he argued that "Perhaps the strongest empirical relationship in the wealth and poverty of nations is the one between ecological zones and per capita income. Economies in tropical ecozones are nearly everywhere poor, while those in temperate ecozones are generally rich."[16] Sachs was writing at the millennium, when scholars and pundits in a wide range of fields were undertaking a global stock-taking of planetary fortunes and ills. His intervention provided the analytical grist for a string of further populist pieces about why the tropical zone in general, and land-locked countries in tropical Africa, Asia and South America in particular, lagged behind the West. He argued that the roots of tropical backwardness and underdevelopment lay in "the combination of physical ecology and societal dynamics," and that it was to the temperate north to which the tropics would need to look for salvation. Other books published around this time and in this millennial spirit – such as Jared Diamond's *Guns, Germs and Steel* (1997); and David Landes's *Wealth and Poverty of Nations* (1998) – painted similar pictures on larger historical canvases, and these three authors agreed that the influence of "physical geography," as Sachs simply put it – meaning latitude and location, and climate, vegetation and terrain – was a much too neglected and maligned factor in how world history and development were understood.[17]

While Sachs insisted that he rejected environmental determinism, as a "false argument that a country's fate is settled by its geography, not merely shaped by it," this literature has breathed new life into a mode of explanation that Gourou had long sought to discredit.[18] Environmental determinism had no place in his outlook and he warned of its consequence to the end of his life. His students took on that responsibility.[19]

Sachs's bold reappraisal of the influence of climate, coastal proximity and disease as a kind of triple lock on tropical underdevelopment legitimised the return of a type of language and thinking that geography, at least, thought it had put to bed, and not least due to Gourou's efforts.[20] It is thus timely to heed Gourou's admonitions about environmental determinism, and of how a purportedly scientific knowledge fell foul of it. Delve beneath the surface of Sachs's stance and it is not hard to find a ground swell of opinion that toys with what Gourou labelled "climatic racism." "Is it not interesting to recall the views of Montesquieu," Gourou wrote in 1963, "for do they not live in us, and still readily come to life, as they lived in him? Have we come to a more correct observation than that in the time of Montesquieu, that the people of the north are taller, calmer, more hardworking, more honest, more enterprising, more trustworthy, more interesting than people from the south?"[21] Gourou may have exaggerated the sway of environmental determinism during the 1960s, when greater faith was being placed in the capacity of science and technology to 'master' an unruly nature. In the longer term, however, it is perhaps prescient to recall Gourou's intimation that there would always be a fine line between dressing up tropical underdevelopment in sophisticated languages of economics, international relations, or geo-politics, and resorting to a more simplistic and sinister language of environmental determinism.

Today, arguments about the relations between society, environment and development are becoming ever more central to ones about the nature of power, governance, and global instability. Climate change and what Andreas Malm terms "a warming world" are now key players in how these relations are understood, and a language of violence, extremes and the desert (the limit of the tropics from Gourou's point of view) has crept its way into the language of order, stability and hierarchy that shaped twentieth-century tropicality.[22] "Extreme weather events and off-kilter weather patterns are causing more humanitarian crises and fueling civil wars . . . [in] that violent and impoverished swath of terrain around the mid-latitudes of the planet," Christian Parenti sets out to show in his 2011 book *Tropic of Chaos*.[23] 'Tropic' is used as a locational metaphor for breakdown, difficult and threat, as it has long been deployed, although the metaphor is now attuned to accelerating climate change rather than just sheer environmental difference. Many of the regions Parenti discusses – Afghanistan, Kyrgystan, Somali – are not in the tropics as Gourou delineated this zone. However, there is in mid-latitude regions a "catastrophic convergence . . . [of] political, economic and environmental disasters," he asserts, and 'the tropics,' thus defined, has become "a belt of economically and political battered post-colonial states . . . [where] we find clustered most of the failed and semifailed states of the developing world."[24]

Cold War militarism, neoliberal economic restructuring, and "banana wars" (guerrilla insurgencies and counterinsurgencies) have played a large part in creating this 'tropic of chaos' – a crime-, corruption-, disease- and famine-ridden zone. In vital respects, however, Parenti's recycles the vision of chronic capitalist and imperialist misadventure found in Joseph Conrad's *Heart of Darkness*, and a much longer association of the tropics with danger. Following 9/11 and the advent of a United States-led 'war on terror,' Greg Bankoff relates, the world became divided up into "regions of risk" (and safety), and the tropical world was configured as a "disease-ridden, poverty-stricken, disaster-prone" hotbed of terrorism and state-failure, and an "extreme environment" of physical and political insecurity.[25] The "climate-induced chaos" dealt with by Parenti has its roots in both this eventuality and a longer history of colonialist rapacity, and he thinks the solution lies in "mitigation," and especially in a redoubling of efforts "to decarbonise our economy."[26] He implies that it is the West's responsibility to sort out the chaos and eschews discussion of what 'tropical' peoples themselves might do.

Wielding a pointed and prescient picture of the concatenation of factors afflicting the tropics – and revolving around what Sachs describes as "the linkages of climatic shocks and extreme social instability" – Parenti rehearses a powerfully disparaging and pessimistic narrative about the tropics as unruly, and his inclination is to alight on and bewail chaos and suffering rather than to investigate the tropics watchfully, in the belief that alternatives and answers might arise from there. The tropical zone becomes a trope – tropic – of Western angst.[27] This "tropic of chaos" might now be included in our characterisation of *les tropiques fantôme*, and as pointing to a new seepage of 'intellectual life' into 'emotional life,' with climate, capital and chaos constituting an existential threat (see Chapter 6).

We also think that over the last 20 years there has been a shift in the texture of tropicality. The idea of neatly separated temperate and tropical zones – the normalised tropicality captured on the cover of Gourou's *Les pays tropicaux*, with these two worlds the home to fundamentally different levels and standards of comfort and wealth, and cycles of progress – has become somewhat scrambled. Gourou wrote remarkably little about the tropical fringes of the temperate north (the American South and northern Australia, for example). However, what were once deemed 'normal' attributes of tropical nature and constitutive of its environmental alterity – constant heat and moisture in some areas, pronounced wet and dry seasons in others, and vulnerability to storms, floods and droughts in yet others – are now permeating the temperate north.[28] It is now argued that natural climatic variations associated with natural *El Niño* and *La Niña* cycles of ocean warming and cooling in the tropical Pacific are become more intense – "supercharged" as a group of atmospheric scientists have recently put it – in a warming global climate, and are exacerbating a range of extreme weather events (drought, flooding, prolonged heatwaves, wildfires) in the Amazon, Arizona, California, Mexico, Australia, and parts of Africa and the Middle East, and with increasingly pronounced land-sea contrasts.[29] Some of the most severe and damaging tropical storms, floods and droughts in recorded weather history have come in the last 20 years are being directly linked with climate change. When such weather 'hits' northern latitudes it is still called 'extreme weather.' But it is now also starting to be talked about as a 'new normal.'

Martin Mahoney avers that "Global climate change . . . has arguably shifted discourses of tropicality from themes of colonial encounter to speculation about tropical invasion."[30] As he and others are now suggesting, human populations the world over "are jointly engaged in the ongoing composition of a common, climate-changed world" in which an older geographical imagination of "tropical nastiness" (disease, disorder, insecurity, scarcity, insalubrities) is supplanted by "mobile climates" in which such nastiness travels and lodges itself in places hitherto deemed distant from and antithetical to the tropics.[31] One no longer needs to travel to the sweltering and beleaguered tropics to experience – revel in, disparage, or suffer the effects of – environmental otherness and excess. Linkages between environmental change and societal discord are not confined to the tropics. The locational marker of a 'tropic of chaos' is becoming more irresolute. Core qualities of the tropicality of yesteryear, and particularly the experience of pronounced seasonality and extreme weather, and the forms of environmental calamity associated with it, have permeated the temperate zone, and are both exposing and creating deep inequalities in the ability of different temperate and tropical actors, constituencies and entities – states, cities, farmers, fire services, forests, river basins, coastlines, crop and irrigation systems – to withstand, adapt to, mitigate, or escape the nature and effects of processes that were once represented as specific to the tropics and merely parroted in the temperate north in the hot houses of botanical gardens, in museum displays, and at world exhibitions.

From a Western vantage point, climate change – although obviously not just climate change – has helped to explode the hoax that danger, difficulty and despair dwell elsewhere, in the Third World or a tropical world. In recent years there has been flurry of work on "global climate anomalies" during the twentieth century, and not least the severe European winters and prolonged *El Niño* in the tropical Pacific in 1940–1942 – at the moment when a *general* circulation model of the tropical and equatorial atmosphere was being devised (see Chapter 5).[32]

"Once we recognize the ecological risks to our economic well-being and even national security, we begin to look much harder for practical approaches to mitigating the pressures that our global society is now placing on the earth's ecosystems," Sachs mused in 2006. "We are all climate refugees now," he has proclaimed more recently – albeit some of 'us' more so than others.[33] A new global division in society is emerging between those who have and can afford air-conditioning and those who cannot. Human cells literally start to boil at 50°C – temperatures now on the horizon in more and more urban and rural environments. Air-conditioned homes and buildings are the new colonial hill stations like Simla and Dalat, which colonists (and Gourou's wife) once coveted as escapes from the burning heat and humidity, and degenerative effects, of the tropics (see Chapter 2).

Les paysans du delta tonkinois redux

In the mid-1980s the Vietnamese regime embarked on its own successful *perestroika*, and in this new context Gourou's work, which has been banished by the communist regime (as was so much other Western research on other areas around the world that assumed the mantle of communism), was rehabilitated. The driving force behind this was the agronomist Dào Thê Tuân. Already influenced by the work of the Soviet economist Alexander Shayanov, and the *Journal of Peasant Studies* (emanating from the University of London), Tuân and his colleagues used Gourou's 1936 thesis as an inspiration and guide for peasant-centred research on Vietnam. These specialists argued that agronomical research in Vietnam had forgotten the main actor in production: the peasant. The revival of interest in Shayanov and peasant studies was also explained by the crisis of socialist (collectivist) agriculture, and Chinese and Vietnamese peasants' clandestine return to family- and village-based peasant production. Tuân created a Department of Rural Systems at the University of Hanoi and launched a programme entitled 'agrarian systems of the Red River basin.' James Scott notes that *Les paysans* figured centrally in the 1970s boom in peasant studies, and 20 years on Tuân realised that Gourou had long before carried out the type of study he envisioned. *Les Paysans* was thus once more lauded as one of the first, and certainly most detailed, studies of rural-village-peasant dynamics.[34]

In 1947 the director of the École française d'Extrême-Orient (EFEO) in Hanoi, Paul Lévy, went on Radio Saigon urging that "the research time machine of our predecessors will start up again," and flagging the importance of Gourou's *Les paysans*.³⁵ But copies of the book were in short supply, and as Tuân reminisced, war and then communist collectivisation brought rural research to a swift end.³⁶ Tuân accepted Gourou's proposal, originally hatched in a colonial context, that the most useful thing that could be done for the peasant was to bring an end to the development of large properties. "Reality," wrote Tuân in 2004, "has demonstrated the rightness of this prophecy."³⁷ Just as the old top-down communist state structures were abandoned, so these agronomists consulted with peasants on how to improve their agrarian *encadrements*. This consultation made an important contribution to the so-called 'Green Revolution' in Vietnam – the success story reported on by Watts, and with Vietnam becoming the second highest rice-grower in the world.

Gourou's name therefore returned to the Tonkin Delta, and a Vietnamese edition of *Les paysans*, translated by Tuân, was published, at last, in 2003.³⁸ To some extent, it had never gone away. His daughter told us how, in 1994, she had received a postcard from anthropologist Georges Condominas mentioning a meeting with General Giáp at a reception in Hanoi: on the mention of Gourou's name, there was 'a flash of recognition' from his former pupil, Condominas reported. The translation of *Les Paysans* into Vietnamese coincided with a warming in Franco-Vietnamese relations: in 1993, President Mitterrand visited Vietnam; in 1995, an office of the EFEO reopened in Hanoi (ironically, in the Avenue Điện Biên Phủ); and in 1997 Vietnam hosted the *Sommet de la Francophonie*. Gourou died before this conference, which was planned to celebrate his hundred years, took place.

In today's Vietnam scholars influenced by Gourou are fighting another rearguard action. In a time of rapid economic growth, the city, once seen by communists and nationalists as alien to the Vietnamese peasant soul, is gaining ground as a development focus and state priority. In his final letters to Jean Gottmann, Gourou expressed regret at neglecting the geography of cities and the challenges posed to landscapes, and especially rice-growing landscapes, by global urbanisation. Contemporary Vietnam would confirm Gourou's late concerns. Despite the successes of Tuân's Gourou-influenced research, agriculture in the Tonkin Delta is being pushed to the margins in state planning and infrastructure initiatives that are literally concreting and asphalting over large parts of the delta. In vain protests against the compulsory purchase of their land, peasants brandish portraits of Ho Chi Minh. The millennia-old irrigation system is becoming fractured, while in Hanoi the Red River still broods behind its tall banks. Vietnamese specialists, as well as those attached to the EFEO, notably the anthropologists Sylvie Fanchette and Olivier Tessier, are united in their fear of an environmental catastrophe in northern Vietnam that is being brought about by break-neck economic growth and global warming, and that is marked by a growing risk of flooding and typhoons.³⁹ Echoing these concerns, General Giáp's last political offensive was to oppose the environmental despoliation of his country by foreign, mainly Chinese, investors. In such a hostile climate, Giáp and Gourou's latter-day Vietnamese acolytes might have agreed with one of Gourou's favourite quotations, from the eighteenth century French writer Bernard Le Bovier de Fontenelle, which he used as an epigraph for *Riz et civilisation*: "You need time to ruin a world, but in the end you only need time."⁴⁰

304 Fin de la tropicalité *(as we knew it)?*

Notes

1 Said addressed this emboldening, in the wake of 9/11, shortly before he died, in his Preface to the twenty-fifth anniversary edition of *Orientalism* (London: Penguin, 2003). On the reassertion of Orientalist imaginative geographies, see Derek Gregory, *The Colonial Present: Afghanistan, Palestine, Iraq* (Oxford: Blackwell Publishers, 2004).
2 Pierre Gourou, *Leçons de géographie tropicale* (Paris: Ecole Pratique des Hautes Études, 1971), 14–16.
3 Aimé Césaire, *Discourse on Colonialism* trans. Joan Pinkham (New York: Monthly Review Press, 1972), 34.
4 Pierre Gourou, *Terres de bonne espérance, le monde tropicale* (Paris: Plon, 1982), 395.
5 Yves Lacoste, "Géographie coloniale et géographie académique: approche epistémologique," in Michel Bruneau and Daniel Dory, eds., *Géographies des colonisations XV-XX siècles* (Paris: L'Harmattan, 1994), 343–348, at 345.
6 Michel and Dominique Inguimberty, Interview with the authors.
7 D. Tabutin, "African population growth: Status and prospects," *Tiers Monde* 32, no. 125 (1991): 159–173.
8 Quotation from Michael J. Watts, "Recombinant capitalism: State, de-collectivisation and the agrarian question in Vietnam," in John Pickles and Adrian Smith eds., *Theorizing Transition: The Political Economy of Post-Communist Transformations* (London and New York: Routledge, 1998), 425–478, at 456. Also see Sophie Devienne, "Red River Delta: Fifty years of change," *Moussons* 9–10 (2006): 255–280.
9 A point emphasised by Michel Bruneau, "Pierre Gourou (1900–1999). Géographie et civilisations," *L'Homme* 153 (2000): 1–25, at 20.
10 Georges Balandier, "General report on the roundtable organised by the International Research Office of the Social Implications of Technological Change (Paris March 1954)," *UNESCO International Social Science Bulletin* 6, no. 3 (1954): 372–384, at 373.
11 For a brief overview of whether genocide today any more fully within geography's ambit, see Jeff Stonehouse, "Genocide and the geographical imagination: Life and death in Germany, China and Cambodia, by James Tyner," *Genocide Studies and Prevention* 8, no. 2 (2014): 85–86.
12 Homi Bhabha, "The humanities and the anxiety of violence," Keynote address to the 'Changing the Humanities/The Humanities Changing' conference (2009), University of Cambridge, http://sms.cam.ac.uk/media/1184691.
13 Pierre Gourou, *Les aventures de Kataboum* (Paris: L'Harmattan, 1997), 23.
14 Baruch Boxer, "Review of *Man and Land in the Far East*," *Journal of Asian Studies* 36, no. 3 (1977): 539–540, at 539.
15 Pierre Gourou, "Sur l'agriculture dans le monde quelques considerations géographique," in *Les sciences sociales: Problems et orientations* (Paris: UNESCO, 1968), 22–31.
16 Jeffrey D. Sachs, "Tropical underdevelopment," NBER Working Paper No. 8119 (2001), 1, http://earth.columbia.edu/sitefiles/file/about/director/documents/nber8119.pdf.
17 Sachs, "Tropical underdevelopment," 1–4. Also see Jeffrey Sachs and Pia Malaney, "The economic and social burden of malaria," *Nature* 415 (2002): 680–685; Jeffrey Sachs, "Climate and refugees," *Scientific American* 296 (2007): 43.
18 Jeffrey D. Sachs, *Common Wealth: Economics for a Crowded Planet* (London: Penguin, 2008), 216–217. For more on this debate, and the bandying about of accusations of environmental determinism, see W.B. Meyers and D.M.T. Guss, *Neo-Environmental Determinism* (London: Palgrave Macmillan, 2017), 40 and *passim*.
19 Henri Nicolaï, for example, took issue with Diamond's account of the swift demise of Mayan Civilisation on account of prolonged drought and concomitant deforestation (partly for wood to fuel the fires that generated the lime plaster used in ornate city constructions). Diamond "refutes, without providing a lot of arguments, the idea that Mayan agriculture was essentially a slash-and-burn agriculture; and while he

Fin de la tropicalité *(as we knew it)?* 305

recognizes that Mayan . . . agriculture was not intensive enough to meet the needs of the cities, he ignores Gourou's argument in *Les pays tropicaux* that the Mayan's relatively sophisticated *techniques d'encadrement* were tied to relatively rudimentary techniques of production." The upshot, he continued, was that when population increased, the response was to curb the extent and duration of extant fallow land, which precipitated soil erosion, the eventual abandonment of land, and the break-up of Mayan cities and society. In short, Diamond constructed a much to narrow environmental explanation. Henri Nicolaï, "Jared Diamond," *Effondrement Belgeo* 4 (2006): 493–497, at 495.
20 Also see Jeffrey D. Sachs, "Institutions don't rule: Direct effects of geography on per capita income," NBER Working Paper No. 9490 (2003), http://citeseerx.ist.psu.edu/viewdoc/download?doi=10.1.1.202.2922&rep=rep1&type=pd
21 Pierre Gourou, "Le déterminisme physique dans 'l'esprit des lois'," *L'Homme* 3, no. 3 (1963): 5–11, at 10.
22 Andreas Malm, *The Progress in This Storm: Nature and Society in a Warming World* (London: Verso, 2018).
23 Christian Parenti, *Tropic of Chaos: Climate Change and the New Geography of Violence* (London: Verso, 2011), 6, 9, 11.
24 Parenti, *Tropic of Chaos*, 6–12.
25 Greg Bankoff, "Rendering the world unsafe: 'Vulnerability' as western discourse," *Disasters* 25, no. 1 (2001): 19–35, at 19; also Greg Bankoff, "Regions of risk: Western discourses on terrorism and the significance of Islam," *Studies in Conflict and Terrorism* 26 (2003): 413–428.
26 Parenti, *Tropic of Chaos*, 10–13.
27 Jeffrey Sachs, "Ecology and political upheaval," *Scientific American* 295 (2006): 37. For example, Parenti, *Tropic of Chaos*, 141, largely accepts Wittfogel's assessment of "water's political impertatives," and without any mention of the streak of environmental determinism that Gourou found in it. That the locations of Bangladesh and Burma are reversed on the map with which he introduces his assessment of 'Asia' does not bode well; nor does the absence of Rwanda – for Gourou and others an epicentre of many of things Parenti seeks to address – in his survey of 'the rise and fall of East African states.'
28 There is also intense debate about whether the increasing preponderance of severe weather – monsoon flooding and drought – in India and other parts of tropical Asia in recent decades can be attributed to climate change (increasing human carbon dioxide emissions) or is being driven by natural variability in regional and global climate systems. For a controversial discussion, see Madhav Khandekar, *Floods and Droughts in the Indian Monsoon: Natural Variability Trumps Human Impact* (London: The Global Warming Policy Foundation, 2014).
29 J.T. Fasullo, B.L. Otto-Bilesner and S. Stevenson, "ENSO's changing influence on temperature, precipitation, and wildfire in a warming climate," *Geophysical Research Letters* (2018), https://agupubs.onlinelibrary.wiley.com/doi/10.1029/2018GL079022.
30 Martin Mahoney, "Picturing the future-conditional: Montage and the global geographies of climate change," *Geo: Geography and Environment* 3, no. 2 (2016): 1–18, at 13–14.
31 Mahoney, "Picturing the future-conditional," 14. Also see Matthias Heymann, "The evolution of climate ideas and knowledge," *WIRES Climate Change* 1 (2010): 581–597; Mike Hulme, "Cosmopolitan climates: Hybridity, foresight, and meaning," *Theory, Culture & Society* 27, nos. 2–3 (2010): 267–276.
32 See, e.g., Stefan Brönnimann, "The global climate anomaly, 1940–1942," *Weather* 60, no. 12 (2005): 336–342.
33 Jeffrey D. Sachs, "We are all climate refugees now," *Project Syndicate*, 2 August 2018, www.project-syndicate.org/commentary/climate-change-disaster-in-the-making-by-jeffrey-d-sachs-2018-2008.
34 James C. Scott, *The Moral Economy of the Peasant* (New Haven, CT: Yale University Press, 1976), 20–25.

35 AEFEO: P74 Paul Lévy (1947–1949) Paul Lévy à Radio Saigon, April 1947.
36 AEFEO: P74 Paul Lévy to Pierre Gourou, 18 December 1948; Dào Thê Tuân, Interview with the authors.
37 D.T. Tuan, "Le pionnier des études rurales vietnamiennes," in Christophe Gironde and Jean-Luc Maurer eds., *Le Vietnam à l'aube du XXIe siècle. Bilan et perspectives politiques, économiques et sociales: mélanges pour commémorer le 100e anniversaire de la naissance de Pierre Gourou* (Paris: Karthala, 2004), 24–27, at 24.
38 Pierre Gourou, *Người Nông Dân Châu Thổ Bắc Kỳ* trans. D.T. Tuan (Hô Chi Minh City: Tre., 2003).
39 Sylvie Fanchette, Interview with the authors; Sylvie Fanchette, "Dynamiques du peuplement, libéralisation économique et décentralisation dans les deltas de l'Asie des Moussons," *Revue Tiers Monde* t. XLV, no 177 (2004): 179–205, at 202; Sylvie Fanchette, "De l'importance des liens géographie physique/géographie humaine pour comprendre les risques de submersion des deltas surpeuplés," *Hérodote* 121 (2006): 6–18; Olivier Tessier, Interview with the authors; Jean-Philippe Fontenelle et Olivier Tessier, "Pression démographique et contraintes politiques: la paysannerie du delta du Fleuve Rouge dans la tourmente du XXème siècle," *Autrepart* 3 (1997): 25–43.
40 Pierre Gourou, *Riz et civilisation* (Paris: Fayard, 1984), 8.

Index

Note: Page numbers in *italics* indicate figures on the corresponding pages.

absentee landlordism 143
Acheson, Dean 43–44
Adams, Frederick 39
Adisa, Jinmi 280
aerial photography 114–119, 178n162
affect 215
affective history 209–210, 212–213; *see also* Febvre, Lucien
Africa: agriculture in 226–227, 295; changing perceptions of 233–234; colonial empire in 207–208; colonial labour in 40; colonial research in 222–223; decolonisation in 226, 247, 252–253; development in 241–242, 285; ecology in 244; famine in 279, 282; French presence in 208, 230; geography in 244, 250–251; Gourou's fieldwork in 217, 224, 226; legacies of colonialism in 245, 277, 278–279; population density in 227, 271, 273, 275, 277, 282; population growth in 295; post-World War II 230; scholarly analysis of 230–232; scientific research in 163; segregation and racial geography in 269–270; UNESCO conference 247; urbanisation in 250; see also *individual African countries by name*
Africa: A Study in Tropical Development (Stamp) 230, 237
agriculture: in Africa 226–227, 295; coffee plantations 270; and colonialism 38; indigenous 239, 267, 269; industrial 14; modernisation of 138, 213, 267, 273; monoculture 14; over-specialisation in 242; rice cultivation 57, 90, 95, 105, 106–108, 144, 227, 240; sugar plantations 270; in the Tonkin Delta 303; traditional 116; *see also* hydraulic projects; irrigation
Algeria 41, 45

Allewart, Monique 17
alterity 13, 119, 301
Amazon, The: A New Frontier? (Hanson) 16
Amazon Valley 228, 292–293
America *see* United States
Amin, Samir 249
Anderson, Benedict 160
André, Max 196
André-Pallois, Nadine 65
Angkor Wat 54, 56, 99, 114, 117
Angola 221, 270
Annales (Febvre) 220
Annam 44, 57, 144
anthropology 242, 266
anti-colonialism 1, 4, 47, 48, 53, 80n71, 140, 142, 215, 216, 230, 292; *see also* colonialism
anti-conquest 37
anti-semitism 188, 254n30
Appadurai, Arjun 280
appropriation 7, 215
Archives du Collège de France, Paris (ACDF) 10
Archives de l'Université Libre de Bruxelles (AULB) 10
Arnold, David 4–5, 9, 10, 11, 12, 13, 16–17, 18, 27, 102, 214, 215, 216, 249, 283, 299
Aron, Raymond 216
art: colonial 65–66, 69; naïve 101; primitive 101; Vietnamese 69
assimilation 45
Associated State of Vietnam 69; *see also* Vietnam
Au, Sokhieng 155
Aufrère, Léon 180
Aymé, Marcel 192

Bach, Ulrich 59
Balandier, Georges 6, 247, 265–266, 295

308 *Index*

Ball, MacMahon 156
Balogh, Brian 145
banana wars 300
Bandung Conference 246
Bankoff, Greg 300
Bao Dai 42, 45, 50
Barnes, Trevor 21, 23
Barnett, Clive 18
Bart, François 249, 278
Basch, Victor 150
Bashford, Alison 56, 154, 168
Bates, Marston 216
Bayly, Susan 110
Beebe, William 16
Beilkin, Jordanna 229
Belgian Congo *see* Congo
Belgian Ministry of Colonies 270
Benveniste, Émile 217
Bernard, Paul 40, 144, 158
Bhabha, Homi 276, 277, 283, 296
Biggs, David 103, 117
Binh, S. E. Pham Van 50
biodiversity 39
biography: and geography 20–21; of Gourou 20–26
bio-politics 139, 141
biopower 141
bio-tropicality 154–160
Blache, Jules 186
Bloch, Marc 183–184, 206, 208–210, 212–213, 224, 296, 298–299
Blum, Léon 147
Boemers, Heinz 202n54
Bonneuil, Christophe 38
Borde, Jean 201n41
Bordeaux Liberation Committee 188–191
Boserup, Ester 227
botanical gardens 38
Bouvier, René 58, 158
Bowman, Isaiah 164, 182, 241, 242, 246
Boxer, Baruch 296
Braudel, Fernand 37–38, 68, 184, 186, 210, 212, 251, 287, 296
Bray, Gilberte Gourou 61, 125
Brazil 164, 227, 248, 286
Bréelle, Dany 98
British Empire Exhibition (1924) 41
British Groundnut Scheme 269
Brocheux, Pierre 53, 143, 148
Broek, Jan 6
Browne, Janet 2
Bruhnes, Jean 114, 124
Bruneau, Michel 21–22, 107, 210, 217, 249, 251, 285
Buchanan, Keith 245–246

Buchanan, Sherry 73
bureaucratisation 15
Burguière, André 213
Burma 56, 62, 229–230, 235, 241
Burrin, Philippe 187
Burundi 268, 270–273

cadastral surveys 109
Caillier, René 189, 202n63
Cambodia 40, 44, 45, 50, 55, 63
capitalism 14, 17, 207; in Indochina 38–40; and tropicality 38
Caribbean 207–8
Carter, Paul 93
Central Africa 150, 274; *see also* Africa
Central African Federation 241
Centre d'Etudes de Géographie Tropicale (CEGET) 217
Centre d'études de politique étrangère (CEPE) 140, 152, 153, 166
Centre médical de l'Université libre de Bruxelles en Afrique centrale (CEMUBAC) 217
Centre national de la recherche scientifique (CNRS) 217
Césaire, Aimé 1–5, 12, 27; anticolonial writings of 9, 244, 246; critique of Gourou 1–4, 17, 18, 87, 169, 210, 224, 242, 296, 297; *Discourse on Colonialism* 1, 5, 9, 216; editing *Tropiques* 185
Chafer, Tony 149
Chakrabarty, Dipesh 251
Chapman, Peter 39
Chevalier, Auguste 228, 269
Chiang Kai-shek 43
China 43–44, 45, 89, 96–97, 166–167, 195, 205n112, 241, 284
Chinh, Truong 143, 159
Chrétien, Jean-Pierre 277
Christen, Xavier de 56
citizenship 45–46
civilization: African 250, 251; agrarian 241, 293; agricultural 156; arrested 137; Asian 220, 227; binaries of 149; colonial 66; determinism of 214; efficiency of 167; Egyptian origins of 3; European 2–3, 38, 47, 212, 234; Far Eastern 6, 212, 240, 293; of France as superior 198; Gourou's assessment of 1–3, 20–21, 91, 104, 122, 126, 220–221, 227; history of 207; in Indochina 50, 53, 65, 94, 102; interaction with geography 181, 214; Khmer 54; landscape moulding by 192; linked to literary

style 24; Mayan 219; in the Orient 11; peasant 124, 156, 297; plant-based 166, 169, 293; pre-capitalist 242; rural 240, 251; superior 2, 99, 220, 227, 234; and technology 93; traditional 91, 125, 144; tropical 1, 214, 219, 220, 227, 242, 292, 293, 297; vegetal 166, 169, 293; Western 12, 92, 98, 252, 294; and Western development 207, 209
civilising mission 2, 41, 70, 137, 143, 159, 271
Claval, Paul 21, 105, 164, 248
Clifford, James 119
climate-induced chaos 300–302
Clout, Hugh 146, 163
Cochinchina (Saigon) 40, 44, 47, 57, 89, 144
Cohn, Bernard 160
Cold War 138, 207, 221, 225, 233, 267, 283, 300
Collège de France 218
colonial discourse 5, 37, 99, 120, 137, 185, 216, 257n98, 275
colonialism 1, 3, 14, 17, 27, 48, 126, 138, 155, 195, 198, 214; in Africa 245; Belgian 278; and diet 42; European 222; French 40, 57–58, 75; historical geography of 242; interventionist 38; justification for 89; legacies of 245, 277, 278–279; 'monsoon' 58–59; problems of 265; in Rwanda 275; and science 163; Western academic engagement with 265; *see also* anti-colonialism; post-colonialism; postcolonial theory
colonial racism 185, 207; *see also* racism
colonial reform 155, 159, 166, 174n68; and international networks 147–154
colonial violence 37, 87, 98, 143, 159, 185, 207
colonisation 11, 15, 41, 76, 155, 188, 209, 215–216, 294; *see also* decolonisation
colonising gaze 9, 16, 27, 297
colour bar 15, 269–270
Comité d'études des problèmes du Pacifique 197
Committee of Amnesty and Defence of the Indochinese People 147
communism 182; in China 43; in France 49, 148, 188, 198, 208; in Indochina 43–4; in Vietnam 171n23, 210
Communist Manifesto (Marx & Engels) 207
Condominas, Georges 107, 198, 238, 286
Congo 266, 270–273, 298; geographical studies of 273–276; maps of 271

Congo Basin 228, 266; *see also* Congo
Conklin, Alice 48, 161, 163
Connell, Raewyn 168
Conrad, Joseph 300
conservation 38, 39, 239–240, 252
Cooper, Frederick 40
Cooper, Nicola 55
Copin, Henri 107
Cordemoy, Pierre 42, 109
Corps Expéditionnaire Français en Extrême-Orient (CEFEO) 61, 64
Courade, Georges 9, 23–24, 107, 249, 285
Critique of Everyday Life (Lefebvre) 207
Cuba 39
Cullather, Nick 225
Cunha, Euclides de 27
Curtin, Philip 238

Dai Nam 44
Dainville, François de 5
Dakar conference 195–196
Dalat negotiations 196–197, 198
Dambaugh, Luella 236
d'Argenlieu, Georges Thierry 196
Dash, Michael 185
Davis, William Morris 182
decolonisation 9, 13, 18–19, 42, 87, 139, 208, 212, 221–222, 228–229, 252; in Africa 226, 247, 252–253; in Asia 268; dialogical side of 251; geographical problems associated with 236–237; in Indochina 36; of knowledge 19; in Malaysia 234; in Rwanda 275
deforestation 39, 165, 227, 304n19
De Gaulle, Charles 183, 188, 191
de Heusch, Luc 282
dehumanisation 1–2
de Lattre de Tassigny, Jean Marie 64
De Lattre Line 64, 70
Delaunay, Gabriel 191
Delavignette, Robert 199n7, 223, 238, 252, 270
Demangeon, Albert 100–101, 104, 126, 140, 146–147, 152, 156, 168, 169, 180, 181, 182, 184, 187, 200n10, 212–213, 297; denunciation of Germany by 181, 182; review of *Les paysans* 146, 158
Democratic Republic of Vietnam (DRV) 36, 42, 43, 44, 73, 196
Descartes, René 24, 296
determinism 207, 214, 223, 278, 287; climatic 223, 228; environmental 104, 207, 220, 228, 253n4, 273, 298–300, 305n27; geographical 59, 234; physical 72, 169, 220; racial 159, 223

310 *Index*

development 19, 266; in Africa 285; agricultural 239, 250; economic 225; post-colonial 266
Deyasi, Marco 54
Diamond, Jared 299
Điện Biên Phủ 44, 72–74, 229, 247
Dieu, Nguyen Thi 45
Dieulefil, Pierre 54
dike building 57, 61, 62, 64, 68, 71, 90, 92, 94, 95, 103, 107, 114–118, 122, 151, 161, 166, 284, 286; American bombing of dikes 284
Dion, Roger 23, 192, 286
Diop, Cheikh Anta 3
Discours sur le colonialisme (*Discourse on Colonialism*) (Césaire) 1, 5, 9, 216
disease: in Africa 235; AIDS 250; beriberi 59; cattle diseases 274; cholera 41, 59, 63; digestive disorders 62; dysentery 41, 59, 62; endemic 59, 225; European 228; factors contributing to 29n14, 255n37; imported 228; in Indochina 38, 59–60, 62–63; intestinal parasites 59; in Jamaica 236; in Japan 240; leprosy 49, 63; malaria 41, 59, 62, 274; in the Philippines 39; respiratory disorders 62; of the sense organs 63; skin diseases 62; sleeping sickness 162; squalor as 157, 160; trachoma 63; tropical 2, 5, 6, 11, 16, 39, 41, 59, 59–60, 62–63, 75, 137, 155, 162, 168, 219, 229, 244, 274, 300; and tropical development 299; tuberculosis 49, 59, 63; typhoid 59; typhus 62; venereal disease 59, 62; in Vietnam 43; yellow fever 41, 196
divination 123
Dobby, E.H.G. 237
Donnadieu Duras, Marguerite 57, 58, 160
Doumer, Paul 53, 160, 162
Dresch, Jean 187, 284
Driver, Felix 11, 15, 17, 18, 27, 90, 117
Dubois, Marcel 96
Duchartre, Pierre-Louis 50, 51, 53–54, 55
Dufaux, Frédéric 108
Dumont, René 158, 242
Duncan, James 16
Dupuy, Paul 126
Duras, Marguerite Donnadieu 57, 58, 160
Dyce, Matt 114

École Coloniale 152
École des beaux-arts de Hanoi 65, *68*, 73
École française d'Extrême-Orient (EFEO) 50, 53, 75, 101, 109, 110, 145, 150, 152, 161, 164, 168, 182, 196

École nationale de la France d'Outre-Mer 181, 182
Edwards, Kathryn 75
El Dorado 215
El Shakry, Omnia 101
emotion, tropical 214–217; *see also* affect; affective history; Febvre, Lucien
empire, ends of 13–20
encadrements 91, 93, 122, 138, 192, 195, 219–221, 226, 231, 239, 240, 247, 250, 279, 283, 293, 296, 303, 304–305n19
Encyclopedia of the Belgian Congo and Ruanda-Urundi 273
Engels, Friedrich 207
Enright, Kelly 99
environmental determinism 104, 207, 220, 228, 254n4, 273, 298–300, 305n27
environmentalism 39
ethnicity, in Rwanda 273, 275–283
ethnocentrism 214
Eurocentrism 14, 19, 169; environmental 2; zonal 214
Europe, population density 105
Evans, Martin 229
exoticism 15, 24, 38, 48, 60, 88, 107, 119, 273, 285, 297
expertise 127, 140, 150, 151, 154, 161, 162, 165, 245; academic 28, 294; comparison-network 139, 160–165; Gourou's 145–146, 196, 208, 224, 252, 266; modern 142; networks of 141; patronage and 145–147; scientific 269; Western 28
Exposition Coloniale 54, 58, 66, 69, 98, 99, 100, 116
Exposition International des Arts Décoratifs et Industriels Modernes 66
Ezra, Elizabeth 99

Fabian, Johannes 160
Fall, Bernard 61, 74
famine 174n72, 300; in Africa 279, 282; in Vietnam 42, 57, 62, 110, 144, 148, 149, 150, 151, 157, 174n72
Fanchette, Sylvie 285, 303
Fanon, Frantz 9, 41, 87, 211, 215, 216, 228–229
Febvre, Lucien 101–103, 104, 123–124, 146–147, 184, 187, 206–216, 220, 237, 242, 297, 298; and decolonization 208, 211–212, 238; geographical conditions 213–214, 237–238, 242; reason vs emotion 212, 238; the tropics as affective domain 102, 146, 206–207, 208, 210, 211, 212, 214, 215, 220,

237–238; World War II 37, 207–208, 210; *see also* affect; emotion, topical
Fire in the Lake (FitzGerald) 111
Fisher, Charles 233–234, 244
FitzGerald, Frances 110
Fletcher, Robert 139
Fontainebleau conference 196–197
Fontenelle, Bernard le Bovier 24, 303
Forde, Daryll 228
forestry 38
Foucault, Michel 125, 141, 154
Fournier-Guérin, Catherine 249
France: in Africa 208, 230; colonialism of 40; colonisation of Indochina 45–50; empire (1919–1939) *71*; liberation of 48; military presence in Indochina 61–65, 69–70, *71*, 72–74, 79n51; and scientific research in the colonies 160
France Illustration 76n1
France Illustration special issue (June 1949) 36–38, 75; article on Bao Dai (by Pham Van Binh) 50; as bid for public support for war 44–45; contemporary writings 38–40; de Christen's article 56; Duchartre's article 50, *51*, 53–55; Gourou's article on École des beaux-arts de Hanoi 65, 68–69, 87; Pignon's article 49–50; Sion's article 55–56, 59; tropicalist imagery *68*; Valluy's article 61; *see also* Vichy regime
France-Liberté 202n54
France-Tireur 202n54
Frank, André Gunder 249
French Africa 208; *see also* Africa
French Antilles 248
French Communist Party (PCF) 49, 148, 198, 208
French India 150
French Indochina *see* Indochina
Freud, Sigmund 141
Freyre, Gilberto 7
Friedman, Susan 209

Gallais, Jean 274
Gallocentrism 214
Gallois, Lucien 96, 126
Gambiez, Fernand 72
Gandhi, Leela 47, 216, 217, 251
Gauguin, Paul 66–67
Geertz, Clifford 242
Gelzer, Gregory 109
General Atlas of the Belgian Congo and Ruanda-Urundi 273
genocide in Rwanda 276–279

Gentleman in the Parlour (Maugham) 56
geodesic triangulation 109
geographers: American 223; anti-colonialist 294; French 27, 186; German 182, 188, 200n10, 223
geographical warfare 284
Géographie de la décolonisation (Isnard) 236
Géographie Universelle projects 178n162
geography: in Africa 250–251; behavioural 235; and biography 20–21; British 248; colonial 20, 181; comparative 297; and the ends of empire 13–20; expansionist 2; in France 114; French 126, 146–147, 170, 246, 249, 266; German 182; ideological differences in 286; original 93, 98; of peace/war 125–126; as performance 17, 120; professionalisation of 18; romantic 107; and the search for order 232–233, 237, 240–241, 246; tropical 170, 181, 186–188, 215, 218–219, 220, 246, 248; and tropicality 13–20, 248; of the tropics 26–27; zonal 6, 170, 237
Geography of Life and Death, The (Stamp) 234–235
geomancy 123
geo-politics 88–89
Geopolitik 181
George, Pierre 180
German Geographical Society 184
Germany: and the Holocaust 280; invasion of Poland 183; occupation of France 213; in Paris 184
Giáp, Trần Văn 182
Giáp, Võ Nguyên 43, 44, 61, 64, 65, 75, 103, 121, 143, 159, 196, 197, 198, 286, 303; meeting with Gourou 72–73
Gibb, Hamilton 17
Gide, André 150
Gilbert, Adrian 64
Gilbert, Etienne 25
Girardet, Raoul 48, 106
Giraude, Marcel 119
global warming 300–302
Godard, Justin 149, 151, 157, 158
Godart, Justin, 151, 157, 158, 159, 168
Golooba-Mutebi, Frederick 278
Gorman, Daniel 140, 180
Goscha, Christopher 45, 63, 73
Gottmann, Jean 9, 183–184, 187, 218, 241, 250, 303
Gould, John 244
Gourou, Barrion Hélène-Georges 61
Gourou, Gilberte (Bray) 61, 125

312 *Index*

Gourou, Pierre: on Africa 224, 226, 252–253; appointment to ULB 180–182, 183; assessment of civilization by 20–21, 91, 104, 122, 126, 220–221, 227; assessment of Rwanda 273–276; biographical focus on 20–26; with CEMUBAC 270–271; Césaire's critique of 1–4, 17, 18, 87, 169, 210, 224, 242, 296, 297; chronology of career 26, 217–221; on colonialism 266; on the 'colour bar' 15, 269–270; conference at Fontainebleau 196–197; on decolonisation and development 267–269; on decolonisation in Africa 247; education 95–97; election to Collège de France 210; as expert on tropics 208–209; fieldwork for *Les paysans du delta tonkinois* 103–104; fieldwork in Africa 217, 225, 226; friendship with Braudel 38; on geography 249; on the Guernut Commission 150–152, 154–160; in Hanoi 61; with IFAN in Dakar 195–196; impressions of his work 18, 242, 244, 245, 249, 251–252, 266–267; indigenist position of 102–103; in Indochina 74–76; on Indochina 287; on the Indochina War 88; influences on 251; at IPR conferences 194–195, 222; on Japan 240–241, 292; late colonial research 221–225; meeting with Giáp 72–73; military service 98–99, 183; missions 217–221; negotiations with Giáp in Dalat 196–197, 198; on peasants of the Far East 214; on peasants of the Tonkin Delta 124–127, 144–145, 156–160; personality of 22–24; photo 25; and the politics of knowledge 298–289; post-war missions and 'system' 217–221, 295; post-war scholarship 211–214; recommended to the Assembly of Professors 206; relationship with Inguimberty 65, 67–69; on the responsibility of the geographer 285–286, 296–297; on the Rwandan genocide 279–283; as scholar 13; on scientific study of the colonies 163–165; Statford conference 197; students of 21–22, 23, 25, 186, 218, 231, 271; 2as teacher 96; teaching at Bordeaux 187; teaching in Hanoi 65, 96; on technology 93; thoughts on writers and writing 24; in the Tonkin Delta 27–28, 74, 140, 146, 152, 283–284; on the Tonkin Delta 68–70, 76, 93–98, 292–293; and the Tonkin Delta peasants 124–127; tropicality of 298–302; on the tropics 246, 292–293; and twentieth century tropicality 27; on urbanisation 251; use of aerial photography by 114–119; use of comparative analysis by 165–170; in Vietnam 72; and the Vietnam War 284–285, 286; wartime research 184, 187–188; work for the Resistance 186, 188–194

Gourou, Pierre, writings of 9–10; *La densité de la population au Ruanda-Urundi* 274; *L'Afrique* 217, 221, 238; *L'Asie* 217, 218, 220; *La terre et l'homme en Extrême-Orient* 166–169, 212–213, 240; *L'avenir de l'Indochine* 197, 198; *Les aventures de Kataboum* 296; *Le Tonkin* (guidebook for Exposition Coloniale) 99–100; *L'utilisation du sol en Indochine française* 154, 160, 166; "My Tropicalist Orientation" 199; *Pour une géographie humaine* 219; "Qu'est-ce que le monde tropical?" 137–138, 142; *Riz et Civilisation* 218, 287, 303; *Terres de bonne espérance* 23, 286–287, 293; *The Tropical World* 7, 230; see also *Les paysans du delta tonkinois* (Gourou); *Les pays tropicaux* (Gourou)

Gregory, Derek 20
Griaulle, Marcel 152
Groslier, Bernard-Philippe 114
Groundnut Scheme 269
Grove, Richard 39
Guernut, Henri-Alfred 150
Guernut Commission 140, 150–152, 154–160, 163, 166, 168
Guha, Ranajit 49–50, 268
Gulick, Anne 3
Gunn, Geoffrey 57, 148, 149

Ha, Marie-Paul 60
habitat destruction 39
Haffner, Jeanne 89, 117
Haiphong 64
Hale, Dana 54
Hall, Stuart 27
Hanoi 59, 64, 94, 96, 149
Hanson, Earl Parker 16
Haraway, Donna 3, 4, 17, 19–20
Hardy, Andrew 144, 158, 159

Hastings, Max 62
Haushofer, Karl 181
Hawaii 39
Healey, Kimberly 55
Hecht, Susanna 7, 27, 252
Heckscher, August 230
Heidegger, Martin 90, 91, 92
Hémery, Daniel 53, 143, 148
Henry, Yves 89
Hérodote (journal) 285
Hiernaux, Jean 281
Hilling, David 236–237
history: Braudel's three levels of 37; colonial 20; and the ends of empire 13–20; and geography 13–20; spatial 93; *see also* affective history; Febvre, Lucien
Ho Chi Minh 44, 47–48, 49, 53, 111
Hodder, Brian 242
Hodder, Jake 20
Holland, William 195
Hoyle, Brian 236–237
Huard, Pierre 181
Huizinga, Johan 212
humanism 22, 98, 100, 125, 149, 150, 165, 166, 198, 279, 293–294, 296–298; colonial 45, 49, 60; French colonial 147; proleptic 276, 296
Hunt, David 126
Huntington, Ellsworth 234, 259n132
Huyên, Nguyễn Văn 182
hydraulic projects 57, 62, 68, 94, 105, 117, 149, 154, 284; *see also* irrigation
Hyman, Paul 183

identity: Caribbean 185; European 10; formation of 277; French 181; geographical 165; of geography 146; Henri Lopes' 'three identities' 277; for Indochina 74; power and 75; regional 104; of the Tonkin Delta 94; tropical 107
Ihde, Don 92
Image of Africa, The (Curtin) 238
imagerie d'Epinal 53–55, 56
imperialism 45, 296; and colonial art 66; French 45–50; and tropicality 38
independence movements 216
India 15; British occupation of 268; population growth in 241
India and Pakistan (Spate) 237, 244
Indian Ocean 248
indirect rule 45, 266, 282
individualism 60, 110

Indochina 27, 41, 150, 241, 298; as Asian 50–55, 74–76; climate and geography of 55–57, 68, 108–110; and colonial art 65–67; demonstrations and marches in 142–143; as French geo-political construct 198; French military in 61–65, 69–70, 71, 72–74, 79n51; French settler community in 47; Gourou's opinion of 88–89; Japanese occupation of 196; Maugham's travels in 56; monsoon environment of 55–61; natural pests in 56; as Oriental 'other' 53–54; popular images of 50–55; travel advertisements 46; as tropical 55–61, 74–76; tropicalisation of 36–37, 38; *see also* North Vietnam; South Vietnam; Vietnam
Indochina War 36, 42–43, 76n2, 185; diseases among troops 62–63; effect on civilians 63; French casualties 64–65, 78n44; and the Tonkin Delta 67–68; *see also* Vietnam War (American)
Indochinese Communist Party (ICP) 142, 159, 171n23
Indochinese Congress 168
Inguimberty, Joseph 57, 65, 66, 67, 69, 70, 75, 204n104, 294, 298
Institut d'Ethnologie 152
Institute for Scientific Research in Central Africa 273
Institute of National Parks of the Belgian Congo 273
Institute of Pacific Relations (IPR) 140, 153–154, 166, 192, 218, 298; Hot Springs Conference 194–195, 222; Kyoto Conference 251; Lucknow Conference 222, 237, 240; Stratford-upon-Avon Conference 222
Institut Français d'Afrique Noire (IFAN) 164, 248; jury in Dakar 195–196
Institut indochinois pour l'étude de l'homme 181, 182
Institut Royal Colonial Belge 218
International Geographical Congresses 181
International Geographical Union (IGU) 181, 184, 218
internationalism 22, 48, 139, 140–141, 153–154, 169, 297, 298
International War Crimes Tribunal 284
irrigation 2, 56, 57, 107, 122, 149, 151, 157, 159, 231, 237, 240, 250, 284, 287, 301, 303; *see also* hydraulic projects
Isnard, Hildebert 236

Jacobs, Jane 20
Jacques, Jules 25
Jamaica 236
Jammes, Ludovic 54, 55
Janes, Lauren 163
Japan 53, 208, 240–241, 292; occupation of Indochina 49, 62, 68, 73, 78n40, 196, 204n91, 212
Jardin d'Agronomie Tropicale 38
Jay, Martin 115
Jennings, Eric 16, 57, 60
jouissance 94
journalists, Vietnamese 80n71
Judt, Tony 193–194, 223
Julien, Charles-André 147
jungle capitalists 39

Kenya 269
Khérian, Grégroire 158
Khoan, Nguyễn Văn 109, 110, 121, 122, 182
Kiernan, Ben 278
Kimble, George 230–231, 233, 235, 237, 245, 259n132
Kingsley, Susan 99
Kirchherr, Eugene 218
Kirk, William 235–236
Kleinen, John 109, 121, 122, 143
knowledge: academic 152, 277; amateur 160; colonial 50, 117; complicity with power 121, 140; expert 4, 53, 152; geographical 62, 153, 222, 231, 233, 259n130, 273, 279; impure 293; indigenous 160, 239; local 39; and location 293; medical 62; Metropolitan 4; politics of 298 298; and power 53, 90, 121, 140, 259n30, 298; production of 11, 50, 141, 222; professional 160; scientific 229, 299; spaces of 108, 109, 121, 147; traditional 110, 127, 269; transnational 143; tropical 3, 14, 185; Western 88, 160–161, 295
Knox, Katelyn 48
Koreman, Megan 193–194
Kyoto Conference (IPR) 251

La 317ème Section (Schoendoerffer) 72
Laborde, Edward 7
labour exploitation 143
Lacoste, Yves 116, 284, 285–286, 287, 292, 293–294
La densité de la population au Ruanda-Urundi (Gourou) 274

La France: Tableau géographique 114
L'Afrique (Gourou) 217, 221, 238
L'Afrique fantôme (Leiris) 249
Lamarchand, René 277
La Méditerranée et le Monde Méditerranéen à l'Epoque de Philippe II (Braudel) 37
land alienation 142
Landes, David 299
Laos 44, 45, 50, 55, 64, 149
Laserre, Guy 218
L'Asie (Gourou) 217, 218, 220
La terre et l'évolution humaine (Febvre) 104, 207, 214
La terre et l'homme en Extrême-Orient (Gourou) 166–169, 212–213, 240
Lattimore, Owen 208
Laval, Pierre 163
L'avenir de l'Indochine (Gourou) 197, 198
La Vérité sur les Colonies 99
League of Nations 138, 163, 182, 189, 270
Lee, Douglas 218, 242
Lefebvre, Henri 101, 102–103, 123, 207
Leiris, Michel 152, 249
L'Empire français (Roques and Donnadieu) 57
Le Réunion 150
le Roux, Hannah 162
Les aventures de Kataboum (Gourou) 296
Les damnés de la terre (Fanon) 211, 228
Les paysans du delta tonkinois (Gourou) 88, 165, 284, 286, 302–303; conclusion to 147; Demangeon's review of 146, 158; English translation of 284; Gourou's relationship with peasants 124–127; Mus's margin notes 110, 119, 135n188; and the objectivity of the geographer 121–122; paternalist view of 123; preface to second edition 98; on the Tonkin peasants 101–106; Vietnamese edition 303
Les pays tropicaux (Gourou) 1, 87, 91–92, 165, 196, 218, 219, 223–225, 228, 237, 250, 266, 286, 297; cover 8; first edition 242, 245, 295; on indigenous agriculture 267; popular reception of 5–10; third edition 226; translation of 7
L'Estoile, Benoît 230
Le Tonkin (guidebook for Exposition Coloniale) 99–100
Lévi-Strauss, Claude 38, 99–100, 139, 217, 247, 251

L'évolution économique de l'Indochine française (Robequain) 154
Lévy, Paul 111–112, 181, 302–303
Lévy, Roger 6, 24, 122, 152, 153–154, 165–166, 167, 178n160, 251
Lévy-Brühl, Lucien 121
Lewis, Arthur 40
L'Homme (journal) 217
Libération-Sud 202n54
Lindsay-Poland, John 16
livestock 274; *see also* water buffalo
living standards 137–138, 156, 157, 208, 214, 221, 224, 240, 267, 273, 274, 294, 295
Livingstone, David 11, 108
Lopes, Henri 277
Loubet, Jean 88, 127, 198
Low, Anthony 143
Ludden, David 12
Luong Xuan Nhi 73
L'utilisation du sol en Indochine française (Gourou) 154, 160, 166
Ly, Abdoulaye 246–247
Lyautey, Marshal Hubert 96
Lycée Albert Sarraut 109

Mabille, Pierre 185
Madagascar 150
Magnella, Paul 282
Mahoney, Martin 11, 301
Malawi 241
Malaya 62
Malaysia 234
male supremacy 17
Malinowski, Bronislaw 121
Malleret, Louis 182
Malm, Andreas 300
Malraux, André 56, 147
Malthusian line 156
Mamdani, Mahmood 277, 280, 281, 282
Mao Zedong 43, 65
Maquet, Jacques 274, 275–276, 281
Marr, David 49
Marseilles, Jacques 105
Martins, Luciana 11, 15, 17, 27, 117
Martonne, Emmanuel de 126, 182, 183, 187, 199n5
Marty, André Louise 65
Marx, Karl 207
Marxism 249, 285
Massignon, Louis 17
Matera, Marc 99
Matisse, Henri 66

Maugham, Somerset 56
Mau-Mau crisis 269
Maurette, Marie-Thérèse 126
Mauss, Marcel 152
Mayan civilisation 219, 304–305n19
McClintock, Anne 59
Medical and Scientific Centre of the Free University of Brussels in Central Africa (CEMUBAC) 270
Meeking, Ludwig 182
Mekong Delta 58, 61, 63, 72, 89, 116, 287
Menil, René 185
Meyer, Angela 66
militarism 17, 300
Mills, Clarence 234–235
Mirzoeff, Nicholas 70
mise en valeur 45–46, 49, 59, 109, 138, 161
Mitchell, Timothy 28, 66, 142
modernisation 230, 297
modernism 99; French 105
modernity 46, 185, 207, 295; in Africa 247; capitalist 207; and the tropical world 297–298
Mogey, John 6
Moïsi, Dominique 215
Monbeig, Pierre 164, 199n5
Monsoon Asia (Dobby) 237
monsoon colonialism 58–59
Montaigne, Michel de 24, 216, 296
Montesquieu 223, 299
Morocco 150
Mouhot, Henri 54, 55, 56
Moutet, Marius 147–148, 149, 150, 151, 157, 159, 180, 196
Mouvements unis de résistance 191, 202n54
Mozambique 270
Müller, Robert 59
Mumford, Lewis 88, 90, 92–93, 125, 129n32
Munger, Edwin 230, 233, 236
Mus, Paul 45, 53, 74, 87, 110–112, 119, 122, 146, 152, 182, 251; margin notes in *Les paysans du delta tonkinois* 110, 119, 135n188
Musée de l'Homme 153, 184
Musée d'Ethnographie du Trocadéro (MET) 151, 153
museums 14, 48, 50, 66, 151, 160, 161, 162, 302; ethnographic/ethnological 151, 153, 174n80; Musée de l'Homme

316 *Index*

153, 184; Musée d'Ethnographie du Trocadéro (MET) 151, 153

Naggiar, Paul Emile 195
Naipul, V.S. 267
National Front for the Liberation of Angola 270
National Institute for the Agricultural Study of Belgian Congo 273
nationalism: African 233, 253, 269, 271, 276, 277, 280; anticolonial 47; Belgian 181; Chinese 43; and colonialism 207; communist 148; and empire 169; in the Far East 233; German 200n10; Vietnam 303; Vietnamese 49, 73, 103, 134, 143, 147, 149, 198, 283
Nations nègres et culture (Diop) 3
négritude 185
Neill, Deborah 162
Neo-Classicism 165
neoliberalism 300
networking 141
Newbury, David 277
Ngoc, Nguyễn Phuong 53
Nguyen, Than 73
Nguyen-Marshall, Van 57, 155
Nicolaï, Henri 25–26, 248, 271, 275, 280, 304–305n19
Norindr, Panivong 74
Nørlund, Irene 144
Northern Rhodesia 241
Nouveaux aspects du problème économique indochinois (Bernard) 144
Nyasaland 241

Oceania 150
Ocitti, Jakayo 253
Office de l'Alimentation Indigène 148
Office de la recherche scientifique et technique outre-mer (ORSTOM) 218, 248, 266, 285
Ogilvie, Alan 232
Orain, Olivier 120, 122
Oriental Despotism (Wittfogel) 220
Orientalism 5, 10–11, 12, 17, 19, 54, 74, 90, 108, 110, 111, 117–118, 121, 127, 163, 165, 216, 244, 292; and colonial art 67; French 50, 74; Orientalist imagery *51*
Orientalism (Said) 10
ORSTOM (Office de la recherche scientifique et technique outre-mer) 218, 248, 266, 285

Osborne, Milton 41
other/Other (othering; otherness) 27, 107, 162, 214, 216, 284; colonial regions as 2; Indochina as 53–54; and native resistance 11
Oudard, Georges 44
overpopulation 140, 142, 144, 154, 155, 156–157, 158, 159, 221, 269, 295; *see also* population density
Overy, Richard 168

Pacific Rim 286
Padrón, Richard 38
Panzer, Wolfgang 182
Papon, Maurice 191
Papy, Louis 201n41, 213, 218, 222, 246
Parenti, Christian 300
Parmentier, Henri 164
Parsons, Talcott 145
Pasquier, Pierre 98, 101
Passarge, Siegfried 182
patronage 142, 145–147, 163, 170, 173n51
peasantism 101
peasants: African 226; and the colonial sciences 162; and labour migration 142; of the Mekong Delta 160; of the Tonkin Delta 144–145, 155, 156–160, 161, 302–303; Vietnamese 151, 302–303; *see also* famine; living standards; poverty; Yên Bái Uprising
peasant studies 142
Pélissier, Paul 23, 186, 192, 198, 201n41, 218, 231, 281
Pelzer, Karl 7, 232
People's Army of Vietnam (PAVN) 43
People's War 43
Perpillou, Aimé 186
Pétain, Marshal Philippe 101, 183, 189
Philippines 39
Phillips, John 242
Phong, Nguyễn Quang 67
Pignon, Léon 44, 49, 50, 60
Pike, Douglas 55
Polynesia 208
population density: in Africa 227, 271, 273, 275, 277, 282; in Asia 6, 68, 91, 100, 107–108, *112*, 114, 118, 121, 127, 140, 144, 154, 159, 224, 242; as cause of disease 29n14, 255n37; in Europe 105
population growth 138, 155, 156, 227, 241, 279, 296; *see also* overpopulation
porno-tropics 59

Portuguese Angola 221, 270
post-colonialism 5, 253, 266, 292; *see also* colonialism
postcolonial theory 18, 27, 46, 149–150
Pourtier, Ronald 282
Pour une géographie humaine (Gourou) 219
poverty 2, 5, 48, 57, 58, 106, 123, 138, 144, 147, 148, 150, 151, 155, 156, 157, 159, 176n129, 212, 214, 217, 225, 236, 239, 240, 267, 275, 276, 278, 299, 300
Power, Marcus 245
power: in Africa 233, 275, 278, 279, 282, 285; Anglo-European 17; bio- 141, 160; British 160; colonial 50, 53, 58, 66, 138, 155, 158, 160, 207, 229, 233, 239, 269; expansionist 2; French 45, 143, 181; and identity 75; imperial 41, 59, 140, 154, 163, 168, 236; industrial 249; international 150; and knowledge 121, 140; knowledge and 53, 90, 121, 140, 259n30, 298; labour 7; linguistic understandings of 125; local 117; metropolitan 87; military 41, 150; national 170; networks of 180; observational 119; theories of 18, 124, 125; Vichy 186; Western 4, 12, 138, 142
Pratt, Mary Louise 37
Price, Grenfell 168, 228
primitivism 15, 99, 101, 185
Prothero, Mansell 230, 251–252
Protschky, Suzie 67
public hygiene 38
Puerto Rico 39
Putnam, Walter 150

"Qu'est-ce que le monde tropical?" (Gourou) 137–138, 142
Quinn-Judge, Sophie 110

racism 4, 20, 58, 80n73, 87, 159, 207; climatic 13, 24, 285, 299; colonial 155, 185, 207; and German fascism 153; sly 4; totalitarian 223
Raison, Jean-Pierre 23, 108, 250
Raj, Kapil 116
Ramaswamy, Sumathi 115
Rancière, Jacques 75
rationalisation 15
rationalism 24, 101, 110, 152, 210, 216, 296
Ratzel, Friedrich 181
realism 140, 211

reason vs. emotion 214–217
Red River 45, 57, 58, 62, 89, 94, 156, 302, 303
René Leys (Segalen) 96–97
resettlement schemes 270
resource extraction 38
retour à la terre 101
Revers, Georges 43, 44
Reynaud, Paul 183
Rhodesia 270
Ribeiro, Orlando 218, 221
Richard-Molard, Jacques 164, 231
Rivet, Paul 6, 122, 146, 147, 151–153, 165, 166, 184, 211
Rivière, Georges-Henri 152–153
Riz et Civilisation (Gourou) 218, 287, 303
Robequain, Charles 47, 74, 106, 122, 146, 150, 151, 156, 157–158, 164, 166, 182; *L'évolution économique de l'Indochine française* 154
Roberts, Priscilla 153–154
Robic, Marie-Claire 114
Rodney, Walter 249
Rogers, Charlotte 56
Rohdie, Sam 164
Roosevelt, Theodore 16
Roques, Philippe 57
Rosenberg, Emily 42, 162–163, 164, 168
Ross, Corey 38, 222, 224, 238–239, 249, 269
Rothman, Adam 11
Rouse, Joseph 125
Rousseau, Henri 14–15, 66
Roussi, Suzanne 185
Royal Academy of Colonial Sciences 273
Royal Geographical Society 18
Royal Geographical Society of Egypt 233
Ruanda-Urundi 266, 268, 270–273, 292, 298; geographical studies of 273–276; population density *272*
Rule of Experts (Mitchell) 142
Russell, John 6, 244
Rwanda 223, 268, 270–273; analysis of genocide 279–283; genocide in 276–279; geographical studies of 273–276; Gourou's assessment of 273–276
Rwandan Truth and Reconciliation Commission 276
Rwanda Patriotic Front (RPF) 276–277, 279
Rydell, Robert 66

Sachs, Jeffrey 299
Sackur, Amanda 149
Safier, Neil 11
Sahlins, Marshall 242
Said, Edward 5, 10–12, 13, 17, 27, 65, 90, 110, 121, 127, 216, 266–267, 292
Saigon 59
Sarraut, Albert 44, 45, 47, 49, 53, 60, 98
Sasaki, Yutaka 195
Sauer, Carl 246
Sautter, Gilles 21, 22–23, 218, 230, 231, 248, 266, 273
Schoendoerffer, Pierre 72
Schulten, Susan 169–170
Schweinfurth, Ulrich 236
science: and capitalist experimentation 139; and colonialism 163
Scott, James 302–303
Seaports and Development in Tropical Africa (Hoyle and Hilling) 237
Sea Wall, The (*Un barrage contre le Pacifique*) (Duras) 58
Section Française de l'International Ouvrière (SFIO) 147, 148
secularisation 15
Segalen, Victor 24, 96–97, 98, 100
segregation 269–270
self-censorship 143
Senegal 269
Service Aéronautique (SA) 109–110
Service Géographique de l'Indochine (SGI) 53, 109–110, 115, 116, 166
Shantz, Homer 232
Shaya, Gregory 48
Shayanov, Alexander 302
Siam 56, 241
Sidaway, James 245
Siderius, Edmund 20
Silvestre, Auguste 151, 158
Singaravélou, Pierre 50, 53, 109, 160, 162
Sion, Jules 55–59, 183
soil erosion 39, 227, 269
Sommet de la Francophonie 303
South Africa 270
Southeast Asia, colonial empire in 207–208
Southern Rhodesia 241, 242
Soviet Union 148, 195, 205n11, 208, 284; nuclear weapons tests by 43–44
Spate, Oskar 229–230, 237, 244, 245–246
Spykman, Nicholas 246

Stamp, Dudley 166, 178n162, 230, 234–235, 237
standpoint epistemology 13
Staszak, Jean-François 67
Steel, Robert 218, 230, 231–232, 251
Stepan, Nancy 11, 37
Stephen, Daniel 41
Stott, Philip 5
Suret-Canal, Jean 7, 9, 40, 88
Sûreté Générale de l'Indochine 143
surrealism 185
Sutter, Paul 11, 161, 250
Sylvest, Casper 225
systematisation 21

Tam, Nguyễn Turòng 196
Tam, Pham Thanh 73–74
Tanganyika 269
Tardieu, Victor 65, 66
taxation 142, 143
Taylor, Nora 67
Taylor, Philip 125
Technics and Civilization (Mumford) 88
techniques d'encadrement 91, 93, 122, 138, 192, 195, 219–221, 226, 231, 239, 240, 247, 250, 279, 283, 287, 293, 296, 303, 304–305n19
technology 92–93, 126; and agriculture 239, 240
Terres de bonne espérance (Gourou) 23, 286–287, 293
Tertrais, Hugues 47
Tesfaye, Facil 281
Tessier, Olivier 303
Tholance, Auguste 143
Thomas, Martin 45, 117, 150, 158–159
Throop, Palmer 211
Tilley, Helen 161, 162, 163, 164, 168
Tonkin (Hanoi) 47, 57
Tonkin Delta 36, 44, 58, 61, 62, 63, 64, 68–69, 72, 74, 76, 116, 287, 292; creation of 89; Gourou's study of 91–92, 292–293; and the Indochina War 67–68, 70–71; maps of 108–110, *113*, 166; peasants of 124–127, 155, 302–303; population density 6, 29n14, 68, 91, 100, 107–108, *112*, 114, 118, 121, 127, 140, 144, 154, 159, 224, 242, 255n37; as tropical environment 94–98; *see also Les paysans du delta tonkinois* (Gourou)
Tønnesson, Stein 196
Topik, Steven 90

topographical studies 109
traditionalism 109, 110, 126, 273
Travel Guide to Belgian Congo and Ruanda-Urundi 271
travelling gaze 15, 16
Trewartha, Glenn 251
Tristes Tropiques (Lévi-Strauss) 99, 139
Tropical Africa (Kimble) 230–231, 237
tropical environments: fear of 16–17; Western fascination with 14–16
tropicalisation 27; of Indochina 36–37, 38
tropicality 4–5, 7, 10, 12, 14, 15, 17, 18, 20, 22–23, 26, 27, 163, 165, 215, 216, 239, 244; aerial 110; affirmative 27–28, 36, 88; and artistic expression 185; blinkered 223; and capitalism 38; construction of 37; and geography 248; and Gourou's biography 21, 293–294; in Gourou's writing 102, 105–106, 297; and imperialism 38; literature on 19; militant 63, 73, 284; negative 28; positive vs. negative 5; as "transactional" process 90; and the tropics as affective domain 28; twentieth-century 292; and war 294–295; zonal 167, 210, 224
tropical knowledge 14
tropical question 155, 219
Tropical Visions in An Age of Empire (Driver and Martins) 27
tropical world (map) *243*
Tropical World, The (Gourou, trans. by Laborde) 7, 230
tropicology 7
tropics: climate and geography of 2; depictions of 14; under-development in 224–225; phantom 223, 246, 248, 249–250, 268, 298, 301; white settlers in 56–60
Tropiques (literary magazine) 184–186
Truman, Harry 43, 138
Tuân, Dào Thê 302
Tucker, Richard 39
Tunisia 150
Twentieth Century Fund 230
Tyner, James 277

under-development 19, 224–225, 239, 249, 297, 299
Unilever 163
Union Française 50, 72, 208
Union Indochinoise 44; *see also* Indochina

United Fruit Company 39, 163
United Nations Educational, Scientific and Cultural Organization (UNESCO) 138, 218, 247, 297
United States: as international leader 138; need for victory over Japan 208; support for France in Indochina 43, 44; and the Vietnam War 111
Université Libre de Bruxelles (ULB) 180–182, 192–193, 218
urbanisation 14; in Africa 250
Uvin, Peter 277, 281, 282

Valéry 296
Valluy, Jean 61, 70
Văn Huyên, Nguyễn 53, 196, 198
van Munster, Rens 225
Vann, Michael 60
Versailles Treaty 181
Veyret, Paul 6
Vichy regime 48, 49, 58, 62, 76n1, 78n40, 101, 164, 183–186, 188, 190, 193, 194, 203n86, 212
Victoir, Laura 60
Vidal, Claudine 281
Vidal de la Blache 104, 114, 124, 297
Viénot, Pierre 147
Viet Minh 36, 42, 44, 61, 63, 64, 70, 72, 73, 185
Vietnam 27, 44, 45, 50, 55, 302–303; communism in 286; famine in 42, 57, 62, 78n40, 110, 144, 148, 149, 150, 151, 157, 174n72; French colonisation of 45–50; and the Indochina War 64; population growth in 295; *See also* Democratic Republic of Vietnam (DNV); Mekong Delta; Tonkin Delta
Vietnamese Nationalist Party 142
Vietnam War (American) 111, 233
violence, colonial 37, 87, 98, 143, 159, 185
Viollette, Maurice 147
Viollis, Andrée 147
Vu, Tuong 111

Walreusse, Jacques 122
war: tropicality and 294–295; *see also* Indochina War; Vietnam War (American); World War I; World War II
water buffalo 94
Waters, Julia 160
Watts, Michael 295
weather events 300–302

Weber, Max 15, 145
Wells, Allen 90
Weulersse, Jacques 146, 182
Where Winter Never Comes (Bates) 216
White, Trumbull 39
White Settlers in the Tropics (Grenfell) 168
white supremacy 207
Whittlesey, Derwent 241, 242
Wilder, Gary 147, 152
Winter, Jay 54
Wittfogel, Karl 220
Wolf, Eric 242
world exhibitions 14, 41–42, 48; *see also* Exposition Coloniale
World War I 212–213
World War II 229, 235, 292, 295; and the Bordeaux Liberation Committee 188–191; and the colonial powers 207–208; effect on academics 186–188; German invasion of Poland 183; German occupation of France 183–184; shortages and rationing 189–191; and the specialist army 62

Xo Viet Nghe-Tinh strikes 142

Yates, Barbara 252
Yên Bái Uprising 54, 142, 143, 147
Yeoh, Brenda 18, 90
Yi Fu Tuan 107
Young, Anthony 232

Zambia 241
Zelinsky, Wilbur 251, 274
Zimbabwe 241
Zimmerman, Maurice 95–96
Žižek, Slavoj 90, 91
zonal fields 127, 139
zonal framing 231, 232
zonal geography 6, 170, 237
zonal imaginary 11, 297
zonality 248
zonal project 213
zonal tropicality 167, 210, 224